Technology Transfer.

Technology and Knowledge

EUROPEAN ASSOCIATION FOR EVOLUTIONARY POLITICAL ECONOMY

General Editor: Geoffrey M. Hodgson, *University of Hertfordshire Business School, UK*

Mixed Economies in Europe: An Evolutionary Perspective on their Emergence, Transition and Regulation
Edited by Wolfgang Blaas and John Foster

The Political Economy of Diversity: Evolutionary Perspectives on Economic Order and Disorder
Edited by Robert Delorme and Kurt Dopfer

On Economic Institutions: Theory and Applications
Edited by John Groenewegen, Christos Pitelis and Sven-Erik Sjöstrand

Rethinking Economics: Markets, Technology and Economic Evolution
Edited by Geoffrey M. Hodgson and Ernesto Screpanti

Environment, Technology and Economic Growth: The Challenge to Sustainable Development
Edited by Andrew Tylecote and Jan van der Straaten

Institutions and Economic Change: New Perspectives on Markets, Firms and Technology
Edited by Klaus Nielsen and Björn Johnson

Pluralism in Economics: New Perspectives in History and Methodology
Edited by Andrea Salanti and Ernesto Screpanti

Beyond Market and Hierarchy: Interactive Governance and Social Complexity
Edited by Ash Amin and Jerzy Hausner

Employment, Technology and Economic Needs: Theory, Evidence and Public Policy
Edited by Jonathan Michie and Angelo Reati

Institutions and the Evolution of Capitalism: Implications of Evolutionary Economics
Edited by John Groenewegen and Jack Vromen

Technology and Knowledge: From the Firm to Innovation Systems
Edited by Pier Paolo Saviotti and Bart Nooteboom

Technology and Knowledge

From the Firm to Innovation Systems

Edited by

Pier Paolo Saviotti

Director of Research, INRA-SERD, Pierre Mendès-France University, Grenoble, France

Bart Nooteboom

Professor, Faculty of Management and Organization, Erasmus University, Rotterdam, The Netherlands

EUROPEAN ASSOCIATION FOR EVOLUTIONARY POLITICAL ECONOMY

Edward Elgar

Cheltenham, UK • Northampton, MA, USA

Published by
Edward Elgar Publishing Limited
Glensanda House
Montpellier Parade
Cheltenham
Glos GL50 1UA
UK

Edward Elgar Publishing, Inc.
136 West Street
Suite 202
Northampton
Massachusetts 01060
USA

A catalogue record for this book is available from the British Library

Library of Congress Cataloguing in Publication Data

Technology and Knowledge: from the firm to innovation systems / edited by Pier Paolo Saviotti, Bart Nooteboom [in association with the] European Association for Evolutionary Political Economy.
Includes bibliographical references and index.
1. Technological innovations—Economic aspects, I. Saviotti, Paolo. II. Nooteboom, B. III. European Association for Evolutionary Political Economy.

HC79 T4 T43916 2000
338.9'26—dc21

99–049007

ISBN 1 84064 224 6
Printed and bound in Great Britain by MPG Books Ltd, Bodmin, Cornwall

Contents

Figures

Tables

Contributors

Wolfgang Becker Institut für Volkswirtschaftslehre, University of Augsburg, Germany

Uwe Cantner Department of Economics, University of Augsburg, Germany

Emmanuelle Conesa International Institute for Applied Systems Analysis (IIASA), Laxenburg, Austria and Groupe d'Analyse et de Théorie Économique, Écully, France

Witold Kwasnicki Institute of Industrial Engineering and Management, Wroclaw University of Technology, Poland

Riccardo Leoncini IDSE-CNR, Milan, Italy

Bart Los Faculty of Economics, University of Groningen, The Netherlands

Maureen McKelvey Department of Technology and Social Change, Linköping University, Sweden

Sandro Montresor IDSE-CNR, Milan, Italy

Bart Nooteboom Professor, Faculty of Management and Organization, Erasmus University, Rotterdam, The Netherlands

Jürgen Peters Deutsche Bahn AG, Corporate Development, Berlin, Germany

Andreas Pyka Department of Economics, University of Augsburg, Germany

Bénédicte Reynaud Director of Research, CNRS-CEPREMAP, Paris, France

Frederico Rocha Adjunct Professor, Institute of Economics, Universidade Federal do Rio de Janeiro, Brazil

Pier Paolo Saviotti Director of Research, INRA-SERD, Pierre-Mendès France University, Grenoble, France

Silke R. Stahl-Rolf Max-Planck-Institute for Research into Economic Systems, Evolutionary Economics Unit, Jena, Germany

François Texier Department of Technology and Social Change, Linköping University, Sweden

Introduction

Pier Paolo Saviotti and Bart Nooteboom

Traces of an evolutionary approach have been found in economics since the beginning of the century. Thorstein Veblen (1919) is perhaps the first author to have adopted such an approach explicitly (see Hodgson 1993). The widely quoted passage by Alfred Marshall (1949) that 'the mecca of the economist lies in economic biology rather than in economic dynamics' (Hodgson 1993, p. 99) proves the attraction that a biological metaphor had for Marshall, in spite of the fact that he never used it in his subsequent work. Joseph A. Schumpeter is, of course, the economist who exerted the greatest influence on the formation of evolutionary economics. His explicit emphasis on non-equilibrium and on innovations constitutes the basis of present-day evolutionary economics. Another economist adopting an explicitly evolutionary approach was Friedrich von Hayek (1982) (see Hodgson 1993).

These economists provided a great wealth of concepts and inspiration, but an explicitly evolutionary approach was developed only after the 1960s, when a series of new studies of innovation and of technical change emerged. Essentially such studies dealt with the factors leading to successful innovations. Their *raison d'être* was the sharply increasing spending on research and development (R&D) after the Second World War and the consequent need to evaluate this type of investment. This change cannot be underestimated, because it amounts to the introduction of a new function in the economic system. It can rightly be considered a revolution (Freeman 1982; Freeman and Soete 1997). Its first and immediate impact was to create the need for the evaluation of this type of investment, although other effects that we shall examine later took longer to manifest themselves.

This need for evaluation could not be satisfied by existing theories. For example, the production function of a firm, even if known, could not be of great help in deciding what can make an innovation successful. The essential problem was that the production function treated technology as a black box. The new studies of innovation were thus very empirical and searching for what could be defined as the internal structure of technology. Simultaneously the heterodox research traditions epitomized by Veblen, Marshall, Schumpeter and Hayek survived within economics. Greatly oversimplifying the situation, we could say that the origin of evolutionary economics lies in case studies of technological innovation and in criticisms of neoclassical economics.

Of course, with such a composition, evolutionary economics had serious intrinsic limitations. On the one hand case studies, while providing a rich and interesting natural history of innovation, were not based on a common methodology and were thus not comparable. The way forward involved creating common methodologies and deriving from the empirical evidence a series of generalizations that would constitute the basis for a theoretical explanation of innovation. Furthermore, these theoretical developments would have to be integrated within a broader theoretical structure that encompassed both innovation and other economic phenomena, thus allowing the transformation of purely critical studies into an alternative theoretical structure and a more systematic understanding of the relationship between evolutionary and neoclassical economics.

In the period since the 1970s, considerable developments have taken place. A series of theoretical generalizations emerged in the form of concepts such as dominant designs (Abernathy and Utterback 1975), technological trajectories and regimes (Nelson and Winter 1977), technological guideposts (Sahal 1981), technological paradigms (Dosi 1982), routines (Nelson and Winter 1982) and new technology systems (Freeman et al. 1982). These concepts, while different, had in common the presence of discontinuities in economic development. For example, a new paradigm establishes concepts, tools and institutional pathways that are qualitatively different and often incompatible with those of the previous paradigm. These concepts can then be considered an extension and an articulation of Schumpeter's emphasis on radical innovations and on their fundamental role in economic development.

If evolutionary economics aimed at being a theory that could provide explanations for wide ranges of economic phenomena and not just for innovation, these first generalizations had to be integrated with other research traditions. For example, innovations are created and adopted mostly by firms. Thus integration with theories of firms and organizations, such as the behavioural theory of the firm or the transaction cost approach, was an important development in evolutionary economics. The beginning of this process of integration is associated with the seminal book by Nelson and Winter, *An Evolutionary Theory of Economic Change*, published in 1982. Similarly, to the extent that science and technology as the bases of innovation could not be developed exclusively by the market but demanded a substantial public involvement, an institutionalist approach was required to understand them. A concept like the national innovation system (Freeman 1987; Lundvall 1988, 1992, 1993; Nelson 1992) combined a systemic and an institutionalist approach.

In this respect we can conceive of two extreme situations: either novelty is exogenously generated and economically selected, thus resembling a Darwinian mode of evolution, or both the generation and the selection of novelty are endogenous to the economic system, thus resembling a Lamarckian mode of evolution. According to the latter mode the generation of variety (variation) is performed with the explicit intention of adapting the innovations to the environment in which they are going to be used. A Lamarckian mode of evolution is more likely to apply to economics than is a purely Darwinian one (Saviotti and Metcalfe 1991; Metcalfe 1998). In a recent reinterpretation of this debate, Nooteboom (1999b) establishes a connection between exploitation and exploration, inspired by the literature on organizational learning. In his view the process of exploitation (application, diffusion) influences that of exploration (discovery, innovation), thus leading not only to an *ex-ante* anticipation of the expected adoption environment, but to a continuous feedback between exploration and exploitation. Once again, socio-economic development seems to be quite different from biological evolution. If indeed the selection process is also a process for the generation of new variety, then the source of variety is neither exogenous nor random, but based on learning from experience in the process of selection, to identify novel combinations with some chance of success because they are based on lessons learned, which feed creative hunches. New variety still has a large chance of failure in selection, but it is no longer random and independent from selection.

Two types of novelty required new empirical and modelling tools. First, new variables that could represent the internal structure of innovative activities had to be identified, taking into account that such activities encompassed a very wide range of institutional set-ups which economists had never studied systematically before. Such variables and the techniques required to treat them were new and thus likely to require a period of learning and maturation. Within this field we can cite the wide utilization of studies of patents and publications. Second, as already mentioned, concepts such as dominant designs or paradigms, following Schumpeter's emphasis on radical innovations, introduced discontinuities and qualitative change in economic development (Saviotti 1996). The analytical tools that could allow us to treat problems involving qualitative change were not available in Schumpeter's time, but they are being developed now. Such tools can be borrowed from several disciplines, the most important of which are biology and non-equilibrium thermodynamics.

Thus we can borrow from biology the concepts of variation, selection and inheritance, and modelling techniques such as replicator dynamics. These concepts and techniques can be very useful because biologists have observed and studied phenomena involving qualitative change long before economists have done so. However, we cannot imagine ever transferring these concepts and techniques from one discipline to another without modifications.

In summary, the emergence of generalizations based on case studies of innovation was followed by the convergence with a number of research traditions and by the development of new empirical and modelling tools. All these developments, while not producing a closed and completely coherent scientific discipline, have considerably enriched evolutionary economics, even if at the price of a temporary fragmentation.

We have already pointed out that innovation studies owed their origin to the sharply increased R&D intensity of highly industrialized countries after the Second World War and to the consequent need to evaluate the effectiveness of this investment in R&D. However, this need for evaluation was only a first-order effect of these phenomena. A longer-term effect has been the creation of a knowledge-based society. This amounts not only to a greatly increased frequency of innovation, but to a general increase of the knowledge intensity of all economic processes. Knowledge creation and utilization becomes the fundamental component of economic activity determining the competitiveness of firms and the capacity of countries to create wealth. Such changes are not only quantitative: sometimes the very nature of innovation processes changes substantially due to the increasing knowledge intensity. Traditionally the creation of new knowledge by means of R&D preceded the creation of prototypes and the introduction of innovations in the market. While the linear models have for a long time been considered an oversimplification, the creation of new basic knowledge through fundamental research was institutionally separate from the commercial exploitation of such knowledge. Fundamental research was carried out mostly in public research institutions while its commercial exploitation was carried out mostly in firms. To this institutional division of labour in the production and use of knowledge corresponds a division of labour in evaluation: the results of fundamental research were evaluated by means of peer review while those of applied research would be evaluated by the market. This mode of knowledge generation and utilization, called Mode 1, is now being challenged by the emergence of a new mode (Mode 2). In Mode 2, knowledge creation and utilization occur simultaneously with continuous feedback processes and they are not as sharply distinguished as in Mode 1 (Gibbons et al. 1994). Changes in the basic mechanisms of knowledge creation and utilization can have wide-ranging economic impacts. For example, industrial organization is affected by the transition to a knowledge-based society. The emergence of interinstitutional collaborative networks in R&D (Chesnais 1998; Mytelka 1991; Coombs et al. 1996) is an example of a new form of industrial organization that, even if it is not yet completely understood, is certainly related to the increasing knowledge intensity of the economy.

The use of R&D statistics, of patents and of publications for the study of innovation could already be considered as a means of studying a knowledge-based economy. Yet other concepts more specific to this type of study have recently emerged. Thus the literature on competencies (Loasby 1998; Cantwell and Piscitello 1999; Powell et al. 1996) provides an interpretation of the behaviour of firms that is based substantially on their knowledge-creating and -using activities. Furthermore, the characterization of knowledge as tacit and codified and concepts such as appropriability and absorptive capacity (Cohen and Levinthal 1989,1990) provide the beginnings of a conceptual apparatus to treat knowledge creation and utilization. Of course, we are talking here about emergent phenomena and explanations: neither the knowledge-based society nor its theoretical understanding are in any sense complete and mature.

In the earlier part of this introduction we gave a very brief sketch of the development of evolutionary economics since the 1960s. Of course, a complete analysis of the subject would deserve a much more detailed treatment. However, the purpose of the description here was simply to place the chapters of this book in the context of the present state of evolutionary economics.

THEMES

As a consequence of the changes described in the previous section, the literature on innovation and innovation systems has begun to pay more attention to the knowledge behind or in technology, and the learning behind or in innovation. That is reflected in the focus of this book: most of the chapters are concerned with the knowledge underlying innovation. An adequate understanding of innovation systems requires insight into system dynamics grounded in a variety of firm competencies and behaviour and a variety of demand (Saviotti 1996). Due to the complexities of the dynamic interactions involved, simulation is often used to model the process. It must be pointed out that the status of simulation in economics, and in particular in evolutionary economics, is changing. While analytical models retain their importance, they can only be developed successfully for very simple systems. Simulations, in spite of their lesser rigour, allow us to treat much more complex situations. This is also reflected in this book: two chapters offer results from simulation (Chapters 1 and 2). In the literature, innovations are seen as the outcome of the interaction between firms: in joint production, exchange or imitation of knowledge (Lundvall 1985, 1988, 1992, 1993; McKelvey 1996). Learning is seen as an interactive phenomenon (Nooteboom 1992, 1999ab). This is also reflected in this book: most of the chapters deal with flows of knowledge and other interactions between and within firms.

We use simulation to analyse how the creation of variety by innovation and the levelling of variety by diffusion interact to produce system-level effects. On the firm level we need to analyse motives and modes of interaction, in terms of flows or joint production of knowledge or technology. Motives for collaboration include the need for flexibility and the need to utilize sources of variety in knowledge and technology. Modes of interaction include tradeoffs between integration, for the internal generation and use of resources, and disintegration, for external collaboration, in activities where others have better capabilities. However, one may need to generate internal resources of 'absorptive capacity' in order to profit from collaboration with others (Cohen and Levinthal 1990). Thus outside and inside research can be substitutes but also complements.

If integrated structures have advantages of scale and scope, and disintegrated structures have advantages of variety and flexibility, how are they to be combined or reconciled? Can integrated structures evade inertia? Can disintegrated structures achieve coordination and efficiency? How can we exploit their dynamic complementarity? How do we balance internal and external coherence? What implications does this balancing and merging of external and internal activities have for the units of analysis: for boundaries, on both the firm level and the level of innovation systems or systems of technology?

This book has four subject areas, in which these themes appear and overlap. All the chapters have a new theoretical or methodological point to make, and most of them (six out of ten) include empirical work. The areas covered are as follows:

1. Simulation of system-level dynamics in the generation and levelling of variety: Uwe Cantner and Andreas Pyka (Chapter 1) and Witold Kwasnicki (Chapter 2).
2. Inertia and radical shifts: empirical chapters by Maureen McKelvey (Chapter 8) François Texier and Emmanuelle Conesa (Chapter 7), and a theoretical reflection by Silke R. Stahl-Rolf (Chapter 10).
3. Interaction between people, firms and institutions: empirical work by Maureen McKelvey and François Texier (Chapter 8), Wolfgang

Becker and Jurgen Peters (Chapter 3), Frederico Rocha (Chapter 6) and Bart Los (Chapter 4), and theoretical papers by Silke R. Stahl-Rolf (Chapter 10) and Bénédicte Reynaud (Chapter 9).
4. Boundaries: Maureen McKelvey and François Texier (Chapter 8) on the boundaries of firms, and Riccardo Leoncini and Sandro Montresor (Chapter 5) on the boundaries of systems.

Chapter 8, by McKelvey and Texier ranges most widely across all themes, dealing with inertia and radical shifts, collaboration and firm boundaries. The theme of interaction between firms draws most of the attention. We do not believe that this is a coincidence. This is where the nexus between system and firms lies, and where understanding is most crucial, as well as most needed, to bring together issues of innovation on the systems level and issues of learning on the firm level, from the perspective that innovations arise from the interactive learning of firms.

System Dynamics

Uwe Cantner and Andreas Pyka, in Chapter 1, simulate the development of an artificial industry constituted by an oligopoly of firms that can adopt one of three strategies: conservative, imitative, or absorptive. Here a conservative strategy is one that relies only on internal R&D, an imitative one does not devote resources to exploratory search but only imitates the most successful methods generated elsewhere, and an absorptive strategy uses a mixture of internal and external knowledge. All firms are satisfying and have investment routines adapted to their strategy. The model assumes different ways of producing products that are not vertically arranged according to quality, but horizontally according to differentiated bundles of characteristics. This variety is reduced by spillover, but enhanced by internal research. The dynamics of such an industry is simulated for an existing trajectory and for changing technological trajectories. Under most circumstances the absorptive strategy is superior, provided that the spillover pool is sufficiently large. Among other things, absorptive firms have lower R&D costs than their conservative or imitative counterparts. Cantner and Pyka's chapter is thus an interesting confirmation of the double role of R&D: the internal creation of new knowledge and the absorption of external knowledge.

In Chapter 2, Witold Kwasnicki presents a general model of the behaviour of firms in an industry, following in the spirit of Nelson and Winter but introducing a number of innovations. The chapter concentrates on a particular application of the general model. Firms produce functionally equivalent products. Innovation is the result of both R&D and of the imitation of competitors. Together with demand, innovation defines product characteristics and price. The operation of the model can be separated into four steps: (i) search for innovation; (ii) firms' decision making; (iii) entry of new firms; and (iv) selling processes. The search for innovation occurs essentially by means of a change in the firm's routines. This can take place by means of four different mechanisms: mutation, recombination, transition and transposition. In their decision making, firms maximize an objective function, but only from year to year. Thus, this is not a once-and-for-all-global optimization. The objective function itself is a combination of long-and short-term objectives. The competitiveness of products in the market is determined by the product characteristics to price ratio, the latter raised to the price elasticity of demand, such a ratio being a measure of the product fitness. Both the product characteristics and the price are determined by the set of routines used by the firm. The model allows different aspects of industry dynamics to be simulated. In this chapter, Kwasnicki concentrates on the importance of what he calls the 'cost ratio', that is, the ratio of the unit cost of production to the total cost of capital. In addition to demonstrating the importance of such a ratio this chapter raises the possibility that, due to a change in the cost ratio, monopolistic firms can start behaving like competitive ones and vice versa.

Inertia and Radical Shifts

Silke R. Stahl-Rolf, in Chapter 10, building on the work of Veblen on institutional change and on social psychology, reflects on how institutions may create inertia, and how they may change. The chapter focuses on institutions in the sense of shared mental models (Johnson-Laird 1983) that direct and shape our thought and action. Inertia may be created by mental models concerning, for example, the merits and value of entrepreneurship, which have taken hold in a culture, and are slow to change. Mental models directing our conduct are socially constructed, on the basis of role models (not to be confused with the mental models) that we employ for guidance. This interactionist perspective is advocated as a way of modifying methodological individualism and taking social influences into account, while maintaining a focus on individual people, in their interaction. As such this chapter also incorporates the third theme of the book: that of interaction. How mental models that guide behaviour are adapted in interaction with potential role models is analysed by means of a theory, derived from Fritz Heider, of balancing positive and negative assessments on two dimensions of relations: emotional and more instrumental or causal. These assessments are produced by mental models, and there is pressure to alter the relations or mental models when assessments are out of balance. Role models that serve as potential sources of mental model, change in weight according to clarity, persuasiveness and prestige. This indicates how novel models may diffuse and take hold in a community, or may be blocked.

Emmanuelle Conesa, in Chapter 4, describes a case study of mission-oriented research. The case is the development of an SSTD: a supersonic, single-stage, transatmospheric craft. It is a cross between a rocket and an aeroplane: a vehicle that takes off and lands as a plane, but which like a rocket goes beyond Mach 8 in speed and flies beyond the atmosphere, but in a single stage, without the waste of jettisoning rocket segments. For mission-oriented research there are basically two approaches. The first is a 'club' system, with concerted exploration of one avenue. The second entails competing projects along alternative avenues. Clearly, the first inhibits variety, so that one may get locked into a suboptimal solution. It also evokes problems of free riding and lack of competition. So, in principle the first is the best, perhaps with some intervention, when the most effective and efficient approach becomes apparent, in order to converge on a dominant design if that is needed for reasons of efficiency and does not by itself emerge by competition, or if such prolonged competition would be excessively wasteful. In this case, however, development requires a radical shift in 'infrastructural' technologies of measurement and experimentation and scientific knowledge to support it. Beyond Mach 8 there are no installations capable of reproducing the combination of speed, pressures and temperatures that would obtain. The solution might be computer simulation, but the scientific knowledge of the relevant formulae was insufficient. The required infrastructural knowledge and technology needs to be shared across many different locations and activities, and thus requires standardization. Thus it is of a public good nature, and the club approach is needed. One of the interesting aspects of this chapter is that it contradicts the received wisdom that for radical innovation, a decentralized, parallel, variety-maximizing approach is always the best.

In a different setting, Maureen McKelvey and François Texier, in Chapter 8, also challenge what they see as the received wisdom concerning integration and differentiation in the case of radical innovation. According to the 'life-cycle' view of technologies and industries, radical innovation is produced in a decentralized mode of small firms in flexible arrangements. Large firms have the advantage of efficient exploitation of an innovation, once it has reached the stage of a 'dominant design' (Abernathy and Utterback 1978). They have difficulty in surviving the creative destruction of competence that a radical innovation entails. McKelvey and Texier use the example of Ericsson's move into mobile communication to falsify that view. Ericsson was able not only to survive but to accomplish that shift successfully, by a combination of a massive increase of internal research and clever alliances with outside partners supplying complementary competencies. In this way, the chapter also belongs in the area of interfirm collaboration, as well as that of boundaries. Interfirm collaboration helped to prevent the inertia that might otherwise have arisen due to lock-in in existing firm perceptions of goals and firm structures. To effect the change, the firm had to make its boundaries fuzzy, to enable novel combinations. In a way, perhaps, this yields a partial confirmation of the thesis the chapter would reject: these steps of disintegration were necessary to prevent the inertia of integration. Furthermore, it is striking that the move into mobile communication at first started outside the scope of official firm policy, and was officially taken up only after it showed success. Interesting also from a theoretical point of view is that in this chapter innovation systems are not seen as something 'out there' which provide the conditions for firms to prosper, but as patterns of dynamically efficient interactions that the firm must itself create. This leads into the theme of interaction.

Interaction

When firms interact, they may jointly produce new knowledge, or they utilize externally generated information. In the case of spillovers, firms can adapt knowledge from others without purchasing the right to do so. In general, the research efficiency of firms depends on the interaction of internal (in-house) R&D activities and the extent to which external knowledge sources can be implemented for own purposes. In Chapter 3, Wolfgang Becker and Jürgen Peters analyse the role of university knowledge as an important external resource for the development of new and improved products. On the basis of theoretical considerations about the interrelation of internal R&D and technological opportunities, the importance of the results of academic research for the innovation activities of 2300 firms in the German manufacturing industry is investigated empirically. Using data of the first wave of the Mannheim Innovation Panel (MIP), the regressions underline that scientific knowledge sources have significant effects on innovative activities of German firms. On the innovation input side, university (scientific) knowledge sources are used as complements. In-house technological capabilities can be expanded with stimulating effects on firms' innovation and R&D intensities as well as on their probability of participating in the R&D process. On the innovation output side, no empirical evidence for significant positive effects of university (scientific) knowledge on the sales shares of new or improved products could be found. In summary, the estimations indicate that university knowledge stimulates the innovation process more indirectly by increasing firms' productivity of in-house research rather than by affecting the outcome of innovation activities directly.

Furthermore, the regressions illustrate the relevance of R&D cooperations as the most systematic form of knowledge transfer between universities and firms and their importance for industrial R&D. Therefore, Becker and Peters investigate the determinants of firms' probability of cooperating with universities. In the German manufacturing industry, firms with insufficient technological opportunities stemming from industrial sources (customers/suppliers) as well as companies whose objective is to create new markets have a higher willingness to cooperate with universities than others. These findings strengthen the assumption that firms try to generate new market potentials by cooperations with universities.

Frederico Rocha, in Chapter 6, studies the issue whether outside research is a substitute or a complement of internal research. Substitution is to be expected when a firm is more specialized in production, and contracts out a greater number activities. Complementarity is expected for absorptive capacity, and when the firm is technologically more specialized. The data used are patents filed at the European Union by 81 high-tech companies. Research collaboration is measured as the share in total firm patents of patents filed jointly with others. Technical specialization is measured as the share of patents in the main activity in total firm patents. Specialization in production (degree of purchasing or contracting out) is measured as the industry average of the percentage of added value in sales. Internal R&D intensity is not found to have an effect on the degree of cooperation, and this fails to confirm the absorptive capacity argument for complementary R&D. The only strongly significant effects are those of percentage added value, confirming the idea of substitutive R&D, and of firm size. Firm size had the expected negative effect: small firms have a narrower scope of specialized capabilities and fewer financial resources for inside research, and must seek external sources more. There is some confirmatory evidence for the hypothesis that technical specialization requires more collaboration. Inclusion of dummy variables for countries showed that the Japanese are more prone to cooperation.

Bart Los, in Chapter 4, also builds on the work of Jaffe (1989), and proposes and tests a new method for measuring technology spillover. In contrast with so-called rent spillovers these are usually not measured on the basis of input–output tables, because trade between sectors i and j does not necessarily represent the relevance of each other's knowledge for their learning processes. Technology spillovers have been measured by other means than input–output tables, but the problem is that the corresponding sources are not as prevalent as input–output tables. Los offers a method of using input–output tables anyway, by using the cosine of input coefficient vectors factors for each pair of sectors (or firms). This is zero when the vectors are perpendicular, and unity when they are completely aligned. In other words, the degree to which input factors are aligned is taken as a proxy for mutual relevance of knowledge. These are then used to weigh R&D in each sector. The performance of this new measure is tested in several ways against the better, that is, more direct but operationally less feasible measures of relevance. One test was to compare effects on productivity. The test was found to perform satisfactorily.

In a theoretical chapter, Bénédicte Reynaud (Chapter 9) explores the notion of 'routines', in the context of change and the coordination of cooperation. The issue of routines was discussed at a meeting at the Santa Fe institute, in 1995. Do routines apply on the level of individuals or organizations, or both? Are they static, dynamic or both? In a corresponding 2 × 2 table she identified skills (static, individual), standard operating procedures (static, firm level), search capabilities (dynamic, individual) and heuristics or meta-routines (dynamic, firm level). Routines are contrasted with procedures and with rules. While procedures are procedurally rational, codified, explicit and calculative, routines are tacit, adaptive and context dependent. While rules yield the official, general, abstract, semantic, 'espoused' regime for coordination, routines constitute the pragmatic, detailed practices 'in use', with *ad-hoc* arrangements and improvisations. They yield a basis for mutual correction and support that could never be foreseen, specified and documented in advance, in some rules that cover all contingencies. This ties in with the situated action (as opposed to the computational–representational) line of thought in cognitive science (Shanon 1990, 1993; Hendriks-Jansen 1996), according to which meaning is inveterately context dependent, and the context yields 'scaffolding' to eliminate ambiguities and supplement gaps.

Boundaries

Most of the studies of interaction deal implicitly with the issue of boundaries. Issues of firm boundaries became more explicit in the study by McKelvey and Texier. Riccardo Leoncini and Sandro Montresor, in Chapter 5, make the boundary issue explicit for systems of innovation and technology. Presently, the issue remains implicit, with the result that different authors draw different boundaries that seem arbitrary and inconsistent. Some draw the boundary around the components of a technical trajectory, others draw national boundaries and yet others cross them. Leoncini and Montresor propose a measure of the spread of internal connections, within a country, and of the spread of external cross-border connections. There is internal coherence when parts of the system are widely linked (little variance in linkages). This indicates a 'system of innovation'. When linkages are concentrated among only few components, this indicates localized innovation, labelled a 'technical trajectory'. When export linkages are spread across sectors of the economy (little variance in the sectoral Balassa index of revealed comparative advantage), this indicates that the system or trajectory is international, otherwise it is national. Thus we obtain a four-way classification: national innovation system, international innovation system, national trajectory, international trajectory. The measures are applied to several countries, in different periods of time. The result is that Japan is the only country that can appropriately be called a national innovation system, while Canada is the only persistent international trajectory across the years. Denmark has moved from a national to an international trajectory.

In summary, all the chapters of this book share an emphasis on the important role of knowledge in the creation of technological change and innovation. However, this common underlying theme relates to a number of trends in evolutionary economics. On the one hand, some chapters use simulations in order to solve models that are too complex to be solved analytically. These chapters help to formulate an important question for economics: how legitimate is the recourse to simulation? How reliable are the results obtained by means of simulation? To the extent that simulation is the only procedure adapted to the study of systems beyond a certain level of complexity, should we rely on simulation or ignore the problems? Of course, the two chapters using simulations cannot give definitive answers to the questions raised. However, they contribute, to the debate about the use of simulation in economics.

Several empirical chapters use methods ranging from input–output analysis to linear regressions. Again, these chapters reflect an important preoccupation in evolutionary economics. The data used are relatively novel and sometimes demand innovative treatment. Thus the tool box of evolutionary economists is being broadened by the use of both new and more-established methods to new types of data. Case studies are still part of the empirical work of evolutionary economists but they have evolved. They are now structured by concepts such as technological paradigms, increasing returns to adoption and national innovation systems, to mention but a few. The role of case studies in exploring new phenomena that are not yet accessible either theoretically or quantitatively remains essential. Finally, some chapters are examples of appreciative theorizing, exploring concepts such as routines and mental models.

This book thus reflects the changes that have taken place in evolutionary economics in the 1990s. The development of new modelling approaches goes hand in hand with the extension and articulation of case studies and of appreciative theorizing. These different approaches are not substitutes but complements of one another. Thus case studies may be required to provide an initial exploration of a completely new subset of reality, to be subsequently analysed by means of appreciative theorizing. In turn, appreciative theorizing can provide an initial articulation for the analysis of some phenomena, subsequently leading to the construction of models or to more advanced and better structured empirical work. The modes of interaction of the different approaches to knowledge generation illustrated here are by no means the only possible ones. The examples given here are intended to show that the continued use and the further development of all these complementary approaches is essential to the intellectual health and creativity of evolutionary economics.

Let us conclude this introduction by stressing how most of the chapters of this book are the work of relatively young members of the European Association of Evolutionary Political Economics (EAEPE). The composition of the book in this sense is not an accident, but the result of a deliberate choice by the editors, to show and emphasize how the development of evolutionary economics is accompanied by a substantial continuity of efforts and achievements among different generations of reseachers.

REFERENCES

Abernathy, W.J. and J.M. Utterback (1975), 'A dynamic model of process and product innovation', *Omega*, **3**(6), 639–56.

Abernathy, W.J. and J.M. Utterback (1978), 'Patterns of industrial innovation', *Technology Review*, **81**, June/July, 41–7.

Cantwell, J. and L. Piscitello (1999), 'The emergence of corporate international networks for the accumulation of dispersed technological competencies', *Management International Review*, **39**(1), special issue, 123–47.

Chesnais, F. (1988), 'Technical cooperation agreements between independent firms: novel issues for economic analysis and the formulation of national technological policies', *STI Review*, **4**, 51–120.

Cohen, W.M. and D. Levinthal (1989), 'Innovation and learning: the two faces of R&D', *Economic Journal*, **99**, 569–96.

Cohen, W.M. and D.A. Levinthal (1990), 'Absorptive capacity: a new perspective on learning and innovation', *Administrative Science Quarterly*, **35**, 128–52.

Coombs, R., A. Richards, P.P. Saviotti and V. Walsh (eds) (1996), *Technological Collaboration: The Dynamics of Cooperation in Industrial Innovation*, Cheltenham: Edward Elgar.

Dosi, G. (1982), 'Technological paradigms and technological trajectories: a suggested interpretation of the determinants and directions of technical change', *Research Policy*, **11**, 147–62.

Freeman, C. (1982), *The Economics of Industrial Innovation*, London: Pinter.

Freeman, C. (1987), *Technology Policy and Economic Performance*, London: Pinter.

Freeman, C., J. Clark and L. Soete (1982), *Unemployment and Technical Change: A Study of Long Waves in Economic Development*, London: Pinter.

Freeman, C. and L. Soete (1997), *The Economics of Industrial Innovation*, London: Pinter.

Gibbons M., C. Limoges, H. Nowotny, S. Schwartzman, P. Scott and M. Trow (1994), *The New Production of Knowledge*, London: Sage.

Hayek, F.A. (1982), *Law, Legislation and Liberty*, 3 vols, London: Routledge.

Hendriks-Jansen, H. (1996), *Catching Ourselves in the Act: Situated Activity, Interactive Emergence, Evolution and Human Thought*, Cambridge, MA: MIT Press.

Hodgson, G.M. (1993), *Economics and Evolution: Bringing Life Back to Economics*, Cambridge: Polity Press.

Jaffe, A. (1989), 'Real effects of academic research', *American Economic Review*, **79**, 957–70.

Johnson-Laird, P.N. (1983), *Mental Models*, Cambridge: Cambridge University Press.

Loasby, B.J. (1998), 'On the definition and organisation of capabilities', *Revue Internationale de Systemique*, **12**(1), 13–26.

Lundvall, B.A. (1985), *Product Innovation and User-Producer Interaction*, Aalborg: Aalborg University Press.

Lundvall, B.A. (1988), 'Innovation as an interactive process – from user–producer interaction to national systems of innovation', in G. Dosi, C. Freeman, C. Nelson, R. Silverberg and L. Soete (eds), *Technology and Economic Theory*, London: Pinter, pp. 349–69.

Lundvall, B.A. (1993), 'User–producer relationships, national systems of innovation and internationalization', in D. Foray and C. Freeman (eds), *Technology and the Wealth of Nations*, London: Pinter, pp. 277–300.

Marshall, A. (1949), *Principles of Economics*, 8th edn, London: Macmillan. (Originally published 1890.)

McKelvey, M. (1996), *Evolutionary Innovations*, Oxford: Oxford University Press.

Metcalfe, J.S. (1998), *Evolutionary Economics and Creative Destruction*, London: Routledge.

Mytelka, L. (ed.) (1991), *Strategic Partnership in the World Economy*, London: Pinter.

Nelson, R.R. (ed.) (1992), *National Innovation Systems: A Comparative Study*, Oxford: Oxford University Press.

Nelson, R. and S. Winter (1977), 'In search of useful theory of innovation', *Research Policy*, **6**, 36–76.

Nelson, R. and S. Winter (1982), *An Evolutionary Theory of Economic Change*, Cambridge, MA: Harvard University Press.

Nooteboom, B. (1992), 'Towards a dynamic theory of transactions', *Journal of Evolutionary Economics*, **2**, 281–99.

Nooteboom, B. (1999a), *Inter-firm Alliances: Analysis and Design*, London: Routledge.

Nooteboom, B. (1999b), 'Innovation, learning and industrial organization', *Cambridge Journal of Economics*, **23**, 127–50.

Powell, W.W., K.W. Koput and L. Smith-Doerr (1996), 'Interorganizational collaboration and the locus of innovation: networks of learning in biotechnology', *Administrative Science Quarterly*, **41**, 116–45.

Sahal, D. (1981), *Patterns of Technological Innovation*, Reading, MA: Addison-Wesley.

Saviotti, P.P. (1996), *Technological Evolution, Variety and the Economy*, Cheltenham: Edward Elgar.

Saviotti, P.P. and J.S. Metcalfe (eds) (1991), *Evolutionary Theories of Economic and Technological Change: Present Status and Future Prospects*, Chur: Harwood Publishers.

Shanon, B. (1990), 'What is context?', *Journal for the Theory of Social Behaviour*, **20**(2), 157–66.

Shanon, B. (1993), *The Representational and the Presentational*, New York: Harvester Wheatsheaf.

Veblen, T. (1919), *The Place of Science in Modern Civilization and Other Essays*, New York: Huebsch. Reprinted with an introduction by W.J. Samuels, New Brunswick Transaction (1990).

1. Investigating innovation strategies in an artificial industry

Uwe Cantner and Andreas Pyka

1 INTRODUCTION

The concept of heterogeneity is central to the population perspective within evolutionary theorizing. The selection process works on the basis whereby most often the fittest alternative, for example the firm with the lowest production costs, dominates (Metcalfe 1994). Besides this selection or competition effect, in social evolution and especially here in economic and technological evolution, heterogeneity is also considered as an additional source of progress. The formal or informal exchange of (technological) know-how (Pyka 1997) leads to cross-fertilization effects, often increasing the probability of further success (Basalla 1988; Mokyr 1990; Sahal 1981; Kodama 1986). Thus, heterogeneity and spillover effects are to be considered as core concepts within an evolutionary approach to techno-economic evolution (Cantner 1996).

Our chapter deals with technological heterogeneity among firms, their innovative activities understood as a cultural evolutionary process, and the resulting characteristic structural developments. Thus we investigate innovation processes which are characterized by intrinsic dynamics and which are difficult or even impossible to tackle using traditional equilibrium-oriented models. Consequently, dealing with heterogeneous agents and evolutionary processes, the frontiers of an analytical solvability are soon reached. Nevertheless, in order to obtain clear-cut results, we choose the so-called 'realistic approach of evolutionary modelling', (Silverberg and Verspagen 1994) which does not focus on an analytic tractability – brought about by heroic assumptions and oversimplifications – but on down-to-earth and plausible relationships with the consequence that numerical simulations are unavoidable. Thus, the model enables us to detect structural developments and turbulences on the intra-industry level which are accompanied by relatively ordered developments on a sectoral level. In addition, our approach also allows for the combination of several dynamic processes that are widely accepted as stylized facts of collective innovation processes. Among these are the exploitation of firms' own technological opportunities, the exploration of external knowledge sources providing for extensive technological potential, and the respective absorptive capacities or receiver competences of agents learning from external know-how sources. Quite important in this respect is a tradeoff between exploiting own and exploring external knowledge whenever the latter is also resource using. Our numerical analysis will identify several characteristic structural developments based on this tradeoff relationship.

Our chapter proceeds as follows. In Section 2 we discuss the theoretical foundations where we explicitly introduce and explain different innovation strategies and the motives behind them. Section 3 describes the simulation model. Section 4 shows the most important results of different simulation runs. We close our discussion with some concluding remarks in Section 5.

2 THE DESIGN OF R&D STRATEGIES

One of the major attempts of modern innovation theory is to provide new insights into the process of technological change by dismissing the often unrealistic assumptions of traditional economic theory. There the assumption of abundant technological opportunities, perfect capabilities, information and foresight combined with only weak uncertainty provide that innovative activities are boiled down to an *optimal* R&D allocation game against competitors. By this, technological progress – banned into a 'black box' – is designed in a way which allows for optimal cost–benefit calculations. Dismissing these assumptions and taking into account that technological progress is also a *game against nature*, it is not at all clear how and along which lines firms design their R&D.

In the following we investigate how different innovation strategies as found in the literature perform in a comparative analysis. For this purpose, first we discuss briefly the technological environment faced by firms. Based on this, the second step provides a characterization of different R&D strategies.

2.1 Supply-side Factors Influencing Innovation

The statement that innovative activities are (also) a game against nature, and thus against the unforseeable, suggests that firms invest resources mainly in order to acquire more information which allows for better decisions and improved performance: they learn and explore their environment, which is not a 'black box', but which has its own structural and dynamic features. Since the 1980s, innovation theory has been engaged mainly in investigating those environments which show the following main features.

Technological uncertainty and the endogenous generation of technological opportunities The search for new technologies and even the improvement of existing technologies are risky and uncertain endeavours. This uncertainty – intrinsic to the innovation process – does not allow the timing, technological features and economic consequences of innovations to be predicted exactly: on the one hand, firms try to find new technological solutions for their production processes with *ex ante* not anticipated consequences; on the other hand, new unforeseen and unexpected discoveries external to a firm may change the current situation. Thus, firm decisions and behaviour are to be seen as *bounded rational.*

The development space within which firms learn and which firms attempt to explore consists of a broad set of technological opportunities providing potentials for progress. Here several regularities can be observed. First, the developmental potential of a specific technology is increasingly exhausted as advances are made on the respective technological trajectory. So-called 'intensive technological opportunities' (Coombs 1988) are becoming depleted step by step. Thereby, technological as well as scientific boundaries come into effect more and more, making further improvements increasingly difficult and sometimes even impossible to achieve.

Second, besides intensive opportunities characterizing a specific technology there are also 'extensive technological opportunities' which arise out of *cross-fertilization* among different technologies (Mokyr 1990). Here, new technical solutions are often actively initiated by firms which then generate new opportunities by the combination of already existing technologies. Sometimes the amalgamation of different – *ex ante* considered as unrelated – technologies leads to totally new technological fields; for example, mechatronic or bionic. As Dahmèn (1990) states, 'structural tensions' between complementary technologies may be resolved in the course of time, providing for new technological opportunities.

Such interdependencies and their combining effects arise out of different sources: besides new ideas and findings in academia the manifold effects between up- and downstream productions among firms within and between industries are potential sources of such cross-fertilization. These mutual influences come into effect mainly by *technological spillovers.*

Appropriability, absorptive capacities, and endogenous spillover pools Spillover effects arise whenever new technological know-how is not a purely private good and thus not entirely appropriable by the innovating firm.[1] Imperfect appropriability conditions are responsible for inventors realistically anticipating that they will receive less than the maximum benefits arising out of an innovation. In mainstream economics – modelling homogeneous agents and single innovation processes – this is a reason for a suboptimal level of innovative activity.[2] New innovation theory does not deny this but emphasizes the *idea-creating* features of knowledge spillovers in the context of heterogeneous agents and different complementary and substitutive innovation activities.

The main reason for imperfect appropriability conditions are the 'latent public good' features of technological know-how. To a large extent this knowledge is only partly excludable and non-rival, making R&D laboratories of firms the potential source of spillovers. Accordingly, this has different impacts on a firm's incentives for R&D. On the one hand, other firms eventually can use its new knowledge, and this will have a negative effect on the incentive to undertake costly innovative endeavours. On the other hand, this leakage of own know-how is often 'compensated' by the opportunity to use the know-how of other firms. The latter argument also underlines the often complementary character of R&D.

The existence of spillover effects necessarily requires both heterogeneity of actors as well as their very ability to understand the respective information content. With respect to the former, the assumption of the *bounded rational behaviour* of agents leads directly to technological heterogeneity resulting from local search processes with specific cumulative experiences, knowledge and competences, as well as lock-in effects. Based on heterogeneity defined in this way, firms require specific (technological) competences or *absorptive capacities* to understand the information content of (technological) spillovers, that is, to transform the information into useful knowledge. These competences may be partly just talents and so on; however, quite often they have to be acquired actively.

Local search processes – and thus heterogeneity – and spillover effects are to be seen in a mutual relationship, and thus, they are *endogenously* determined. Keeping first spillovers to one side, local search processes often lead to increasing heterogeneity allowing for a larger pool of potential spillovers. The feedback of spillovers on heterogeneity, however, is twofold. First, with substitutive know-how stocks, spillovers tend to reduce heterogeneity because local search activities become more similar (*negative feedback*). Second, by cross-fertilization between complementary know-how, spillovers may lead to further heterogeneity because new local opportunities are opened up (*positive feedback*.)

Within a context where innovative activities are considered as local search and exploration, one may ask what strategies firms design in order to cope with the uncertainty envolved? We take up this issue next.

2.2 Firm Strategies to Cope with Innovation

The rate of technological progress is not God-given but determined by the specific behaviour of firms which try to improve or to introduce new technologies. The restrictions arising from technological heterogeneity, uncertainty and rationality constraints are – as just mentioned – indeed responsible for an abandonment of the global optimization principle with a clear-cut optimal strategy derivable. However, this does not imply that there are no longer regularities in firm behaviour. In their decisions, firms do not randomly allocate R&D budgets and select certain research directions just by chance. Instead they are guided in a cumulative manner by their past experiences and the capabilities they have already built up. Consequently, the resulting behaviour is neither unique among actors nor optimal and can be described by the concept of *routines*.

Strong regular patterns in the innovative activities of firms suggest that innovative behaviour be described as 'routinized' (Nelson and Winter 1982). Firms operate in environments characterized by a spectrum of market and technological possibilities providing opportunities for overcoming current constraints. Despite the prevalence of technological uncertainty with respect to the results of their actions, firms have expectations. Hence, within the above restrictions and constraints firms are able to design different innovation strategies (Freeman 1982) in which they decide how to use and employ their technical skills and resources.

These strategic decisions are guided by *procedural rationality* (Simon 1976). Abstract questions of how not to measure the marginal productivity of R&D expenditures are not on the agenda of firms; rather, questions on reasonable procedures for fixing these amounts have to be addressed. Therefore, we simply regard R&D decision rules as behavioural patterns which cannot be explained by optimization, but by reference to historical circumstances, experiences and evolutionary development. Firms design and adjust their routines by means of learning and adapting to changing environments. In this perspective, firms are learning organizations, constrained by their cognitive capabilities (Heiner 1988). They do not know the complete set of actual and future opportunities open to them, and therefore they cannot choose the globally best alternative; they are rather constrained to local opportunities.

The introduction of new technologies, the improvement of existing ones, learning how to adjust behaviour, and imitation of other successful actors are the most important components for improving the firms' (technological) performance and strengthening their competitive advantage. Within this context, modern organizational theory points to a tradeoff dilemma between the *exploitation* of existing, and the *exploration* of new, opportunities (Winter 1971; March 1991). Typically included in exploration are behaviours such as search, variation, risk taking and experimentation, whereas exploitation means refinement, production, efficiency and execution. This is also reflected in the returns from exploration which are, compared to the returns from exploitation, less certain and more remote in time. In the following we introduce three general behavioural patterns and show how exploration and exploitation find a variety of expressions in different routines.

The first strategy considered is the so-called *conservative strategy*[3] where all innovative efforts concentrate exclusively on own research. The conservative attitude neglects the change in the nature of modern innovation processes with its increasing interdependencies; it nevertheless has an overall innovation orientation.[4] External technological developments are neglected as long as only investment in the refinement of own specific opportunities is undertaken. Thus, the technology and know-how required for growth and competitiveness are generated in (technological) isolation. Innovative efforts are directed into exploitation of the existing technology, that is, process improvements, and into exploration of new technologies, that is, product innovation.

Firms applying the *imitative strategy*[5] do not devote resources to explorative search. By introducing new technologies they attempt to imitate only the most successful methods generated elsewhere rather than innovating by themselves (Winter 1986). Thus, this strategy exploits only external knowledge and opportunities. Firms acting in this way are not willing to explore risky new opportunities. They want to avoid failure and, even more, learn from the failure of competitors. Therefore, they are satisfied with not being technological leaders. According to Freeman (1982) the imitative strategy is a kind of insurance. It enables firms to react and adapt to technical change introduced by competitors.

Imitation becomes possible whenever new technological know-how is not completely appropriable by the innovating firm. In a regime of total non-appropriability of the know-how, imitative firms have the advantage of knowing and learning *ex ante* that the aim of their imitative efforts is a workable solution. However, technological knowledge is typically characterized as specific, tacit as well as cumulative. 'In such cases "technology transfer" may be as expensive and time consuming as independent R&D' (Nelson 1990, p. 197). Therefore, it is the very nature of technological knowledge that also makes imitation a costly endeavour. Thus, the imitative strategy also requires R&D expenditures and imitative firms may also be research intensive. Moreover, in cases of limited access to the technology to be imitated, it is very unlikely that imitation yields the same technological results as the original innovative technology.[6]

Firms applying the so-called *absorptive strategy* decide to employ a combination of internal and external know-how. With respect to the former they undertake research endeavours aimed at two goals which are also targeted by conservative firms: they exploitatively improve their production processes and exploratively introduce product innovations. Additionally, they exploit external knowledge sources, not in order to imitate, but to achieve cross-fertilization effects which allow them to extend the opportunity space.

However, it would be unrealistic to believe that these spillovers can be integrated without cost into the knowledge stock of a receiving firm. On the contrary, firms with an absorptive strategy have to invest a share of their R&D budget for scanning the general (external) technological development.[7] Thus, they expect synergistic benefits which help to overcome limited technological opportunities.

In contrast to imitative firms, absorptive firms do not simply copy a successful technology but try to integrate external knowledge in order to *create* additional technological opportunities; and in contrast to conservative firms, they do not spend their total R&D budget on improving their in-house technology. Absorptive firms rather attempt to find a balance between the short-term benefits of exploiting own specific opportunities with a mixture of a long-ranged exploring and exploiting of external technological possibilities. In this respect, they try to avoid – at least in a dynamical perspective – the dilemma that 'exploitation of existing knowledge reduces the capabilities and the speed with which new alternatives can be explored' (Levinthal and March 1981, p. 312).

Following the above characterization of the innovative process, investment is made in absorptive capacity in technological heterogeneous environments in order to exploit extensive technological opportunities by understanding the content of knowledge spillovers. Therefore, investing in absorptive capacity is not immediately targeted towards a specific well-described research purpose. In a way, it is done as a precaution, in the event of some unforeseen technological developments. In this context Cohen and Levinthal (1994) refer to the words of Louis Pasteur 'Fortune favours the prepared mind'.

In reality a clear distinction between these stylized strategies is obviously impossible. This applies especially to the building up of absorptive capabilities, which is often seen as a kind of byproduct of 'normal' R&D activities. In these cases conservative as well as imitative firms indirectly acquire the capabilities to understand and use know-how applied in close-neighbour technologies. With increasing technological distances, however, these absorptive capabilities become less effective and direct efforts to acquire complementary knowledge become necessary. Despite this, however, for analytical purposes we shall make a clear-cut distinction between the three strategies.

2.3 Market Competition

Innovative and imitative activities of firms are undertaken in an economic environment which is characterized by a certain degree of competition among firms. Concerning a technological breakthrough that is made collectively by several firms the issue of competition attains special importance, because the situation where firms are in a competitive relation is different from one in which there is no competiton. Thus, in the following we distinguish the kind and the intensity of competition.

Firms compete in terms of price and quality, which in a dynamic context are influenced by process and/or product innovation. In general, one would expect that a successful innovator will be able to divert demand from its competitors because consumers can substitute between several goods. These substitution effects are due to price and quality changes which are the results of the following actions and reactions:

- A process innovation allows the innovator to charge a lower price, attracting additional demand.
- As a reaction, non-innovators are forced to lower their prices in order to counteract the loss of market share.
- A product innovation in the sense of a quality improvement provides for additional demand, allowing the innovator to charge a higher price.
- As a first reaction to this quality-induced substitution effect, non-innovators lower their prices.

Within such an economic environment, competition between firms is not the textbook-perfect price competition with a homogeneous output. Competition rather has to be considered in an oligopolistic setting with differences in unit production costs and different product qualities.

3 THE SIMULATION MODEL

In this section we present a model of a dynamic oligopoly in which firms not only compete in the market, but also may influence each other by their innovative activities. The firms under consideration belong to three 'camps' , each characterized by a fixed strategy: the *conservative* strategy, the *imitative* strategy, and the *absorptive* strategy.

3.1 Market

Modelling the economic interdependence of firms, we suggest a setting which allows price and quality competition among *heterogeneous* firms to be represented. In order to represent the four actions/reactions mentioned in Section 2.3 it would be necessary to construct respective functions for each firm in the model. To keep the presentation of this part as simple as possible we suggest using a model of *heterogeneous oligopoly* which by its very structure represents the kind of competitive relationships described.[8] One obviously might argue here that a model of this type is an equilibrium model based on static optimization which is not suitable for modelling evolutionary dynamics. However, we apply this model in a way that such equilibrium states will never be reached. First, the reaction functions are based on *past* prices (as the only available market information) so that, starting in a situation of disequilibrium, the attainment of the equilibrium state takes several periods.[9] Second, those tendencies are continuously disturbed and influenced by upcoming innovations. Thus, what we observe here is the interplay between *equilibrating* and *disequilibrating* forces.[10]

Within the proposed heterogeneous oligopoly every firm faces an individual linear demand function:[11]

$$p_{it} = a_{it} - \eta\, x_{it} + \frac{h_{it}}{n-1} \cdot \sum_j p_{jt-1}; \quad i, j \in \{1, 2, \ldots, 10\}; \quad i \neq j; \qquad (1.1)$$

p_{it} = price of firm i at time t;
η = slope of demand function;
a_{it} = prohibitive price of firm i's product;
h_{it} = demand switch variable;
n = number of firms;
x_{it} = output of firm i at time t.

The prohibitive price a_{it} will be used later on to model the consumers' assessment of product quality. Consequently a_{it} is to be considered as dependent on the innovative activities in the sector aiming for quality improvement.

Regarding production processes we assume constant returns to scale. Thus, unit costs c_{it} are independent of the produced output. Besides mere production firms periodically devote investments r_{it} to R&D. Therefore, the profit function π_{it}, of firm i reads as follows:

$$\pi_{it}, = (p_{it} - c_{it}) \cdot x_{it} - r_{it}. \qquad (1.2)$$

With respect to market behaviour we rely on the Bertrand assumption. In our context this has two implications: first, firms believe that there will be the same number of competitors in the market as in the preceding period; and second, firm i will expect the prices of its competitors j to be unchanged in the current period.

Based on this we assume profit-maximizing behaviour for short-term decisions, that is, for one period. Under this assumption it is straightforward to develop the firms' reaction functions

$$p_{it} = \frac{a_{it} + c_{it}}{2} + \frac{h_{it}}{2(n-1)} \cdot \sum_j p_{j,\, t-1} \tag{1.3}$$

and the corresponding output level

$$x_{it} = \frac{a_{it} - c_{it}}{2\eta} + \frac{h_{it}}{2\eta(n-1)} \cdot \sum_j p_{j,\, t-1}. \tag{1.4}$$

3.2 Technological Progress I: Process and Product Innovations

To secure and even to enlarge their market shares, firms try to improve on their technologies. They are engaged in R&D endeavours aiming at two goals. First, these innovative efforts are directed towards process innovations which make production techniques more efficient. Consequently, process innovations are represented by unit cost reductions. The innovative success is transformed to an economic one by the enlargement of production due to reduced prices. Second, innovative efforts of firms are directed to the creation of new products. These product innovations attract additional demand by quality competition.

In order to accomplish technological progress firms have to invest in R&D activities. Since the development of a new technology, even the improvement of an existing technology, is risky and uncertain, R&D decisions of firms are not guided by the maximization principle. Instead entrepreneurs apply certain routines for their decisions, which are influenced mostly from past experience but are shaped by future expectations. As an approximation of an entrepreneurial R&D routine, firms invest a share γ_{it}, $\gamma_{it} \in [0, 1]$, of their turnover[12] in the preceding period:

$$r_{it} = \gamma_{it} \cdot (p_{i,t-1} \cdot x_{i,t-1}). \tag{1.5}$$

The behaviour of firms with respect to the determination of the concrete share can be described best with the help of *satisficing behaviour*. In an earlier paper (Cantner et al. 1998) we compare the impact of different orientations with respect to behavioural adaptations, a technological and an economic one. In the present chapter, the satisficing behaviour is focused solely on the technological side. R&D activities are used as strategic devices in order to reach a better relative position with respect to process technologies RP^{PC}_{it} by process innovations or to reach a relative better quality position RP^{PD}_{it} by new products in order to cause substitution effects on the demand side. The relative positions RP^{PC}_{it}, RP^{PD}_{it} are in the interval [10; 1], 1 indicating the respective leading position.[13] These figures are determined by the following equations:

$$RP^{PC}_{it} = \frac{c_{\min}}{c_{it}}; \quad c_{\min} = \min_i \{c_{it}\}; \tag{1.6}$$

$$RP^{PC}_{it} = \frac{PDI_{it}}{PDI_{\max}}; \quad PDI_{\max} = \max_i \{PDI_{it}\}; \tag{1.7}$$

$c_{\min}$ = minimal unit costs in the oligopoly;
PD_{it} = counter of product innovations of firm *i*;
$PDI_{\max}$ = *latest* product innovation in the oligopoly.

A technologically lagging firm ($RP^{PC}_{it} \cdot RP^{PD}_{it} < 1$) tries to catch up by raising its R&D efforts. Instead, a technologically leading firm ($RP^{PC}_{it} \cdot RP^{PD}_{it} = 1$) slightly reduces R&D (0.5 per cent per period) in order to raise profits by reducing R&D costs. Formally this kind of satisficing behaviour is described by equation (1.8):

$$\gamma_{it} = \begin{cases} \gamma_0 \cdot [1 + (1 - RP^{PC}_{it} \cdot RP^{PD}_{it})], & \text{for } RP^{PC}_{it} \cdot RP^{PD}_{it} < 1 \\ \gamma_{i,t-1} - 0.005, & \text{for } RP^{PC}_{it} \cdot RP^{PD}_{it} = 1; \end{cases} \tag{1.8}$$

γ_0 = initial value of the R&D share.

Section 2 identified cumulativeness as another feature of technical progress along a technological trajectory. To reach a certain technological level all preceding levels have to be passed through, because otherwise the relevant technological understanding cannot be achieved. This feature is reflected in the R&D capital stock R_{it}, representing the accumulated technological know-how. This stock is built up by *direct* R&D expenditures which are a share $(1 - \sigma_{it})$, $\sigma_{it} \in = [0, 1]$ of the periodical R&D investments r_{it},[14] The determination of σ_{it}, the share of investment in absorptive capacities, will be explained below.

$$R_{it} = \sum_{t}(1-\sigma_{it})\cdot r_{it}. \tag{1.9}$$

Besides R&D activities, the rate of technological progress depends also on the degree of exhaustion of the intensive technological opportunities. In order to take account of this we assume positive but decreasing potential innovative success ie^{pot}_{it}. The pace of exhaustion is further influenced by the parameter α describing the bending of this function. To represent technological uncertainty, innovative success ie_{it}, is determined stochastically. An equally distributed random number ψ_t, $\psi_t \in =[0, 1]$, reflects the uncertainty inherent in process innovations.

$$ie^{pot}_{it} = 1-\exp(-\alpha \cdot \kappa^{PC}_{it} \cdot R_{it}) \quad \text{and} \quad ie_{it} = \begin{cases} ie^{pot}_{it}, \text{ for } f(R_{it}) \geq \psi_t \\ ie_{i,t-1}, \text{ for } f(R_{it}) < \psi_t; \end{cases} \tag{1.10}$$

$$\frac{\partial f}{\partial R_{it}} > 0; \ \frac{\partial^2 f}{\partial R^2_{it}} < 0;$$

κ^{PC}_{it} = weight determined by the capability to integrate external process know-how.

Firms applying the imitative strategy try to improve their production techniques by imitating the most successful process innovations $ie^{\max}_{t}$ of their competitors. Therefore, the aim of the imitative endeavours is an already workable technical solution, so imitative firms are not confronted with uncertainty in the same way. However, new innovation economics with its knowledge orientation is denying the possibility of a perfect imitation because of the only latent public good properties of technological know-how: ‘When imitation is attempted under conditions that permit only limited access to the thing being imitated, it becomes very similar to innovation and of course is unlikely to yield an exact copy’ (Winter 1984, p. 292). Equation (1.11) describes the imitative success ie^{im}_{it}, which also depends on an equally distributed random number μ^{PC}_{t}.

$$ie^{im}_{it} = ie^{\max}_{t-1} \cdot \mu^{PC}_{t} \cdot [1 - \exp(-\alpha \cdot R_{it})]; \tag{1.11}$$

ie^{im}_{it} = imitative success of firm i;
$ie^{\max}_{t-1}$ = maximum of innovative success in the oligopoly; $ie^{\max}_{t} = \max\{ie_{it}\}$;
μ^{PC}_{t} = equally distributed random number; $\mu^{PC}_{t} \in \{\mu^{\min}; \mu^{\max}\}$.

Since the innovative success comes into effect only with a time lag, the R&D efforts of period $t-1$ influence unit costs of period t which develop as follows:

$$c_{it} = c_0 \cdot (1 - ie_{i,\,t-1}), \tag{1.12}$$

c_0 = initial value of unit costs.

Besides improving production processes firms are assumed to engage in product innovations. The uncertainty involved in those endeavours is quite different from the one we assume for process innovations. Whereas the direction and impact of process innovations along certain trajectories can be roughly expected, this does not apply to product innovations. In the literature this context is described with the notion of *intrinsic* uncertainty. In order to model this quite different feature of product innovations we assume a poisson-distributed random number ρ_t.[15]

The R&D efforts devoted to product innovations are again represented by the stock of R&D capital R_{it}. In the course of time, firms accumulate a success probability pr_{it}, which approximates asymptotically the mean value of the poisson-distributed random number ρ_t. The growth of the success probability is characterized by positive but decreasing rates:

$$pr_{it} = 1 - \exp(-\alpha \cdot \kappa_{it}^{PD} \cdot R_{it}) \text{ and } PDI_{it} \begin{cases} PDI_{it} + 1, & \text{for } pr_{it} \geq \rho_t \\ PDI_{it}, & \text{for } pr_{it} < \rho_t; \end{cases} \tag{1.13}$$

κ_{it}^{PD} = weight determined by the capability to integrate external product know-how.

If the accumulated probability pr_{it} in period t is equal to or even larger than the poisson-random number ρ_t, then firm i introduces successfully a new product. Whenever a firm succeeds with a product innovation, the knowledge to master the old technology is assumed to become irrelevant. Therefore, the old stock of R&D capital will be totally depreciated every time a product innovation occurs.[16] The new technology shows full technological opportunities and consequently a large potential for new process innovations.

Again imitative firms are confronted with less severe uncertainty, which is reflected in the normal distributed random number μ_t^{PD}. Their imitative endeavours pr_{it}^{im} are directed on that product of competitors which technologically ranks superior:

$$pr_{it}^{im}(PDI_{jt} > PDI_{it}) = 1 - \exp(-\alpha \cdot R_{it}); \tag{1.14}$$

$$PDI_{it} = \begin{cases} PDI_{it} + 1, \text{for } pr_{it}^{im} \geq \mu_t^{PD} \\ PDI_{it} \text{ else;} \end{cases} \tag{1.15}$$

pr_{it}^{im} = probability of imitation of firm i;

μ_t^{PD} = normal distributed random number; $\mu_t^{PD} \in \{\mu^{\min}; \mu^{\max}\}$.

The product innovations of our simulated industry are characterized by quality improvements which are acknowledged on the demand side. Here, they have a twofold effect. First, the demand switch variable h_{it} will be influenced. With higher heterogeneity among products the substitutability decreases and with it also the demand switch, thereby affecting the competitors by different degrees. The effect will be the largest for those firms whose quality distance increases measured by the absolute aggregated (normalized) distance $|Q_i|$[17] of firm i to the other firms.

$$h_{it} = h_0 \cdot (1 - \frac{|Q_{it}|}{n}); \tag{1.16}$$

h_0 = initial value;
Q_{it} = relative quality deviation with respect to the average quality.

Second, a successful product innovation of firm i changes the prohibitive price a_{it}. Here the innovating entrepreneur produces a higher quality and the consumers' assessment of his/her product will increase. Other firms j experience a decrease in their a_{jt} ($j \neq i$) value, because the product innovation decreases their respective measure Q_{it} of the aggregated (normalized) distances[18] of firm i in quality space. Consequently, we get:

$$a_{it} = a_0 \cdot n_{Q_{it}}; \tag{1.17}$$

a_0 = initial value;

where in the present chapter, n the number of firms simply is a constant because we do not consider entry and exit of firms.[19]

3.3 Technological Progress II: Spillovers and Absorptive Capacity

As already mentioned at the beginning, an essential feature and determinant of a collectively pushed forward research process are technological spillovers. In order to use the information content of spillovers, a minimal amount of understanding the other's technologies is necessary. This understanding of the knowledge created outside the firm has to be acquired actively and is institutionalized in a firm's so-called 'absorptive capacity'.

In the model we explicitly distinguish between process and product spillovers. The first category affects the technology used by the firm and helps to improve this technology. The latter category of spillovers increases the probability of product innovations. They contain either technical information about new product opportunities or information concerning the combination of different knowledge elements.[20] Effective spillovers necessarily can only arise when there is some heterogeneity among firms.[21] In this respect we assume that the relevant spillover pool increases with increasing heterogeneity in the industry.

Spillover effects are generated *endogenously*. For process spillovers, the variance of the unit costs s_{ct}^2 of the different firms is taken as a proxy for spillover potentials. For product spillovers, the variance s_{at}^2 of the quality measure a_{it} serves as the relevant measure for product or quality heterogeneity.

The firms applying the absorptive strategy do not invest their whole R&D budget in direct R&D but also develop absorptive capacities in order to understand external knowledge. They invest the share σ_{it} in building up absorptive capacity. Similar to the R&D decision above, firms use their relative technological positions to determine their investment in absorptive capabilities. The technologically leading firm is investing only a minimum share σ_{min} of its R&D budget, because the possibilities of learning endogenously are quite low. Instead, technologically lagging firms are increasing their respective share to not higher than σ_{max} in order not to neglect their own research activities. Intermediate firms, however, orientate according to their relative technological position: the more a firm is lagging behind its competitors the higher the efforts to learn from its competitors' technologies.

$$\sigma_{it} = \begin{cases} \sigma_{max}, & \text{for } \sigma_{it} \geq \sigma_{max} \\ 1 - RP_{it}^{PC} \cdot RP_{it}^{PD}, & \text{for } \sigma_{min} \geq \sigma_{it} \geq \sigma_{max} \\ \sigma_{min}, & \text{for } \sigma_{it} \leq \sigma_{min} \end{cases} \tag{1.18}$$

σ_{max} = maximum share of R&D investment in absorptive capacity;
σ_{min} = minimum share of R&D investment in absorptive capacity.

The absorptive capacity of a firm, such as the stock of R&D capital, has to be accumulated:

$$ac_{it} = \sum_{t} \sigma_{it} \cdot r_{it};$$ (1.19)

ac_{it} = absorptive capacity of firm *i*.

Besides the cognitive aspect of integrating external knowledge, empirical investigations hint at the importance of technological aspects reflected in the technological gap G_{it} of a firm to the technological frontier which influences the possibilities for a firm to integrate external knowledge according to an inverted u-form relationship.[22] Firms technologically falling back too far are not able to keep up with the pace of the leaders; the respective knowledge is too specific. Also, firms near to the technological frontier cannot benefit very much from the spillover pool, because for them spillovers contain little that is new. But for firms in-between, spillovers have a large impact because these firms are technologically able to integrate this frontier know-how. In the model we differentiate between a technology gap G_{it}^{PC} in process technologies (PC) and G_{it}^{PD} in product technologies (PD).

$$G_{it}^{PC} = \sum_{\substack{j \\ j \neq i}} \frac{RP_{jt}^{PC}}{n-1} - RP_{it}^{PC}.$$ (1.20)

$$G_{it}^{PD} = \sum_{\substack{j \\ j \neq i}} \frac{RP_{jt}^{PD}}{n-1} - RP_{it}^{PD}.$$ (1.21)

The spillover functions in (1.22) for $F[G_{it}^{PC}]$ and in (1.23) for $F[G_{it}^{PD}]$ reflect the interplay between technological and cognitive factors with respect to a firm's ability to integrate technological spillovers. This is shown in Figure 1.1. For quite small and quite large technological gaps at the borders absorptive capacities yield only a small impact, whereas for intermediate gaps we find with increasing absorptive capacities an increasing impact of spillovers.

$$F[G_{it}^{PC}] = [\delta - (G_{it}^{PC})^2] \cdot ac_{it};$$ (1.22)

$$F[G_{it}^{PD}] = [\delta - (G_{it}^{PD})^2] \cdot ac_{it};$$ (1.23)

$F[G_{it}^{PC}]$ = spillover function of process innovation;
$F[G_{it}^{PD}]$ = spillover function of product innovation;
δ = scaling parameter.

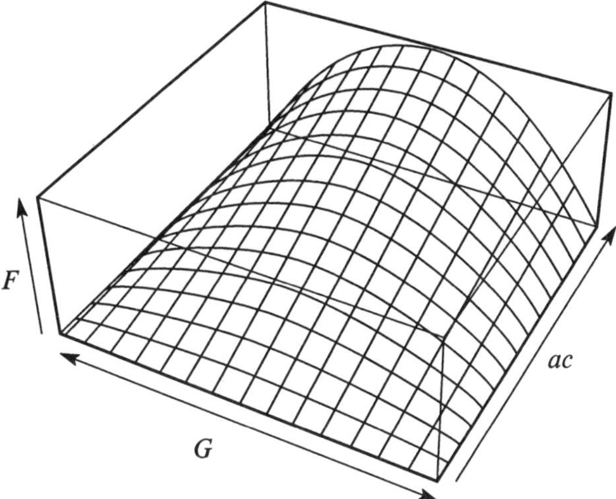

Figure 1.1 Spillover function

Equation (1.10) for the potential innovative success of process innovations is modified by the weight κ_{it}^{PC} representing the spillover pool s_{ct}^2 and the spillover function $F[G_{it}^{PC}]$ as a weight for the R&D capital stock:

$$\kappa_{it}^{PC} = \begin{cases} 1 + \dfrac{s_{ct}^2}{1 + \mathrm{Exp}(\tau - F[G_{it}^{PC}])}, & \text{for } \sigma_{it} > 0 \\ 1, & \text{for } \sigma_{it} = 0 \end{cases} \tag{1.24}$$

τ = difficulty in building up absorptive capacity.

The probability of a product innovation is also positively influenced by the possibility of exploiting product spillovers. This is expressed in the weight κ_{it}^{PD} representing the spillover pool s_{at}^2 and the spillover function $F[G_{it}^{PC}]$ as a weight for the R&D capital stock in equation (1.13):

$$\kappa_{it}^{PD} = \begin{cases} 1 + \dfrac{s_{ct}^2}{1 + \mathrm{Exp}(\tau - F[G_{it}^{PD}])}, & \text{for } \sigma_{it} > 0 \\ 1, & \text{for } \sigma_{it} = 0 \end{cases} \tag{1.25}$$

For an absorptive firm, a successful product innovation bears an additional consequence: the absorptive capacity, such as the stock of R&D capital, falls victim to the *competence-destroying technical change* and becomes obsolete, that is, it will be depreciated with every new technological trajectory a firm adopts.

Figure 1.2 summarizes the basic structure and order of events. Starting with the heterogeneous oligopoly at period t, firms make their decisions in the boxes characterized as rhombs. On the left-hand side we find market decisions which finally determine profit and turnover of the current period. On the right-hand side R&D decisions are arranged, whereby the firms determine their investment in R&D and in building up absorptive capacities, leading to cost and quality improvements. The endogenous character of the spillover pool is marked with a circle.

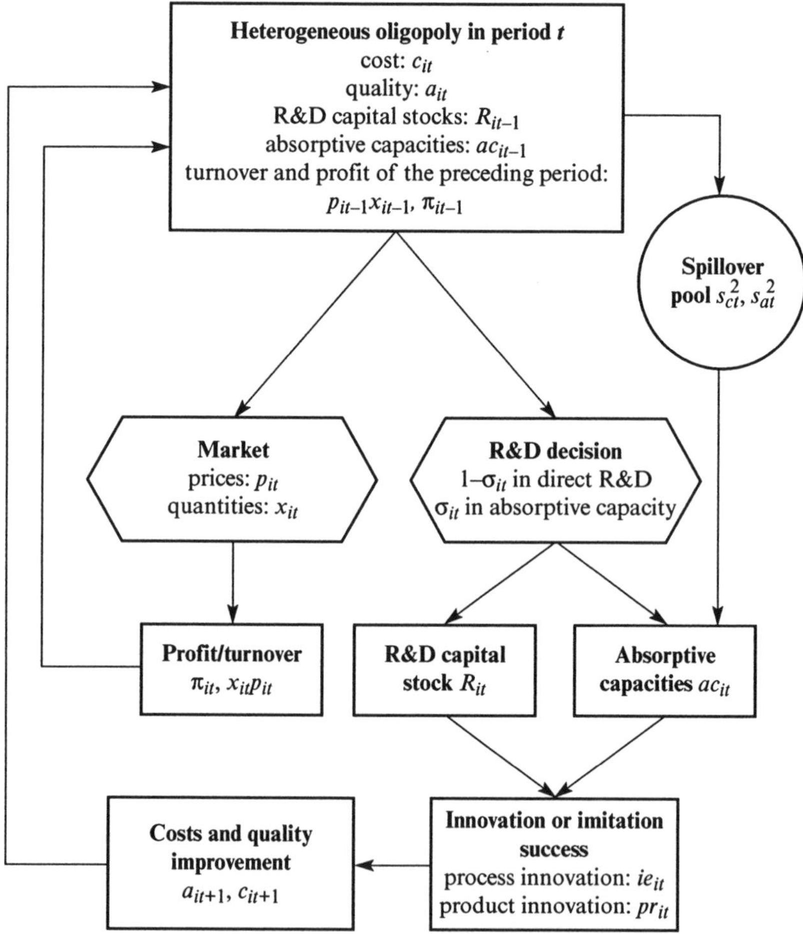

Figure 1.2 Structure of the model

4 SIMULATION RESULTS

Before starting with the numerical experiments, some comments on the simulations are necessary. In the oligopoly, 10 or 15 firms compete which can be classified according to the respective innovation strategies applied. We are comparing three different 'camps' of firms of the same size: absorptive, conservative, and imitative. In the beginning, unit production costs and product quality of all firms are identical. The supposed time scale of 250 and 1000 iterations is artificial and was chosen in order to work out the specific developments. Finally, we perform so-called 'Monte Carlo' simulations, meaning that we try to eliminate any random disturbances by performing 100 runs with different pseudo-random numbers and computing the respective averages.

4.1 Development Along a Single Trajectory

For the first simulation experiments we exclude the possibility of product innovations and only investigate the development along a single technological trajectory. Consequently, we have to notice that in this scenario the development will finally reach asymptotically an equilibrium state where technological opportunities are exhausted and the technical development comes to rest. Figure 1.3 shows the development of unit costs in a scenario where the imitative camp is excluded[23] and only absorptive and conservative strategies compete. We summarized the different firms by showing camp averages.

In phase I we find only minor differences in unit costs between the conservative (dotted line) and the absorptive (solid line) camp. High

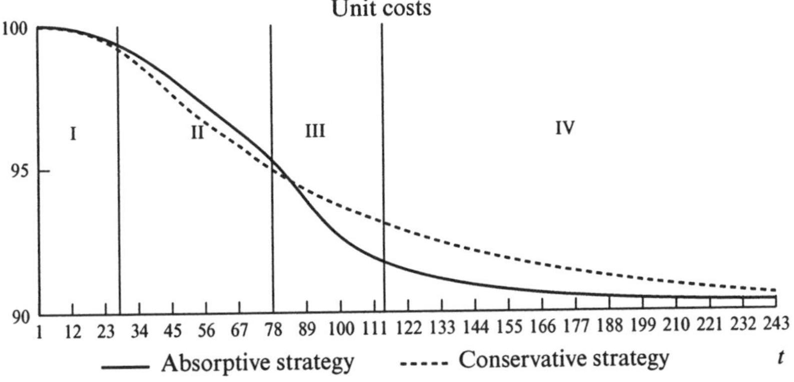

Figure 1.3 Development of unit costs along a single trajectory

uncertainty at the beginning of the technological development together with still small R&D capital stocks are responsible for only minor cost reductions at this early state of the technological trajectory. In phase II it becomes obvious that the conservative camp is more successful in its innovative endeavours. Because conservative firms invest their whole budget in direct R&D, they are faster in accumulating R&D capital stocks responsible for an early exploitation of high technological opportunities. During this period they produce with lower unit costs than absorptive firms. In phase III the first technological bottlenecks become visible for conservative firms, thus leading to a lower rate of cost reduction. Compared with that, two reasons are responsible for a significantly different performance of absorptive firms. On the one hand, the innovative advances of conservative firms have raised heterogeneity in the oligopoly, leading to a steadily increasing spillover pool (Figure 1.4a).

On the other hand, meanwhile, absorptive firms have accumulated a significant stock of absorptive capacity (Figure 1.4b) enabling them to understand and integrate external knowledge. Both developments cause a kind of threshold effect between the 80th and 115th iterations, which reflects the exploration of additional potentials of cost reduction by absorptive firms offered by spillover effects and the accompanying extensive technological opportunities. Ultimately, drawing back on external knowledge allows the absorptive camp to leapfrog their conservative competitors at about period 85, and to produce with lower unit costs. Phase IV represents the final exploitation of the remaining opportunities and the already assumed convergence of unit costs. To show the economic impacts of this technological development we show the respective market shares in Figure 1.5.

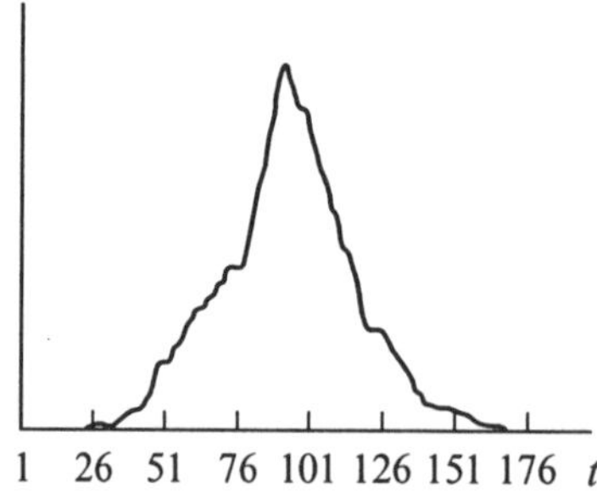

Figure 1.4a Spillover pool (variance of costs)

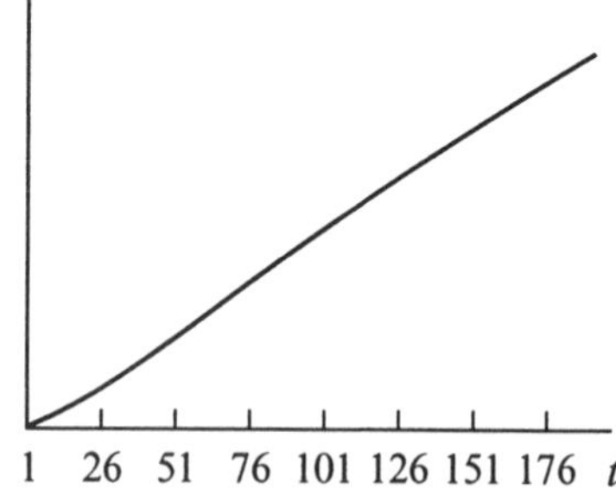

Figure 1.4b Accumulated absorptive capacities

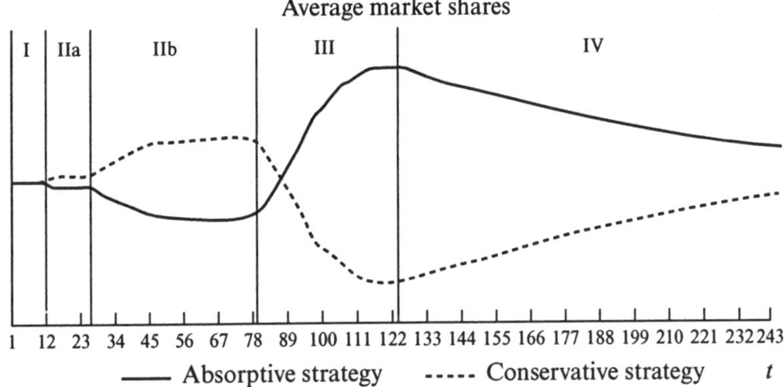

Figure 1.5 Development of market shares

In phase I the market splits up nearly equally between both camps because no remarkable cost reductions can be realized in this early period. However, the first R&D successes of the conservative camp allow higher cost reductions in phase II and, consequently, they achieve the leading position with respect to market shares. The conservative firms are able to offer lower prices and thus attract additional demand. In phase IIa this development is first slowed down because of the routinized decision making. While absorptive firms increase their engagement in R&D due to their technologically falling behind, conservative firms in the leading position begin to decrease their R&D budgets in order to increase periodic profits. However, at this early state of development neither spillover pool nor absorptive capacities are large enough, so that in phase IIb the advantage of concentrating on only own R&D becomes effective for conservative firms. They are in the position to break away decisively with respect to market shares.

Market shares exert a direct impact on periodic profits as can be seen in Figure 1.6. In phase II the conservative camp is able to transpose its leading technological position into higher profits. However, this development changes abruptly when the absorptive capacity effects come into action. Visible progress in cost reduction and finally even undercutting the conservative competitors allows also for reversing the economic development. So after 70 to 80 iterations, absorptive firms also occupy the economic leading position. At the final phase IV we again find an asymptotic development of both camps.

Already on a single technological trajectory going hand in hand with only restricted opportunities, the simulation experiments show character-

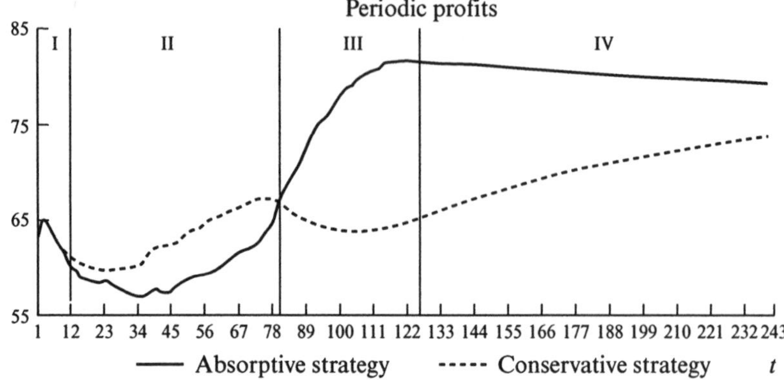

Figure 1.6 Development of periodic profits

istic structures of a collective innovation process. While conservative strategies are quite successful in an environment of rich technological opportunities, strategies emphasizing the collective character of technological development do relatively badly. In such a framework there is no need for cognitive abilities to integrate external knowledge; the respective means could be invested more effectively in direct R&D. However, restricted opportunities are responsible for the increasing difficulties conservative firms have to cope with in order to maintain their pace of development. Now, external knowlege gains importance as it allows firms to explore new extensive technological opportunities. However, without having invested in the absorptive capabilities necessary to understand external know-how, these potentials are inaccessible. Absorptive firms that have started to accumulate the respective capabilities in good time are now able to outperform their competitors by exploiting these extensive opportunities. So, despite a quite ordered technological development on a sectoral level, we find changing technological and economic leaderships between firms.

4.2 Development with Changing Technological Trajectories

The second scenario includes the possibility of product innovations which have a twofold effect. On the one hand, a new technological trajectory with yet unexploited opportunities is opened up by a successful innovator. On the other hand, product innovations go hand in hand with quality improvements, leading to a higher assessment and with it to the possibility of charging higher prices on the demand side. Now, by also investigating a longer time horizon, the imitative camp can be included. Figure 1.7a shows the average results of introducing new products. It becomes obvious that the absorptive firms are producing a higher quality in the long run, whereas the conservative and imitative camps on the second and third position behave in shifts in the medium term until finally the imitative camp is successful in holding on to the better-quality level. When looking at the development of prohibitive prices in Figure 1.7b, the clear falling behind of the conservative firms in the long run becomes quite evident. Up to the 650th iteration they can at least catch up with the imitative camp; however, afterwards the quality lag compared to their competitors increases more and more.

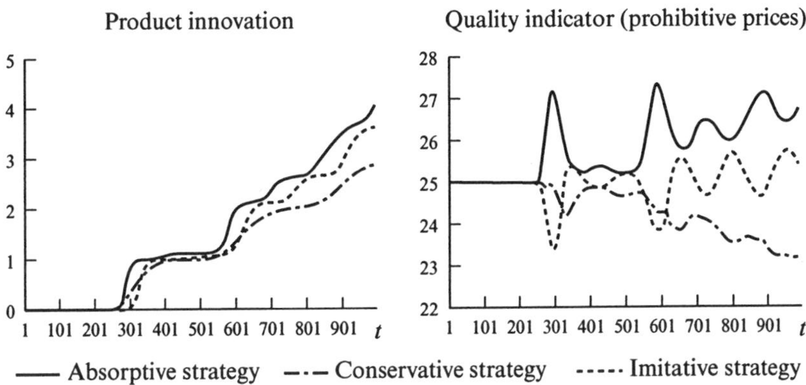

Figure 1.7a Product innovation *Figure 1.7b Relative quality*

But how does this technological development transpose in the economic realm? Figure 1.8 shows the respective development of average market shares. In phase I, again we find the previous scenario along a single technological trajectory, now also including the imitative camp, which performs badly at this stage. However, after the absorptive firms introduce their first product innovation at the beginning of phase II, the situation changes dramatically. In the first place, because of their new product technology, absorptive firms are confronted with high unit costs of production forcing them to raise prices. Despite their higher quality they now lose demand, attracted to the cheaper products of conservative and imitative firms. Because the imitative endeavours of the latter were directed towards lower unit costs of absorptive firms at the end of the first trajectory, the imitative camp attracts the bulk of this demand and reaches the leading market position for a short period. Also, on a somewhat lower scale, conservative firms were able to attract additional demand. However, a new technological trajec-

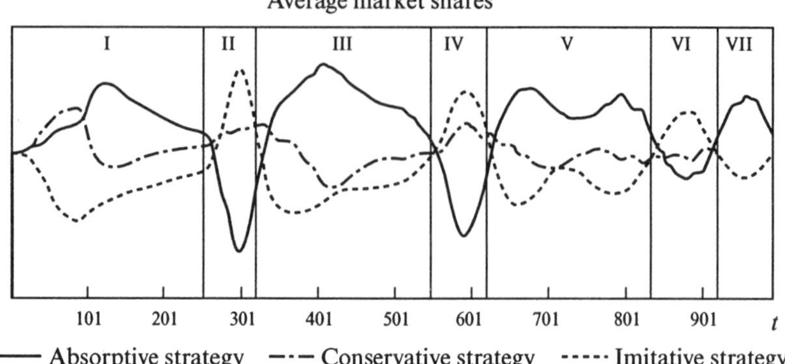

Figure 1.8 Development of market shares including product innovations

tory with nearly undepleted technological opportunities offers high potentials for cost reduction. As soon as the absorptive firms are able to exploit these potentials and realize high cost reductions, they gain back market shares step by step, and finally in phase III they are back to the leading market position. When the imitative firms introduce a new technology, even before the conservative camp, high unit costs also lead to a loss of demand in the first place for the benefit of the conservative competitors. In phase IV the temporary disadvantage of introducing a new technology is already smaller for absorptive firms even if they have to surrender their market leadership again for some periods. With respect to the technological developments characterized above, in the long run a better economic performance of imitative firms would have been expected compared to the conservative firms. As imitative firms do not allocate resources to own R&D, they are not able to realize the same cost reductions as their competitors. So, as a result of on-average higher unit costs of production, their advantages in quality competition are compensated for by unfavourable high prices. Therefore, in the economic sphere a clear-cut dominance of neither the imitative nor the conservative camp can be observed, even if technologically imitative firms clearly leapfrog their conservative competitors.

An alternative to comparing different strategies is to investigate the respective 'R&D effectiveness'. Here, a clear distinction between success in process and that in product innovation must be made. Concerning the latter, firms applying the absorptive strategy are on average the first to introduce a new product. With respect to process innovations, Figures 1.9a, b and c show for the three strategies the relationship between periodic R&D budgets and the resulting cost reductions. To interpret these

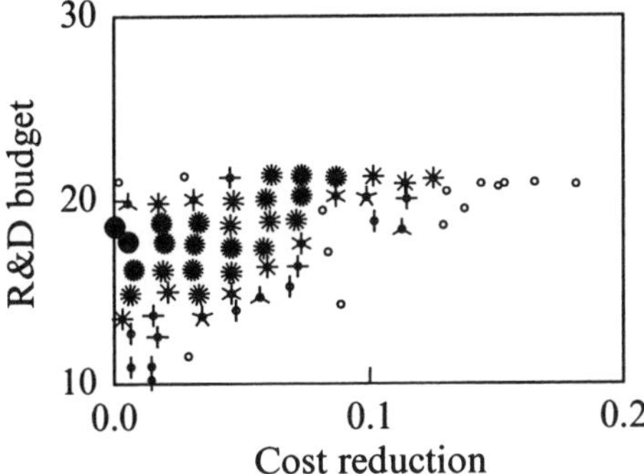

Figure 1.9a R&D effectiveness of absorptive firms

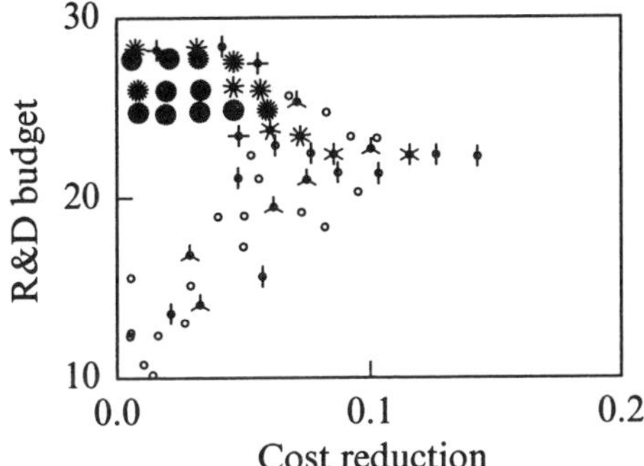

Figure 1.9b R&D effectiveness of conservative firms

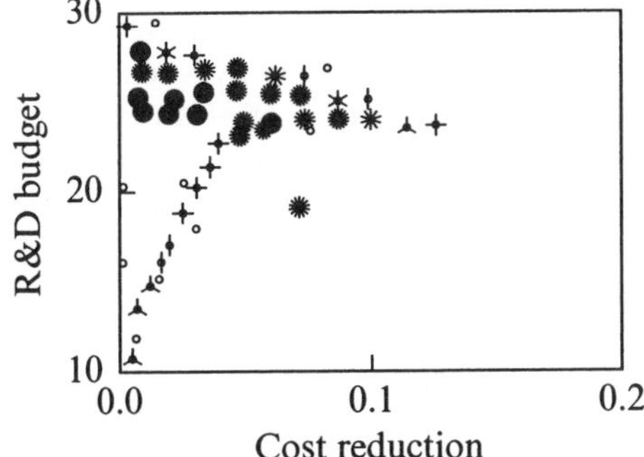

Figure 1.9c R&D effectiveness of imitative firms

figures, observations are represented by 'sunflowers'. The number of petals to each sunflower reflects the number of observations falling in a certain interval.

The following results are interesting. First, the average R&D endeavours of the absorptive firms are smaller than those of their imitative and conservative competitors. The absorptive firms are more frequently in leading technological positions, causing them not to increase or even to decrease the resources allocated to R&D. At the same time, their R&D successes are at least partly above those of the other camps. There are two reasons for this. First, absorptive firms are able to explore external know-how which endogenously rises with heterogeneity in the oligopoly. Thus, they can draw back not only on the intensive opportunities of a single trajectory but also on extensive opportunities resulting from cross-fertilization effects of different technologies. Second, absorptive firms reveal and adopt new technological trajectories more often, which is why they are not confronted with depleted technological chances of a single trajectory as severely as their competitors.

5 CONCLUSIONS

This chapter has presented a theoretical discussion and a simulation analysis of an artificial industry allowing the interplay of endogenous know-how and spillover pools and the absorptive capacity of heterogeneous agents to be investigated. More specifically, the simulation experiments aimed at comparing three different kinds of firm strategies designed to accomplish technological advance: the conservative strategy, the imitative strategy and the absorptive strategy.

The experiments refer to two groups of results. First, looking at a single technological trajectory, where only process innovations are accomplished, the correspondence between high spillover pools and high absorptive capacities shows up clearly. Only when these two conditions fit does the advantage of the absorptive strategy over the other strategies become apparent. Finally, this also translates into comparatively higher economic success, measured by market shares or profits, which then works its way by a success-breeds-success mechanism. A second group of results refers to the appearance of a new technological trajectory and the succession of several trajectories. The success-breeds-success effects are why absorptive firms are generally able to open up a new technological trajectory. This time technological and economic success go hand in hand with some non-negligible time lag. Any survival of firms with other strategies is due to price competition where only the imitative strategies are able to keep pace – however, on a considerably lower economic and technological level. A final interesting result concerns R&D efficiency. Absorptive firms on average spent less R&D to achieve a specific level of innovative success.

Of course, the results found and the characteristic patterns detected can be accepted only on the basis of some qualifications. First, whenever the spillover pool is only of small magnitude, the absorptive strategy will never be the most successful (Cantner and Pyka 1998). Second, firms are designed in such a way that they stick to their strategy independently of success or failure. Pyka (1999) shows in a very long time horizon including entry and exit of firms that strategy switches lead to an almost absolute dominance of absorptive strategies, only sometimes slightly threatened by conservative successes. Third, the success of gradual new technological trajectories also depends on the consumers' assessment of quality. This is modelled in a straightforward and simple way: any strong substitutability or complementarity effects are not dealt with. Fourth, we presented our results as average developments and structures; sensitivity and variance analyses have been performed in order to check robustness. For more details, see Cantner and Pyka (1998).

NOTES

1. See, for example, Winter (1989).
2. The basic reference is Arrow (1962).
3. In the literature, this is also known as the 'go-it-alone strategy'. See Fusfield and Haklish (1985).
4. The conservative strategy, therefore, clearly differs from Freeman' s (1982) 'traditional strategy' where the respective firms do not see any reason to change products because neither demand nor competition compels them to do so.
5. Freeman (1982) alternatively uses the notion 'defensive or dependent strategy'.
6. See Winter (1986).
7. Cohen and Levinthal (1989) state in this respect: 'When a firm wishes to acquire and use new knowledge that is unrelated to its ongoing activity, then the firm must dedicate exclusively to creating "absorptive capacity" (i.e. "absorptive capacity" is not a by-product)' (p. 129).
8. The heterogeneous oligopoly setting is also applied in a simulation study by Meyer et al. (1996).
9. For our model, simulation runs excluding any innovation processes show that starting in a disequilibrium situation it takes about 30 periods to achieve equilibrium.
10. See Witt (1992, p. 42).
11. See Kuenne (1992).
12. See, for example, Nelson and Winter (1982, p. 132).
13. In the model, the firms move in a discrete quality space where introducing a product innovation is reflected in reaching the next higher-quality level.
14. Along a specific trajectory we assume no obsolescence of know-how, which would lead in our case just to a slowdown of the developments observed in the model. Of course, for firms which are engaged only in own R&D, $\sigma_i = 0$.

15. This probability distribution, which in the literature is often called 'the distribution of the low probability for happenings with a low probability' seems to be adequate with respect to product innovations.
16. This reflects the extreme case of a totally embodied innovation. Tushman and Anderson (1986) introduce the notion of 'competence destroying technical change' in this respect.
17. The larger $|Q_{it}|$, the larger is the aggregate quality distance to the competitors' products and, consequently, the lower are the substitutive effects due to h_{it}.
18. With a positive (negative) Q_{it} firm i is on the aggregate in a more leading (lagging) position in the quality space. Correspondingly, consumers' assessment is above (below) the initial value a_0.
19. For a consideration of exit and entry, see Pyka (1999).
20. '[S]uccessful product development requires two types of knowledge. First, it requires component knowledge, or knowledge about each of the core design concepts and the way in which they are implemented in a particular component. Second, it requires architectural knowledge or knowledge about the ways in which the components are integrated and linked together into a coherent whole' (Henderson and Clark 1990, p. 11).
21. See Metcalfe (1994).
22. See Verspagen (1992) and Cantner (1995).
23. Simulation runs containing the imitative camp have shown that in the short run (without product innovation) these firms are not able to keep pace with the technological leaders. In these scenarios imitative firms perform at the last position when compared to the other strategies.

BIBLIOGRAPHY

Arrow, K.J. (1962), 'Economic welfare and the allocation of resources for invention', in R.R. Nelson (ed.), *The Rate and Direction of Inventive Activity*, Princeton, NJ: Princeton University Press, pp. 104–19.

Basalla, G. (1988), *The Evolution of Technology*, Cambridge: Cambridge University Press.

Cantner, U. (1995), 'Technological dynamics in asymmetric industries – R&D, spillovers and absorptive capacity', Institut für Volkswirtschaftslehre der Universität Augsburg, Volkswirtschaftliche Diskussionsreihe 143.

Cantner, U. (1996), 'Heterogenität und Technologische Spillovers: Grundelemente einer ökonomischen Theorie des technologischen Fortschritts', (Heterogeneity and technological spillovers: basic elements of an economic theory of technological progress), Habilitationsschrift, Universität Augsburg.

Cantner, U., H. Hanusch and A. Pyka (1998), 'Routinized innovations: dynamic capabilities in a simulation', in G. Eliasson and C. Green (eds), *Microfoundations of Economic Growth, A Schumpeterian Perspective*, Ann Arbor, MI: Michigan University Press, pp. 131–55.

Cantner, U. and A. Pyka (1998), 'Absorbing technological spillovers, simulations in an evolutionary framework', *Industrial and Corporate Change*, **9**(1), 369–96.

Cohen, W. M. and D.A. Levinthal (1989), 'Innovation and learning: the two faces of R&D', *Economic Journal*, **99**, 569–96.

Cohen, W.M. and D.A. Levinthal (1994), 'Fortune favours the prepared firm', *Management Science*, **40**(2), 227–51.

Coombs, R. (1988), 'Technological opportunities and industrial organization', in Dosi et al. (eds) (1988), pp. 295–97.

Dahmèn, E. (1990), 'Development blocs in industrial development', in B. Carlsson (ed.), *Industrial Dynamics*, Dordrecht: Kluwer Academic, pp. 109–21.

Dasgupta, P. and J.E. Stiglitz (1980), 'Industrial structure and the nature of innovative activity', *Economic Journal*, **90**, 266–93.

Dosi, G., C. Freeman, R. Nelson, G. Silverburg and L. Soete (eds) (1988), *Technical Change and Economic Theory*, London: Pinter.

Dosi, G. and M. Egidi (1991), 'Substantive and procedural rationality', *Journal of Evolutionary Economics*, **1**, 145–68.

Dosi, G., D.J. Teece and S. Winter (1992), 'Towards a theory of corporate competence: preliminary remarks', in G. Dosi, R. Giannetti and P.A. Toninelli (eds), *Technology and Enterprise in a Historical Perspective*, Oxford: Clarendon, pp. 1–27.

Freeman, C. (1982), *The Economics of Industrial Innovation*, 2nd edn, London: Pinter.

Fusfield, H.I. and C.S. Haklish (1985), 'Cooperative R&D for competitors', *Harvard Business Review*, **6**, 66–76.

Heiner, R. (1988), 'Imperfect decisions and routinized production: implications for evolutionary modelling and inertial technical change', in Dosi et al. (eds) (1988), pp. 147–69.

Henderson, R. and K. Clark (1990), 'Architectual innovation: the reconfiguration of existing product technologies and the failure of established firms', *Administrative Science Quarterly*, **35**, 9–30.

Kodama, F. (1986), 'Technology fusion and the new R&D', *Harvard Business Review*, July–August, 70–78.

Kuenne, R.E. (1992), *The Economics of Oligopolistic Competition*, Cambridge, MA: Blackwell.

Levinthal, D.A. and L.G. March (1981), 'A model of adaptive organizational change', *Journal of Economic Behaviour and Organization*, **2**, 307–33.

March, J.G. (1991), 'Exploration and exploitation in organizational learning', *Organization Science*, **2**(1), 71–87.

Metcalfe, J.S. (1994), 'Competition, Fisher's Principle and increasing returns to selection', *Journal of Evolutionary Economics*, **4**, 327–46.

Meyer, B., C. Vogt and R. Vosskamp (1996), 'Schumpeterian competition in heterogeneous oligopolies', *Journal of Evolutionary Economics*, **6**, 411–24.

Mokyr, J. (1990), *The Lever of Riches: Technolological Creativity and Economic Progress*, Oxford: Oxford University Press.

Nelson, R.R. (1990), 'What is public and what is private about technology?', CCC (Copyright Clearance Center) Working Paper 90-9, University of California at Berkeley.

Nelson, R.R. and S. Winter (1982), *An Evolutionary Theory of Economic Change*, Cambridge, MA: Harvard University Press.

Pelikan, P. (1988), 'Can the innovation system of capitalism be outperformed', in Dosi et al. (1988), pp. 370–88.

Pyka, A. (1997), 'Informal networks', *Technovation*, **17**, 207–20.

Pyka, A. (1999), *Der kollektive Innovationsprozeß: Eine theoretische Analyse informeller Netzwerke und absorptiver Fähigkeiten* (*Collective Invention: A Theoretical Analysis of Informal Networks and Absorptive Capacities*), Berlin: Duncker & Humblot.

Reinganum, J.E. (1985), 'Innovation and industry evolution', *Quarterly Journal of Economics*, **100**, 81–99.

Sahal, D. (1981), *Patterns of Technological Innovation*, Reading, MA: Addison-Wesley.

Schumpeter, J.A. (1942), *Capitalism, Socialism and Democracy*, 5th edn, London: Allen & Unwin.

Silverberg, G. and B. Verspagen (1994), 'Collective learning, innovation and growth in a boundedly rational, evolutionary world', *Journal of Evolutionary Economics*, **4**, 207–26.

Simon, H.A. (1976), 'From substantive to procedural rationality', in S.J. Latsis (ed.), *Method and Appraisal in Economics*, Cambridge and London: Cambridge University Press, pp. 129–48.

Tushman, M.L. and P. Anderson (1986), 'Technological discontinuities and organizational environments', *Administrative Science Quarterly*, **31**, 439–65.

Verspagen, B. (1992), *Uneven Growth between Interdependent Economies*, Maastricht (Aldershot: Avebury, 1993).

Winter, S.G. (1971), 'Satisficing, selection, and the innovating remnant', *Quarterly Journal of Economics*, **85**(2), 237–61.

Winter, S.G. (1984), 'Schumpeterian competition in alternative technological regimes', *Journal of Economic Behavior and Organization*, **5**, 287–320.

Winter, S.G. (1986), 'Schumpeterian competition in alternative technological regimes', in R. Day and G. Eliasson (eds), *The Dynamics of Market Economics*, Amsterdam: Elsevier, North-Holland, pp. 199–232.

Winter, S.G. (1989), 'Patents in complex contents: incentives and effectiveness', in V. Weil and R. Snapper (eds), *Owning Scientific and Technical Information*, New Brunswick, NJ: Rutgers University Press, pp. 41–59.

Witt, U. (1992), 'Überlegungen zum gegenwärtigen Stand der evolutorischen Ökonomik' (Considerations of the present state of evolutionary economics), in B. Bievert (ed.), *Evolutorische Ökonomik: Neuerungen, Normen und Institutionen* (*Evolutionary Economics; Novelties, Norms and Institutions*), Frankfurt am Main: Campus Verlag, pp. 23–55.

2. Monopoly and perfect competition: there are two sides to every coin

Witold Kwasnicki

1 INTRODUCTION

It is a well-documented fact that the profit gained by firms and market prices are related to the market structure. It is generally acknowledged that a greater concentration of the market causes higher prices and more profit for firms. A pure competition case relates to the situation of a large number of firms competing on the finite market; in the course of market development the possibilities of making a positive profit disappear and at the equilibrium state the profit of competitive firms is equal to zero, the only profit made by firms being the 'normal' profit embedded into the supply function in the form of the opportunity costs. Monopoly is the other side of a market spectrum. A monopolist feels secure on the market and tries to increase prices to a relatively high level in order to make substantial profits. Some theoretical considerations based on the evolutionary model of industrial dynamics suggest that this is true in most situations but there are some specific situations (industrial regimes) which cause a monopolist to behave similarly to the firms competing on the pure market and vice versa; there are situations which allow competitive firms to apply the monopolists' strategy and thereby make huge profits.

In Section 2 of the chapter a description of an evolutionary model of industrial dynamics is presented. Sections 3 and 4 contain results of a simulation study of the model. The model's analysis presented in Section 3 is focused on equilibrium analysis. Behaviour of the model for a wide spectrum of the model's parameters is investigated. Study dynamics of industrial development related to the main subject of the chapter is presented in Section 4. In both these sections it is shown how different modes of industrial development emerge for different industry concentrations and different values of the unit cost of production and the productivity of capital. Section 5 concludes.

2 THE EVOLUTIONARY MODEL OF INDUSTRIAL DYNAMICS

The model is described in detail in (Kwasnicki and Kwasnicka 1992, 1994; Kwasnicki 1996a). Because of space limitations, the presentation of the model here will be confined to a general description without going into the mathematical details. The model describes the behaviour of a number of competing firms producing functionally equivalent products. The decisions of a firm relating to investment, price, profit and so on are based on the firm's evaluation of the behaviour of other, competing firms, and the expected response of the market. The firm's knowledge of the market and knowledge of the future behaviour of competitors is limited and uncertain. Firms' decisions can thus only be suboptimal. The decisions are taken simultaneously and independently by all firms at the beginning of each period (for example, annually or quarterly). After the decisions are made, the firms undertake production and put the products on the market. The products are evaluated by the market, and the quantities of different firms' products sold in the market depend on the relative prices, the relative value of the products' characteristics and the level of saturation of the market. In the long run, a preference for better products, that is, those with a lower price and better characteristics, prevails.

Each firm tries to improve its position in the industry and in the market by introducing innovations in order to minimize the unit costs of production, maximize the productivity of capital, and maximize the competitiveness of its products on the market. The general structure of the model is presented in Figure 2.1. The product's price depends on the current technology of the firm, on market structure and on the assumed level of production to be sold on the market. The two arrows between 'price' and 'production' indicate that the price is established in an interactive way to fulfil the firm's objectives (that is, to keep relatively high profits in the near future and to ensure further development in the long run). Modernization of products through innovation and/or initiating new products by applying radical innovation depends on the investment capacity of the firm. Thus, in managing innovation, each firm takes into account all economic constraints, as they emerge during the firm's development. It thus frequently occurs that economic constraints can prevent a promising invention from being put into production.

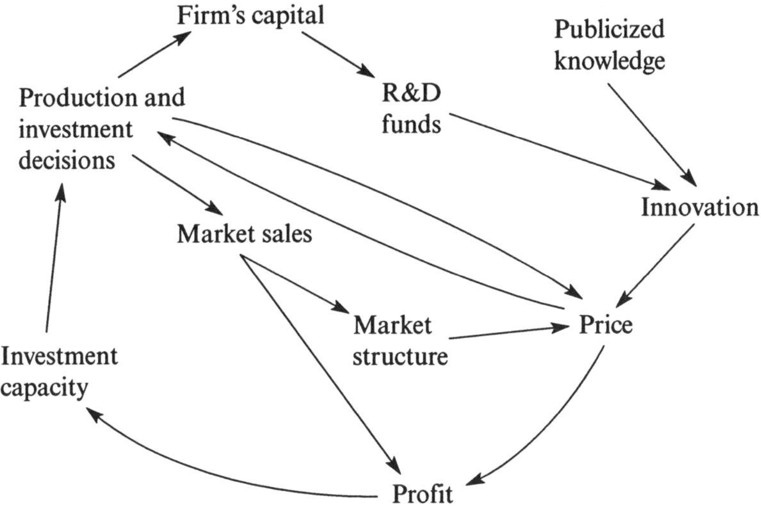

Figure 2.1 General structure of the evolutionary industrial model

One of the distinguished features of the model is the coupling of technological development and economic processes. Current investment capacity is taken into account by each firm in the decision-making process. The success of each firm in the search for innovation depends not only on the amount of R&D funds spent by each firm in the search for innovation, but also on the extent to which firms make private knowledge public. Making the private knowledge of a firm public can in some cases speed up industrial development, but also diminishes a firm's incentives to spend more funds on R&D projects. We may therefore expect only a certain part of private knowledge to be made public.

Firms' investment capacity depends on firms' savings and available credits, and also, indirectly, on the firm's debt. Production and investment decisions are based on the firm's expectations about the future behaviour of its competitors, its market structure, its expected profit and the past trend of the firm's market share. Current technical and economic characteristics of products offered for sale and the technology used to manufacture the products are taken into account in price-setting decisions, investment and production. Because of the inevitable discrepancies between a firm's expectation and the real behaviour of the market, the firm's production offered for sale on the market is different from market demand (it can be either smaller or larger than demand).

We distinguish between invention (that is, a new item being considered for production) and innovation (an invention introduced into the production process). There are two ways in which firms search for inventions: autonomous, in-house research, and imitation of competitors. Public knowledge allows not only for imitation of competitors, but may also concern the research process (in Figure 2.1, the arrow from publicized knowledge to innovation indicates this influence). From all inventions only a small fraction are actually developed. An innovation may modernize current production but it can also initiate new, radical ways of production, that is, by introducing essentially new technology. In general, each innovation may reduce unit costs, increase the productivity of capital, and improve product performance. However, it frequently happens that an improvement of one factor is accompanied by a deterioration of the other two. Firms therefore face the problem of balancing the positive and negative factors of each invention. An invention will only become an innovation if the positive factors prevail.

In the model, each firm may simultaneously produce products with different prices and different values of the characteristics, that is, the firm may he a multi-unit operation. Different units of the same firm manufacture products by employing different sets of routines. Multi-unit firms exist because of searching activity. New technical or organizational solutions (that is, a new set of routines) may be much better than the ones in use but immediate full modernization of production is not possible because of investment constraints on the firm. In such situations the firm continues production using the old routines and tries to open a new unit where production applying the new set is started on a smaller scale. Subsequently, old production techniques may be phased out gradually.

Simulation of industry development is done in discrete time in four steps:

1. *Search for innovation* – search for new sets of routines which potentially may replace the old set currently employed by a firm.
2. *Firms' decision-making process* – calculation and comparison of investment, production, net income, profit, and some other characteristics of development which may be attained by employing the old and the new sets of routines. Decisions of each firm on: (a) continuation of production by employing old routines or modernizing production, and (b) opening (or not) of new units.
3. *Entry of new firms*.
4. *Selling process* – market evaluation of the offered pool of products; calculation of firms' characteristics: production sold, shares in global production and global sales, total profits, profit rates, research funds and so on.

2.1 The Search for Innovation

The creative process is evolutionary by nature, and as such its description should be based on a proper understanding of the hereditary information (see Kwasnicki 1996a, ch. 2). According to the tradition established by Schumpeter (1934, 1950), and Nelson and Winter (1982), we use the term 'routine' to name the basic unit of the hereditary information of a firm. The set of routines applied by the firm is one of the basic characteristics describing it. In order to improve its position in the industry and in the market, each firm searches for new routines and new combinations of routines to reduce the unit costs of production, increase the productivity of capital, and improve the competitiveness of its products in the market. Nelson and Winter (ibid. p. 14) define routines as 'regular and predictable behavioral patterns of firms' and include in this term such characteristics as 'technical routines for producing things . . . procedures of hiring and firing, ordering new inventory, stepping up production of items in high demand, policies regarding investment, research and development, advertising, business strategies about product diversification and overseas investment'. A large part of research activity is also governed by routines. 'Routines govern choices as well as describe methods, and reflect the facts of management practice and organizational sociology as well as those of technology' (Winter 1984, p. 292).

Productivity of capital, unit costs of production, and characteristics of products manufactured by a firm depend on the routines employed by the firm (examples of the product characteristics are reliability, convenience, lifetime, safety of use, cost of use, quality and aesthetic value). The search activities of firms 'involve the manipulation and recombination of the actual technological and organizational ideas and skills associated with a particular economic context' (ibid., p. 292), while the market decisions depend on the product characteristics and prices. We may speak about the existence of two spaces: the space of routines and the space of product characteristics.[1]

We assume that at time *t* a firm is characterized by a set of routines actually employed by the firm. There are two types of routines: *active*, that is, routines employed by this firm in its everyday practice, and *latent*, that is, routines which are stored by a firm but not actually applied. Latent routines may be included in the active set of routines at a future time. The set of routines is divided into separate subsets, called segments, consisting of similar routines employed by the firm in different domains of the firm's activity. Examples are segments relating to productive activity, managerial and organizational activity, marketing and so on. In each segment, either active or latent routines may exist. The set of routines employed by a firm may evolve. There are four basic mechanisms for generating new sets of routines, namely: *mutation*, *recombination*, *transition* and *transposition*.

The probability of discovering a new routine (mutation) depends on the research funds allocated by the firm for autonomous research, that is, in-house development. It is assumed that routines mutate independently of each other. The scope of mutation also depends on funds allocated for in-house development.

The firm may also allocate some funds for gaining knowledge from other competing firms and try to imitate some routines employed by competitors (recombination). It is assumed that recombination may occur only between segments, not between individual routines, that is, a firm may gain knowledge about the whole domain of activity of another firm, for example, by licensing. A single routine may be transmitted (transition, see Figure 2.2) with some probability from firm to firm. It is assumed that after transition a routine belongs to the subset of latent routines. At any time a random transposition of a latent routine to the subset of active routines may occur (Figure 2.3). It is assumed that the probabilities of transition of a routine from one firm to another and the probabilities of transposition of a routine (from a latent to an active routine) are independent of R&D funds, and have the same constant value for all routines.

In general, the probability of transposition of a routine for any firm is rather small. But randomly, from time to time, the value of this probability may increase abruptly and very active processes of search for a new combination of routines are observed. This phenomenon is called 'recrudescence'. Recrudescence is viewed as an intrinsic ability of a firm's research staff to search for original, radical innovations by employing daring, sometimes apparently insane, ideas. This ability is connected mainly with the personalities of the researchers and random factors play

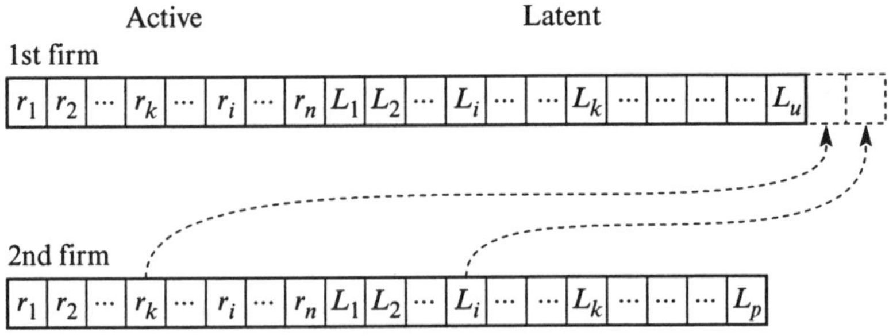

Figure 2.2 Routines transition

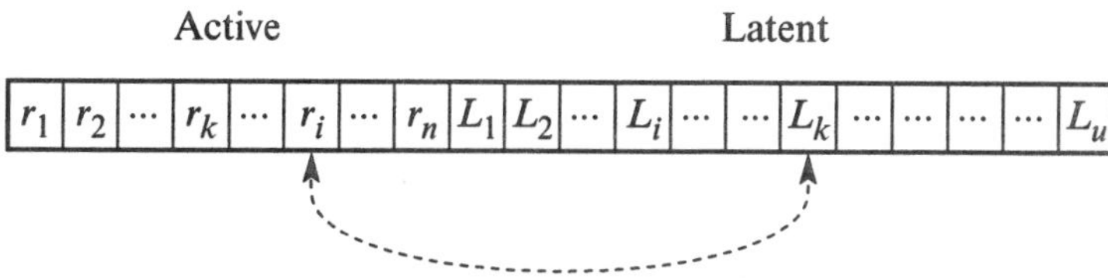

Figure 2.3 Routines transposition

an essential role in the search for innovations by recrudescence, so that its probability is not related to R&D funds allocated by a firm to 'normal' research. It is assumed that recrudescence is more probable in small firms than in large ones which spend huge amounts on R&D, although it is possible to assume that the probability of recrudescence does not depend on firm size.

As a rule, mutation, recombination and transposition on a normal level (that is, with low probabilities in long periods) are responsible for small improvements and, during the short periods of recrudescence, for the emergence of radical innovations.

2.2 Firms' Decisions

It seems that one of the crucial problems of contemporary economics is to understand the process of decision making. Herbert Simon states that 'the dynamics of the economic system depends critically on just how economic agents go about making their decisions, and no way has been found for discovering how they do this that avoids direct inquiry and observations of the process' (Simon 1986, p. 38).

The background of the decision-making procedure adopted in the model is presented in detail in Kwasnicki (1996a). It is assumed that each firm predicts future developments of the market (in terms of future average price and future average product competitiveness), and on the basis of its expectations about future market development and expected decisions of its competitors, each firm decides on the price of its products, investment and the amount of production which it expects to sell on the market. Current investment capability and the possibility of borrowing are also considered by each firm.

The decision-making procedure allows diversified situations faced by different firms to be modelled, for example, the power of a small firm to influence the average price is much smaller than that of a large firm. So, small firms are, in general, 'price takers' in the sense that they assume that the future average price will be very close to the trend value, while large firms generally play the role of 'price leaders' or 'price makers'.

Price, production and investment are set by a firm in such a way that some objective function is maximized. Contrary to the neoclassical assumption, it is not a maximization in the strict sense. The estimation of values of the objective function is not perfect and is made for the next year only. In other words, it is not a global, once-and-for-all, optimization, but rather an iterative process with different adjustments taking place from year to year.

Different price-setting procedures (based on different objective functions and the mark-up rules) have been scrutinized, the results of which are presented in Kwasnicki and Kwasnicka (1992), and Kwasnicki (1996a). In many simulation experiments, firms were allowed to select different price-setting procedures. The results of these experiments suggest that firms applying the objective O_1 function (presented below) dominate on the market and in the long run supersede all others. This objective function has the following form:

$$O_1(t+1) = (1 - F_i)\frac{\Gamma_i(t+1)}{\Gamma(t)} + F_i\frac{Q_i^s(t+1)}{QS(t)}\,,$$

$$F_i = a_4 \exp\left[-a_5\frac{Q_i^s(t+1)}{QS(t)}\right], \qquad (2.1)$$

where F_i is the magnitude coefficient (with values between 0 and 1), Q_i^s the supply of firm i, Γ_i the expected income of firm i at $t+1$ (defined by equation (2.2) below), QS is the global production of the industry in year t and Γ the global net income of all firms in year t. $\Gamma(t)$ and $QS(t)$ play the role of constants in equation (2.1) and ensure that the values of both terms in this equation are of the same order.

The expected income of firm i (Γ_i) and the expected profit of this firm (Π_i) are defined as

$$\Gamma_i = Q_i^s(t)\,\{p_i(t) - Vv\,[Q_i^s(t)] - \eta\}, \qquad (2.2)$$

$$\Pi_i = \Gamma_i - K_i(t)\,(\rho + \delta), \qquad (2.3)$$

where V is unit production costs, $v(Q_i^s)$ is the factor of unit production cost as a function of the scale of production (economies of scale), η is the constant production cost, $K_i(t)$ the capital needed to obtain the output $Q_i^s(t)$, ρ the normal rate of return and δ the physical capital depreciation rate (amortization).

The function O_1 expresses short- and long-term thinking of firms during the decision-making process (the first and second terms in equation (2.1), respectively). Plausible values for the parameters are $a_4 = 1$ and $a_5 = 5$, implying that the long run is much more important for survival and that firms apply a flexible strategy, that is, the relative importance of short- and long-term components changes in the course of a firm's development (the long-term one is much more important for small firms than for the large ones).

The decision-making procedure presented above, with the search for the 'optimal' price-setting procedure based on the objective function concept, constructs a formal scheme for finding the proper value of the price and expected production to be sold on the market. Naturally this scheme is only an approximation of what is done by real decision makers. They, of course, do not make such calculations and formal optimization from year to year; rather, they think routinely: 'My decisions should provide for the future prospects of the firm and also should allow income (or profit) to be maintained at some relatively high level'. Decisions on the future level of production and the future product price depend on the actual investment capabilities of the firm.

2.3 Entry

In each period $(t, t + 1)$ a number of firms try to enter the market. Each entrant enters the market with assumed capital equal to *InitCapital* and with the initial price of its products equal to the predicted average price. The larger the concentration of the industry, the greater the number of potential entrants (that is, firms trying to enter the market). The value of *InitCapital* is selected in such a way that the initial share of an entrant is not larger than 0.5 per cent.

In general, any firm may enter the market, and if a firm's characteristics are unsatisfactory then it is quickly eliminated from (superseded in) the market. But because of the limited capacity of computer memory for simulations, a threshold for potential entrants is assumed. It is assumed that a firm enters the market only if the estimated value of objective 0_1 of that firm is greater than an estimated average value of the objective 0_1 in the industry. It may be expected that a similar (rational) threshold exists in real industrial processes.

2.4 Product Competitiveness on the Market

The productivity of capital, variable costs of production and product characteristics are the functions of routines employed by a firm (see

Figure 2.4). Each routine has multiple, pleiotropic effects, that is, it may affect many characteristics of products, as well as productivity, and the variable costs of production. Similarly, the productivity of capital, unit costs of production and each characteristic of the product can be a function of a number of routines (polygeneity). We assume that the transformation of the set of routines into the set of product characteristics is described by m functions F_d,

$$z_d = F_d(r), \quad d = 1, 2, 3, \dots, m, \tag{2.4}$$

where z_d is the value of characteristic d, m the number of product characteristics, and r the set of routines. It is assumed also that the productivity of capital $A(r)$ and the unit cost of production $V(r)$ are also functions of a firm's routines, where these functions are not firm specific and have the same form for all firms.

Attractiveness of the product on the market depends on the values of the product characteristics and its price. The competitiveness of products with characteristics z and price p is equal to

$$c(p, z) = \frac{q(z)}{p^{\alpha}}, \qquad z = (z_1, z_2, z_3, \dots, z_m), \tag{2.5}$$

where $q(z)$ is the technical competitiveness, z a vector of product characteristics, and α price elasticity.

In the presence of innovation, technical competitiveness varies according to the modification of routines made by each firm, or because of introducing essentially new routines. Technical competitiveness is an explicit function of product characteristics. As explained above, each routine does not influence the product's performance directly, but only indirectly through the influence on its characteristics. We assume the existence of a function q enabling calculation of technical competitiveness of products manufactured by different firms. We say that q describes the adaptive landscape in the space of product characteristics. In general, this function depends also on some external factors, varies in time, and is the result of co-evolution of many related industries. The shape of the adaptive landscape is dynamic, with many adaptive peaks of varying altitudes. In the course of time some adaptive peaks lose their relative importance, others become higher.

Because of the ongoing search process, at any moment each firm may find a number of alternative sets of routines. Let us denote by r the set of routines actually applied by a firm and by r^* an alternative set of rou-

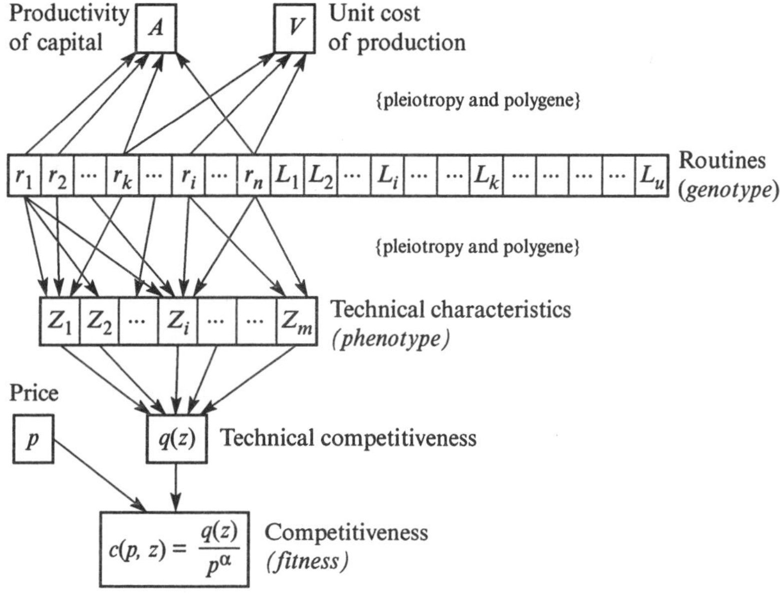

Figure 2.4 From routines to competitiveness, productivity of capital and unit cost of production

tines. Each firm evaluates all potential sets of routines r^* as well as the old routines r by applying the decision-making procedure outlined in section 2.2. For each alternative set of routines the price, production, investment (including the modernization investment) and value of the objective function are calculated. The decision of firm i on modernization (that is, replacing the r routines by r^* routines) depends on the expected value of the firm's objective function and its investment capability. Modernization is undertaken if the maximum value of the objective function from all considered alternative sets of routines r^* is greater than the value of the objective function possible by continuing the actually applied routines r, and if the investment capability of the firm permits such modernization. If the investment capability does not allow modernization, then the firm:

1. continues production employing the 'old' routines r, and
2. tries to open a new small unit where routines r^* are employed; production is started with an assumed value of capital equal to *InitCapital*.

To modernize production, extra investment is necessary. This 'modernization investment' depends on the discrepancy between the 'old' routines r and the 'new' routines r^*. For simplicity, it is assumed that modernization investment IM is a non-decreasing function of distance between the old routines r actually applied by a firm and the new set of routines r^*.

All products manufactured by the entrants and the firms existing in the previous period are put on the market and all other decisions are left to buyers; these decisions primarily depend on the relative values of competitiveness of all products offered, but quantities of products of each firm offered for sale are also taken into account. It is assumed that global demand $Q^d(t)$ for products potentially sold on a market is equal to an amount of money, $M(t)$, which the market is inclined to spend on buying products offered for sale by the firms divided by the average price, $p(t)$, of the products offered by these firms,

$$Q^d(t) = \frac{M(t)}{P(t)} . \tag{2.6}$$

$M(t)$ is assumed to be equal to

$$M(t) = N\exp(\gamma t)\,[p(t)]^{\beta} \tag{2.7}$$

where N is a parameter characterizing the initial market size, γ the growth rate of the market, and β the (average) price elasticity. The average price of all products offered for sale on the market is equal to

$$p(t) = \sum_i p_i(t)\,\frac{Q_i^s(t)}{Q^s(t)} , \tag{2.8}$$

where $Q^s(t)$ is global supply and is equal to

$$Q^s(t) = \sum_i Q_i^s(t). \tag{2.9}$$

Global production sold on the market is equal to the smaller value of demand $Q^d(t)$ and supply $Q^s(t)$,

$$QS(t) = \min\{Q^d(t), Q^s(t)\}. \tag{2.10}$$

The selection equation describing competition among firms (products) in the market has the following form (f_i is the market share of products manufactured by firm i):

$$f_i(t) = f_i(t-1)\,\frac{c_i(t)}{c(t)}\,, \tag{2.11}$$

where $c(t)$ is the average competitiveness of products offered for sale,

$$c(t) = \sum_i f_i(t-1)\, c_i(t). \tag{2.12}$$

This means that the share (f_i) of firm *i* in global output increases if the competitiveness of its products is higher than the average of all products present on the market, and decreases if the competitiveness is lower than the average. The rate of change is proportional to the difference between the competitiveness of products of firm *i* and average competitiveness.

Finally, the quantity of products potentially sold by firm *i* (that is, the demand for products of firm *i*) is equal to

$$Q_i^d(t) = QS(t) f_i(t). \tag{2.13}$$

The above equations are valid if the production offered by the firms exactly fits the demand of the market. This is a very rare situation and therefore these equations have to be adjusted to states of discrepancy between global demand and global production, and discrepancy between the demand for products of a specific firm and the production offered by this firm. The details of this adjustment process is presented in Kwasnicki (1996a). Equations (2.11 and 2.13) describe the market demand for products of firm *i* offered at a price $p_i(t)$ and with competitiveness $c_i(t)$. In general, however, the supply of firm *i* is different from the specific demand for its products. The realization of the demand for products of firm *i* does not depend only on these two values of demand and supply, but on the whole pool of products offered for sale on the market. The alignment of supply and demand of all firms present on the market is an adaptive process performed in a highly iterative and interactive mode between sellers and buyers. In our model, we simulate the iterative alignment of supply and demand in a two-stage process in which a part of the demand is fulfilled in the first stage, and the rest of the demand is, if possible, fulfilled in the second stage. If there is no global oversupply of production, then in the first stage of the supply–demand alignment process all demand for production of specific firms, wherever possible, is fulfilled, but there is still the shortfall in production of firms which underestimated demand for their products. This part of demand is fulfilled in the second stage of the supply–demand alignment process. At this stage, the products of the firms which produce more than the specific demand are sold to replace the shortfall in production by the firms which underestimated the demand for their products.

The supply–demand alignment process is slightly different if a global oversupply of production occurs. It seems reasonable to assume that in such a case the production of each firm sold on the market is divided into (i) the production bought as the outcome of the competitive process (as described by equations (2.11) and (2.13)), and (ii) the production bought as the outcome of a non-competitive process. The latter part of production does not depend directly on product competitiveness but depends primarily on the volume of production offered for sale, that is, random factors play a much more important role in the choice of relevant products to be bought within this part of the production. In general, the division of production of each firm into these two parts depends on the value of global oversupply. The higher the oversupply, the larger is the part of production of each firm which is sold on the basis of non-competitive preferences.

Usually global oversupply, if it occurs, is small, so the major part of production is distributed under the influence of competitive mechanisms and only a small part is distributed as a result of non-competitive distribution. But to clarify the necessity of distinguishing the two proposed stages of the selling–buying process, let us consider the following, albeit artificial, situation. Except for one firm, the production of all other firms exactly meets the demand for their products. The atypical firm produces much more than the demand for its products. It could be assumed that the production sold by all firms is exactly equal to the specific demands for their products, which is equivalent to the assumption that the volume of overproduction of the atypical firm does not influence the behaviour of the market. In an extreme case, we may imagine that the volume of production of the atypical firm is infinite and the rest of the firms continue to produce exactly what is demanded. Does this mean that the excessive production would go unnoticed by the buyers and that they would remain loyal to firms producing exactly what is demanded? A more adequate description would require the incorporation of the assumption that the future distribution of products sold on the market depends on the level of overproduction of all firms, and particularly the level of overproduction of the atypical firm. In the case of the overproduction of one firm its share in the global production sold will increase at the expense of all firms producing exactly what is demanded. In the extreme case, when overproduction of the atypical firm tends to infinity, the only products sold on the market belong to that firm, and the shares of all other firms will be zero. But it does not mean that producing more than is demanded is an advantageous strategy for the firm and that it is an effective weapon to eliminate the competitors. In fact, the bulk of overproduction is not sold on the market and is lost by the firm. In effect, the atypical firm's profit is much smaller than expected, or even may be negative. After some time the firm's development stops and in the end the firm will be eliminated from the market.

3 SIMULATION – EQUILIBRIUM ANALYSIS

One of the general validity tests of any model, used especially at the initial phases of a model's development, is to check whether some characteristics (for example, rates of changes of some important variables) of the model's behaviour do not depend on used units of measurement. While testing the validity of our model from this point of view we have observed that the development of the model is exactly the same for the same value of a factor equal to the multiplication of the productivity of capital and the unit cost of production, that is, the AV factor. The preliminary observations on this phenomenon are presented in Kwasnicki (1996a, pp. 122–7). In this chapter I should like to investigate this phenomenon more closely to build a platform for further, envisaged, empirical research of selected industries.

By definition, productivity of capital A is equal to quantity of production Q divided by the capital used in production process K, and unit cost of production is equal to total cost of production C divided by the production volume Q. Therefore, the AV factor is equal to the total cost of production to capital, that is, C/K. Let us call this factor the 'cost ratio'.

One of the aims of introducing innovations is increasing the productivity of capital and reducing the unit cost of production. Therefore, introduced innovations also change the value of the cost ratio. It can either increase (if improvement of productivity of capital is greater than cost reduction), decrease (cost reduction is greater than productivity improvement), or become constant (if ratios of changes of productivity of capital and cost reduction are exactly the same). There is some empirical evidence that in the last one hundred years the productivity of capital has remained almost at the same level but the unit cost of production of most industries has been reduced significantly. Also, simulation results focused on a search for so-called 'innovation regimes' (see Kwasnicki 1996b) suggest that innovation allowing for reduction of the unit cost of production is more eagerly accepted by firms than innovations leading to an increase in the productivity of capital. The view that in the course of modern industry development the cost ratio is significantly reduced seems to be justified. What, then, is the significance of AV reduction on the modes of development of modern industries?

Of course, the economic process is extremely dynamic, with a large number of emerging innovations and bounded rational decisions of managers (related to prices, investments and level of production) influencing the development of any industry. Because of the dynamic nature of the economic process, reaching equilibrium is very difficult, and the vision of trying to attain equilibrium seems to be more likely in real industrial development. Using our simulation model we can investigate the dynamics of industrial processes but nevertheless it is difficult to arrive at general conclusions concerning industrial development looking only at the dynamic trajectories of the development of different industries. An attempt to investigate the dynamic properties of the model in the presence of innovation and economy of scale is presented in the next section. A comparison of the simulation results presented in both sections allows us to see the differences between the static and dynamic views of the economic process.

For a general view of industrial development let us start by investigating the equilibrium properties of the model, keeping in mind that this is a dynamic process. Making a number of simulations for different values of the cost factor and also for other parameters of the model (for example, the number of firms and their size, that is industry concentration) we are able to generate a 'surface' describing the equilibria of industry development for different values of the cost ratio and industry concentration. As has been said, industry development can be seen as the pursuit of equilibrium. If we assume that in the course of development the cost ratio is reduced and also the concentration of industry changes, then, as a first approximation, trajectories of industry development can be seen as lines on the previously generated surface of equilibrium values.

A large number of simulation experiments were carried out for different values of the cost ratio (A/V), industry concentration (that is, the number of competing firms n), rate of industry growth (γ), normal rate of return (ρ), price elasticity in the demand function (β), and many others. Equilibrium values of one such series of simulation experiments are presented in Tables 2.1 and 2.2 (in some of these experiments, industry does not reach stable equilibrium but fluctuates around the steady state – these are shown in bold in the tables). It was assumed that there is no innovation. Some parameters are constant in these experiments (for example, $\beta = -0.3$, $\gamma = 0$, $\rho = 0.05$). The simulation approach allows for a relatively large spectrum of change of the cost ratios (from 0.01 to 100) and also for a wide spectrum of change of industry concentration (from pure monopoly, $n = 1$, to pure competition, $n = 24$). Looking at the results presented in Table 2.1 we can see that there are two zones – the first where equilibrium profit is equal to zero, and the second where there is positive profit. The surface of equilibrium profit to capital ratio is presented in Figure 2.5,

Table 2.1 Profit to capital for different values of the cost ratio and the industry concentration

	Number of firms (*n*)							
AV	1	2	4	8	12	16	20	24
0.1	0.00	0.00	0.00	0.00	0.00	0.00	0.00	0.00
0.02	0.00	0.00	0.00	0.00	0.00	0.00	0.00	0.00
0.05	0.00	0.00	0.00	0.00	0.00	0.00	0.00	0.00
0.075	4.29	0.00	0.00	0.00	0.00	0.00	0.00	0.00
0.1	12.38	0.00	0.00	0.00	0.00	0.00	0.00	0.00
0.2	44.76	5.08	0.00	0.00	0.00	0.00	0.00	0.00
0.5	141.10	42.69	12.10	0.00	0.00	0.00	0.00	0.00
0.75	228.86	74.04	28.14	4.75	0.00	0.00	0.00	0.00
1	303.80	105.38	44.19	12.10	1.28	0.00	0.00	0.00
2	627.60	230.77	108.38	44.19	22.57	11.08	5.27	1.73
5	1 599.00	606.90	300.96	140.50	86.43	59.50	**39.29**	**78.57**
7.5	2 409.00	920.40	461.44	220.75	139.64	99.26	**26.25**	
10	3 218.00	1 233.80	621.92	301.00	192.85	139.01	**43.43**	
20	6 456.00	488.00	1263.80	622.00	405.70	298.02	**112.00**	
50	16 170.00	6 249.20	3 189.60	1 585.00	1 044.30	775.04		
75	24 266.00	9 383.80	4 794.40	2 385.50	1 576.40	1 172.60		
100	32 361.00	12 151.80	6 399.20	3 190.00	2 108.60	1 570.10		

Table 2.2 Price to unit cost for different values of the cost ratio and the industry concentration

	Number of firms (*n*)							
AV	1	2	4	8	12	16	20	24
0.01	21.00	21.00	21.00	21.00	21.00	21.00	21.00	21.00
0.02	11.00	11.00	11.00	11.00	11.00	11.00	11.00	11.00
0.05	5.00	5.00	5.00	5.00	5.00	5.00	5.00	5.00
0.075	4.24	3.67	3.67	3.67	3.67	3.67	3.67	3.67
0.1	4.24	3.00	3.00	3.00	3.00	3.00	3.00	3.00
0.2	4.24	2.25	2.00	2.00	2.00	2.00	2.00	2.00
0.5	4.24	2.25	1.64	1.40	1.40	1.40	1.40	1.40
0.75	4.24	2.25	1.64	1.32	1.27	1.27	1.27	1.27
1	4.24	2.25	1.64	1.32	1.21	1.20	1.20	1.20
2	4.24	2.25	1.64	1.32	1.21	1.16	1.13	1.11
5	4.24	2.25	1.64	1.32	1.21	1.16	**1.13**	**1.11**
7.5	4.24	2.25	1.64	1.32	1.21	1.16	**1.09**	
10	4.24	2.25	1.64	1.32	1.21	1.16	**1.09**	
20	4.24	2.25	1.64	1.32	1.21	1.16	**1.09**	
50	4.24	2.25	1.64	1.32	1.21	1.16		
75	4.24	2.25	1.64	1.32	1.21	1.16		
100	4.24	2.25	1.64	1.32	1.21	1.16		

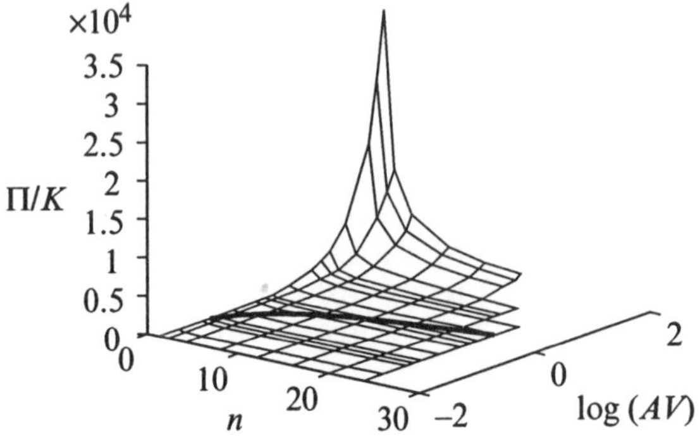

Figure 2.5 Profit as a function of the cost ratio and the industry concentration

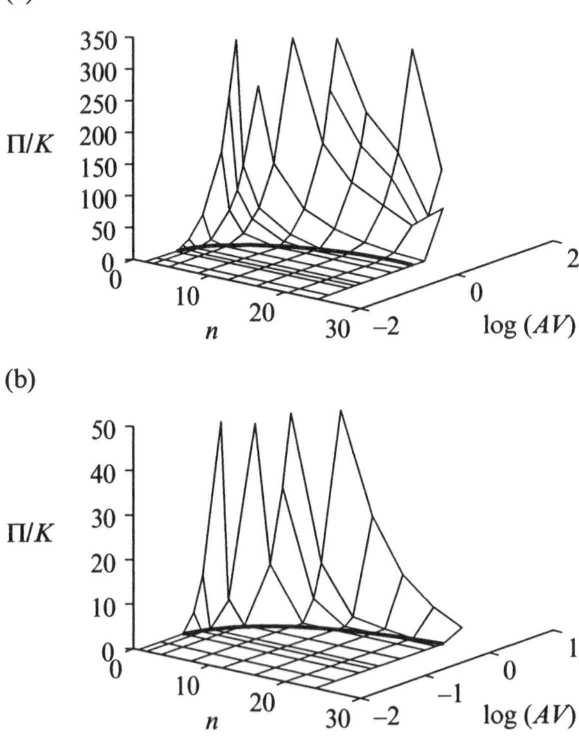

Figure 2.6 Profit as a function of the cost ratio and the industry concentration (enlargements)

which is a graphical presentation of the results in Table 2.1. As we can see, values of the profit to capital ratio for some simulations are very large; it is unlikely that such high values would be observed in actual industrial processes but one advantage of the simulation approach is that we are able to create some extreme conditions and observe the hypothetical development of the system under those extreme conditions. We present all simulation results in order to give a general (although in some cases purely theoretical) view of industrial development. The border separating the two zones is drawn in Figure 2.5 as a solid, thick line. In Figure 2.6 the same surface is enlarged so that more details can be observed around the border. To see the rate of growth of the profit, the surface presented in Figures 2.5 and 2.6 can be 'sliced' for constant values of the cost ratio or constant values of the industry concentration. The relevant functions are presented in Figures 2.7 and 2.8 (Figures 2.7b and 2.8b correspond to the

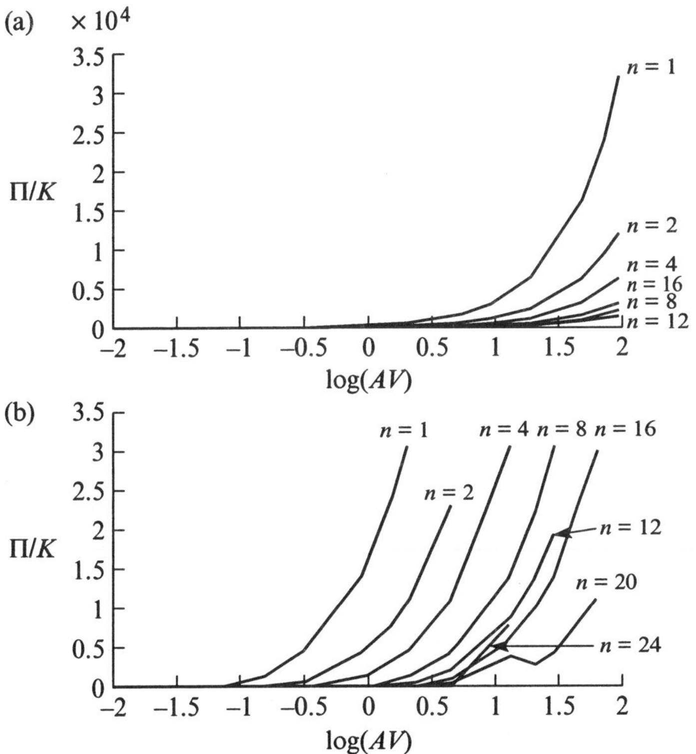

Figure 2.7 Profit as a function of the cost ratio (industry concentration is a parameter)

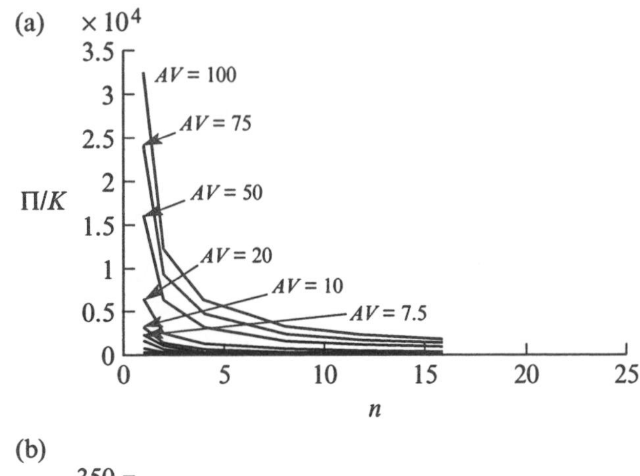

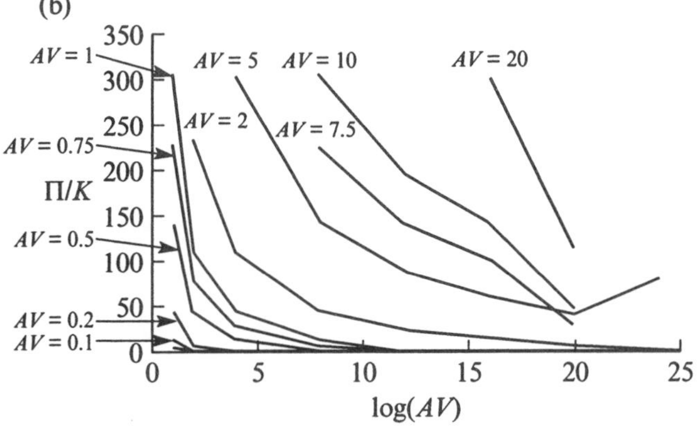

Figure 2.8 Profit as a function of the industry concentration (cost ratio is a parameter)

surface presented in Figure 2.6a). From these figures, it can be seen that after crossing the border from the zero profit zone, profit increases relatively quickly with increasing values of the *AV* ratio or with an increasing concentration of the market. The irregular behaviour observed for a relatively large number of firms (small concentration) is caused by fluctuations of development, as mentioned earlier.

An analogous surface can be drawn using the simulation data from Table 2.2. The shape of the surface of the price margin is slightly different from that of the profit ratio surface (see Figure 2.9 and its enlargement in Figure 2.10). To see more details of the price margin surface we must look at it from a different perspective, and therefore the order of the *AV* and *n* axes is different from that on Figures 2.5 and 2.6.

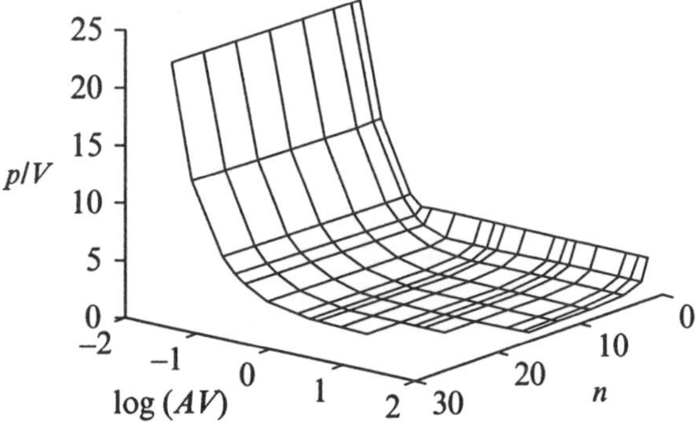

Figure 2.9 Price margin as a function of the cost ratio and the industry concentration

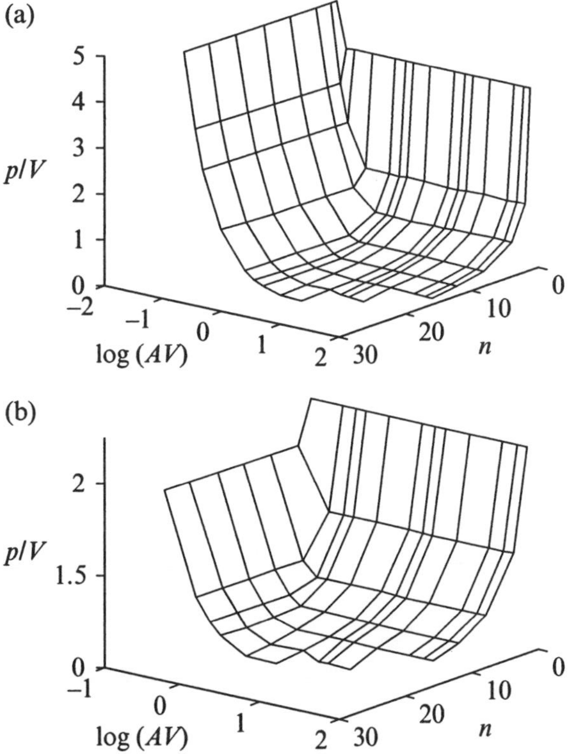

Figure 2.10 Price margin as a function of the cost ratio and the industry concentration (enlargements)

In the zero profit zone the price margin hyperbolically decreases (see also Figures 2.11 and 2.12, where the price margin surface is sliced in a similar way as in Figures 2.7 and 2.8 with the profit ratio surface). After reaching the border separating the two zones, the price margin is constant but this constant value depends on industry concentration (and the price margin is smaller the smaller is the concentration and the smaller is the value of the cost ratio). It can be said that the shape of the price margin surface reverses the shape of the profit ratio surface – in the zero profit zone the price margin decreases hyperbolically, and vice versa, n the hyperbolical growth of the profit ratio, the surface of the price margin is relatively flat.

From a qualitative point of view the shapes of both the profit ratio and the price margin surfaces are almost the same for different values of the other model's parameters. For example, for different kinds of the market (defined by the values of the price elasticity β in the demand func-

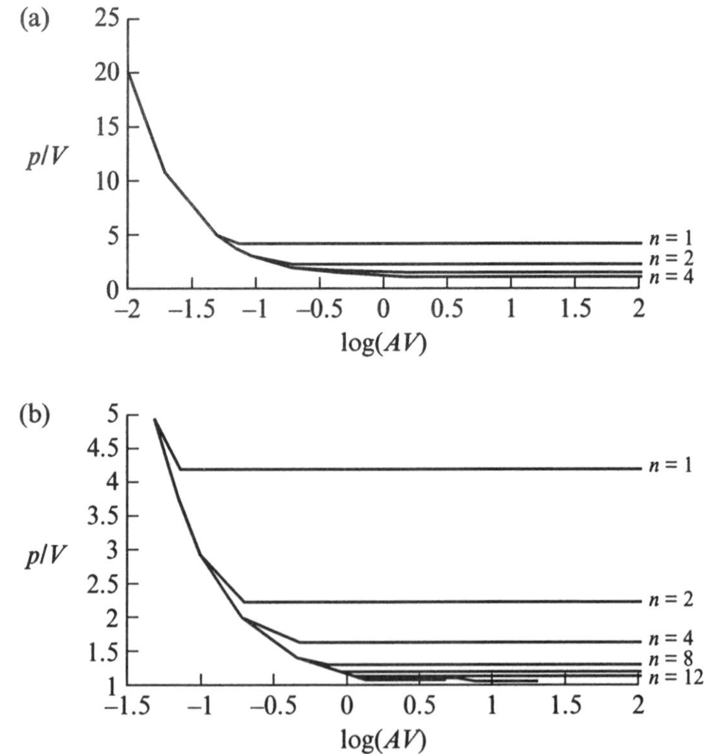

Figure 2.11 Price margin as a function of the cost ratio (industry concentration is a parameter)

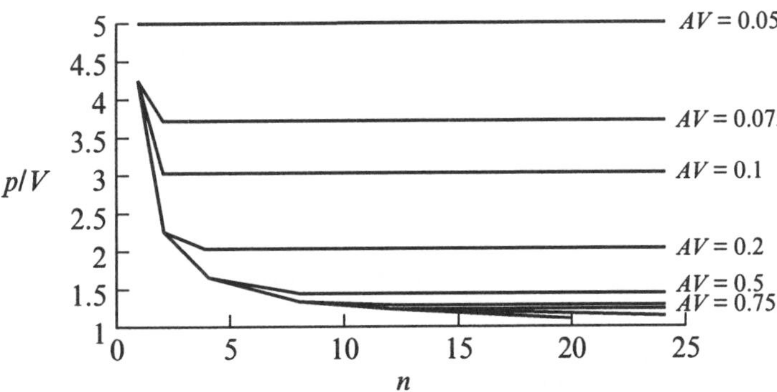

Figure 2.12 Price margin as a function of the industry concentration (cost ratio is a parameter)

tion) we observe the same two zones but the border between these two zones has shifted slightly. The β parameter is greater than zero for commodities fulfilling primary needs and is negative for commodities fulfilling higher-order needs. The shift is greater the smaller is the cost ratio AV and the greater is the concentration of the industry. The borders for four kinds of markets are presented in Figure 2.13, where it can be seen that even for the primary goods markets ($\beta > 0$) and for high concentration, there are industry states where profit is equal to zero.

For a growing market (that is, the rate of the market growth γ is greater than zero) the shape of the surfaces is almost exactly the same (and also the border does not change), the only difference is that instead of a zero

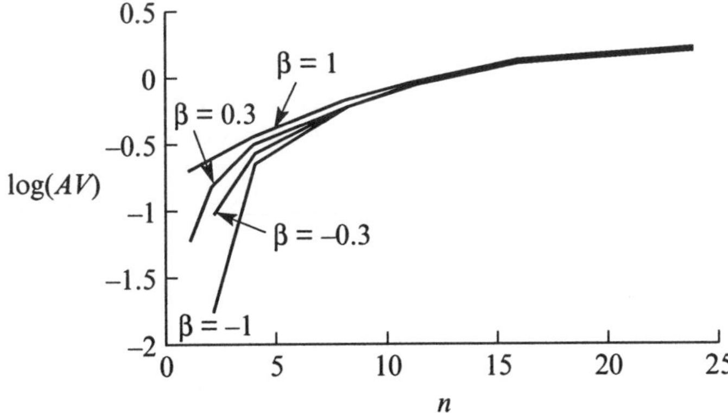

Figure 2.13 Border between zero profit and positive profit regions for four different kinds of industries

profit zone we have a flat area of positive profit, for example, for $\gamma = 0.05$ the equilibrium value of the profit to capital ratio in this zone is equal to 4.88 per cent. Discussion on the impacts of the price elasticity β and the rate of market size growth γ on industry development is presented in Kwasnicki (1996a, pp. 136–41).

4 SIMULATION – DYNAMIC ANALYSIS

The results presented in the previous section can wrongly suggest that modes of development of real industry processes are smooth and seemingly deterministic. Because of innovation and firms' interaction, and also 'natural' errors made by decision makers (bounded rationality of industry actors), industry development is far from being a smooth process. In this section we present results of three simulation runs to illustrate the theoretical findings presented in the previous section. At least at the level of stylized facts, the simulation results are very close to real industrial processes. Of course, the best way to illustrate the theory would be to collect relevant data of real industrial processes and to compare the simulation results with industrial records, and the author hopes that this will be possible in the future.

In the first simulation run it was assumed that innovation can be introduced by firms and that there is no economy of scale. Innovation can lead to a unit cost (V) reduction and a productivity of capital (A) increase. The changes in these two factors can cause either an increase or a decrease in the cost factor (AV). But concurrently with changes in costs and productivity there occur improvements in products' technical performance, that is, an increase in technological competitiveness. Therefore, each innovation can be considered as complex. Each firm trying to introduce innovation faces the problem of considering positive and negative factors of each innovation. It frequently happens that accepted innovation allows an increase in technical competitiveness and also an increase in the unit cost of production or a decrease in the productivity of capital; but it also happens that a reduction in the unit cost of production is accompanied by a decrease in technical competitiveness and a decrease in the productivity of capital.

In the second experiment it was assumed that there are no innovations and the only way to reduce the unit cost of production is by exploitation of the economy of scale. The third experiment is a combination of the first two simulations – innovations and economy of scale are both present, and in a sense cooperate in the process of cost reduction.

Let us call these three experiments 'innovation', 'economy of scale' and 'innovation and economy of scale', respectively. These terms are used in all the following figures. The results of the basic characteristics of development in all these experiments (namely, unit cost of production, productivity of capital and technical competitiveness) are presented in Figures 2.14 and 2.15. In the first experiment, innovation allows the average value of the unit cost of production to be reduced five times (from the initial value of 5 to 0.986). Economy of scale in the second experiment allows the average cost to be reduced almost nine times (from the initial value of 4.86 to 0.53 in the last year of simulation). In the second experiment there are rather smooth curves of characteristics and fluctuating development in the presence of innovation (which is caused mainly by a random process of innovation emergencies). Because there is no innovation in the 'economy of scale' experiment, values of productivity of capital and technical competitiveness are constant and equal to their initial values during the whole simulation period (0.1 and 0. 14, respectively). Cost reduction in the third experiment is accelerated because of the cooperation between innovation and economy of scale – the average unit cost of production is reduced more than 300 times (from the initial value of 4.86 to 0.014). In year 49, one small firm introduced innovation which led to a significant improvement in its products' technical performance (technical competitiveness of that innovation was almost 40 per cent better than the average technical competitiveness in the industry) and also improved the productivity of capital (by about 10 per cent). As is frequently observed in real situations, the technical performance and productivity of capital improvements are accompanied by an increase in the unit cost of production (in our simulation experiment this is about seven times above the average cost – which increased because of economy of scale exploitation by large firms in the period just before introducing the innovation). In spite of this disadvantage, the innovation was accepted by the market. Therefore in the next decade an almost tenfold increase in the unit cost of production was observed, but because of both further innovations and also economy of scale exploitation the cost was reduced to the final value of 0.014. At the same time, improvement in technical competitiveness was observed, and the final value of average competitiveness is 2.36. The second radical innovation emerged at the beginning of the ninth decade. This innovation allowed for a further improvement in technical competitiveness of almost 30 per cent, but it was accompanied by a small increase in the unit cost of production and a slight deterioration in the productivity of capital (compare Figures 2.14 and 2.15). The improvement in technical performance was considered as more advantageous than the deterioration in the cost of production and productivity of capital.

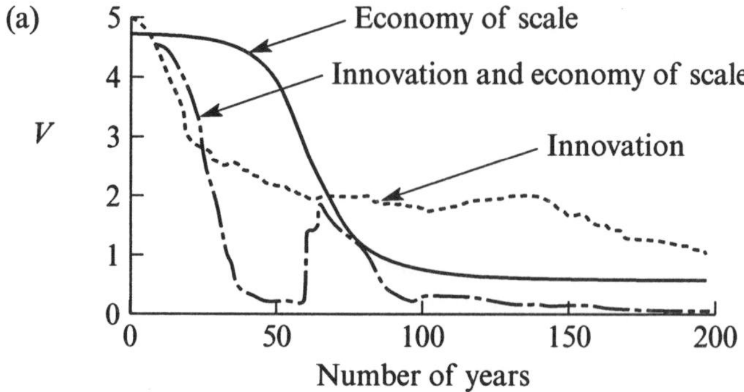

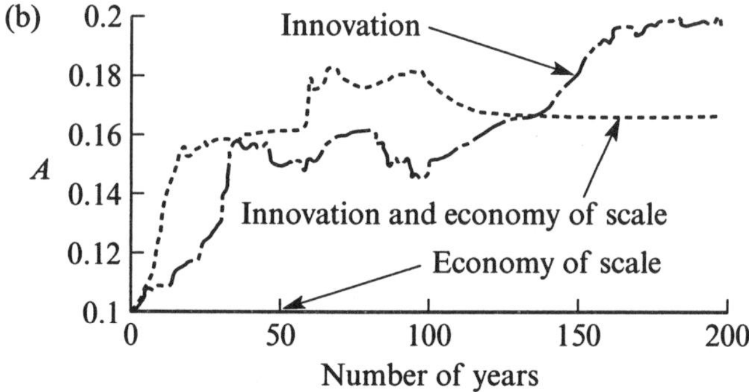

Figure 2.14 Variable cost of production and productivity of capital (average values)

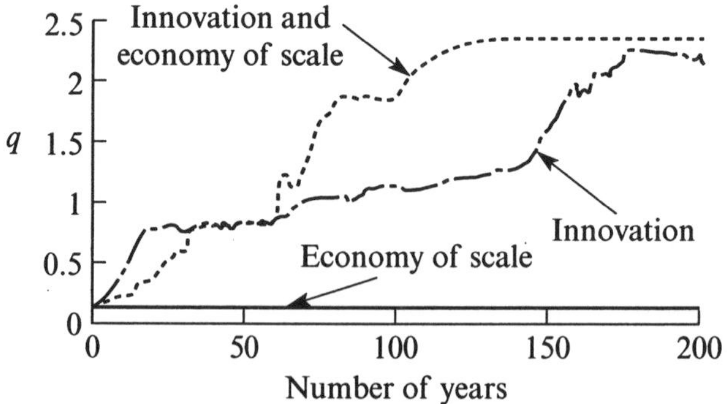

Figure 2.15 Average values of technical competitiveness

Changes in the cost ratio (*AV*) are presented in Figure 2.16a (note the logarithmic scale of the *AV* axis). In the presence of innovation, the cost ratio was reduced 2.6 times (from the initial value of 0.5 to the final value in the last year of 0.19). Economy of scale allowed the cost ratio to be reduced more than nine times in the second experiment (to the final value of 0.053). In the third experiment the cooperation of innovation and economy of scale allowed the cost ratio to be reduced more than 200 times (to the final value of 0.0023). We can see that the synergic effect of innovation and economy of scale is much greater than the simple multiplication of the separate influences of each factor.

The differences in development in all three experiments are also visible in the changes in industry concentration. In Figure 2.16b, the Herfindahl

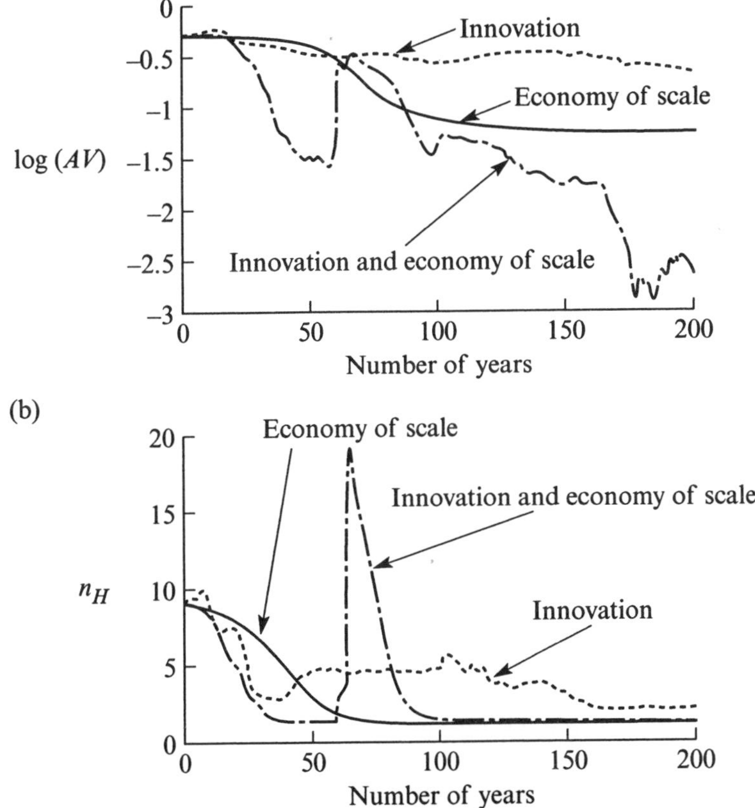

Figure 2.16 The cost ratio (AV) *and Herfindahl firm number equivalent*

firm's number equivalent (n_H) is presented. All three experiments started with 12 different-sized firms (the largest firm has a 16.7 per cent share of the market and the smallest a 0.5 per cent share, which means that the initial industry was equivalent to almost nine equal-sized firms – $n_H = 8.93$). Each year a number of new firms try to enter the industry. In all the experiments, because of either innovation or economy of scale, the industry becomes more and more concentrated (although in the second experiment in the first decade after the emergence of the first radical innovation industry concentration decreased). In the 'innovation' experiment in the final year of the simulation there are 18 firms but most of them are rather small and the market is equivalent to duopoly (n_H = 1.82). Economy of scale causes much greater concentration. When there is no innovation and only economy of scale influences concentration, the smaller firms are very quickly eliminated and since year 90 only one firm has been operating in the market (out of the original 12 firms). Allowing firms to innovate causes a fluctuation in industry concentration. In the third experiment at the end of the simulation five firms are operating in the market, although four of them were rather small ($n_H = 1.013$).

Comparing the fluctuation in industry concentration with parallel changes in other characteristics of industry development we can observe that the fluctuations are correlated mainly with the emergence of innovation, and the more radical the innovation, the higher the increase in the number of firms entering the market. As mentioned earlier, in the third experiment two radical innovations emerged around years 50 and 90, but only the first innovation caused a reduction in industry concentration (there are more than 50 firms operating on the market in the decades following the emergence of the first radical innovation and the maximum value of n_H is equal to 18.9). The first radical innovation was made by small firms and the large firms were unable to copy it within a relatively short period. This enabled other small firms to enter the market and therefore by the end of the sixth decade the Herfindahl equivalent number of firms was equal to almost 20 firms. The second radical innovation was found by the largest firm and therefore we do not observe a similar reduction in industry concentration in the ninth decade.

The results presented in Figures 2.14 to 2.16 allow trajectories of development of profit to capital ratio and the margin of price over the plane $n_H \times \log(AV)$ to be drawn. The best way is to draw these trajectories together with the equilibrium surfaces (as presented in Figures 2.6 and 2.10); however the clarity of such figures would be very poor, so we shall present only the trajectory of the profit/capital ratio (Figure 2.17). In Figure 2.17a all three trajectories are drawn but the picture is dominated by the trajectory of the third experiment (where profit to capital ratio is much greater than in the other two experiments). Therefore, in Figure 2.17b we present the trajectories for the first two experiments. We can see that the trajectories fluctuate and are placed above and also below the 'equilibrium' surface. Fluctuations are much more significant in the presence of innovation. In all three trajectories we can distinguish two phases: the first is characterized by the domination of an increasing concentration of industry and the second by the reduction in the cost factor *AV*. The trajectory marked 'innovation' in Figure 2.17b is a typical trajectory in the presence of innovation – with rather large fluctuations, turns and reverses during the course of development. A similar picture would be produced by showing the trajectory of a price margin, but rather than doing this, in order to observe more details we shall look at the changes in these two characteristics as a function of time. In Figure 2.18, the average value of the profit to capital ratio is presented. When there is no

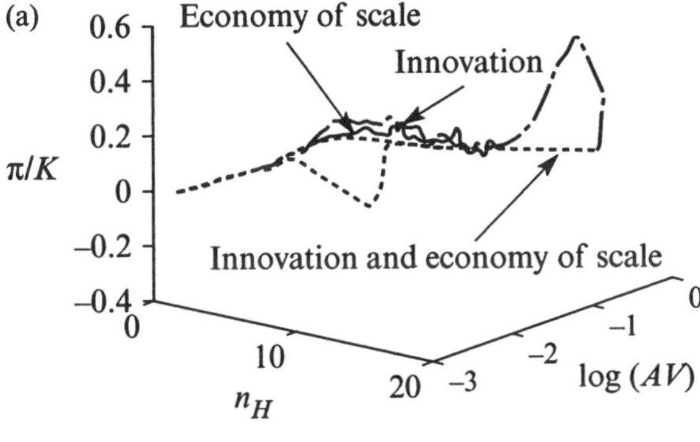

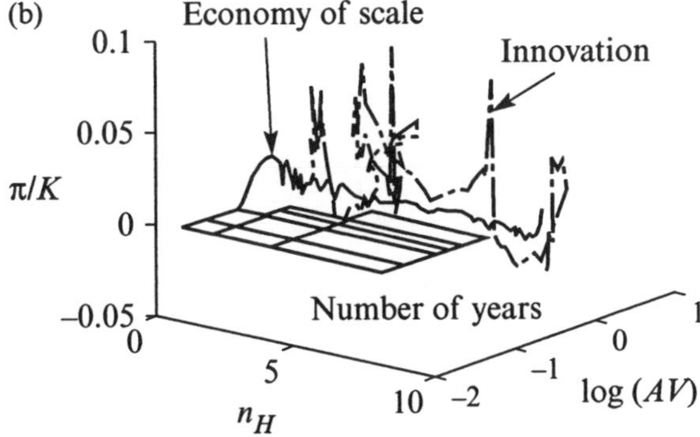

Figure 2.17 Profit to capital ratio in the space of AV *and* n

economy of scale and only innovation processes influence industry development, the fluctuations of profit to capital ratio and price margins are significant. Introducing (incremental) innovation results in a profit of about 8 per cent but also, in some periods, causes the average profit to decrease to –2 per cent (nevertheless, even at this stage of development, some leading firms are still able to produce a rather high positive profit). Small fluctuations in the 'economy of scale' simulation (and no innovation) are caused by discrepancies in firms' expectations about the future development of the industry and real industry development (wrong decisions can be ascribed to firms' bounded rationality). The same kind of fluctuation is observed in the third experiment (see Figure 2.18b, and also a more detailed chart in the top right-hand corner of that figure). Radical

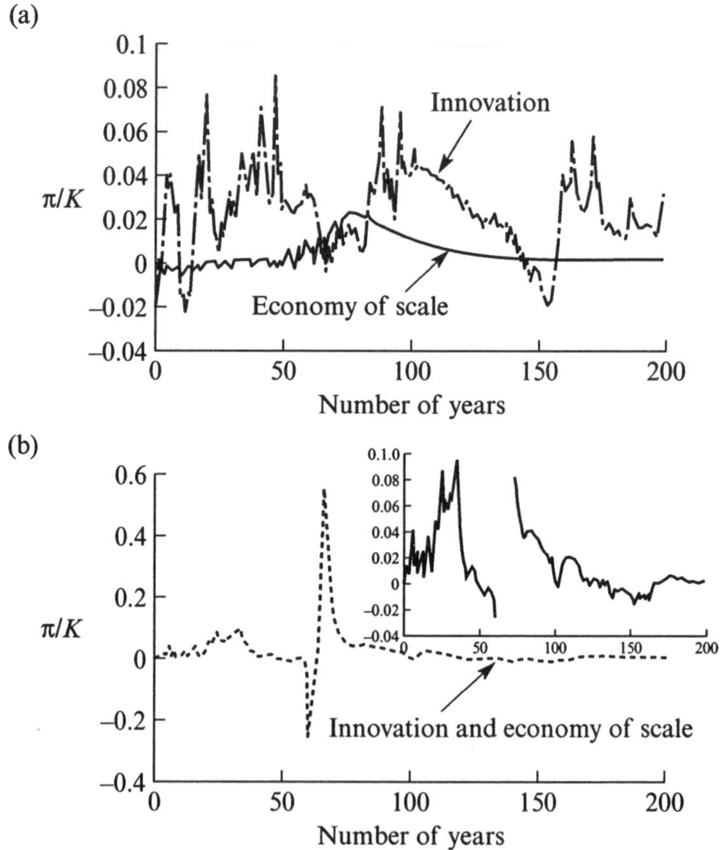

Figure 2.18 Profit to capital ratio (average values)

innovation which emerged in year 49 caused a relatively deep reduction in the average profit in the fifth decade (to –27 per cent) but within the next few years the industry witnessed high growth and the average profit increased to 55 per cent (but very quickly dropped to the relatively small value of 4 per cent). The second radical innovation (in the ninth decade) was made by the largest firm. When that innovation was introduced, there were 32 firms in operation, but most of them were very small (n_H = 1.97). The largest firm was not able to exploit the advantages of this radical innovation because at this stage the capital ratio AV was relatively small (about 0.03). The maximum average profit after the emergence of this innovation was less than 2 per cent.

Development of a price margin is presented in Figure 2.19. In the presence of innovation (and no economy of scale) the price margin was no

(a)

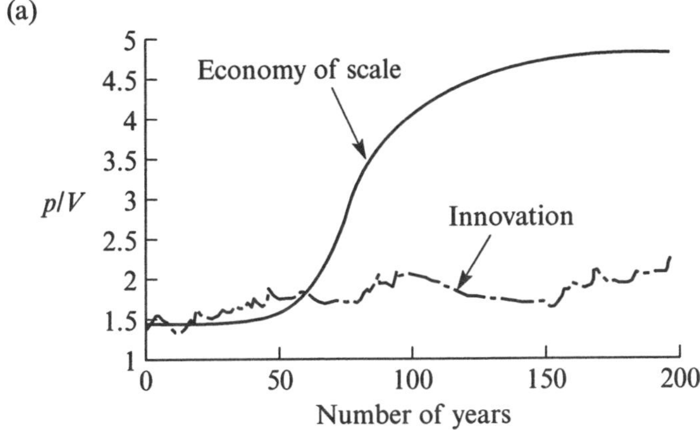

(b)

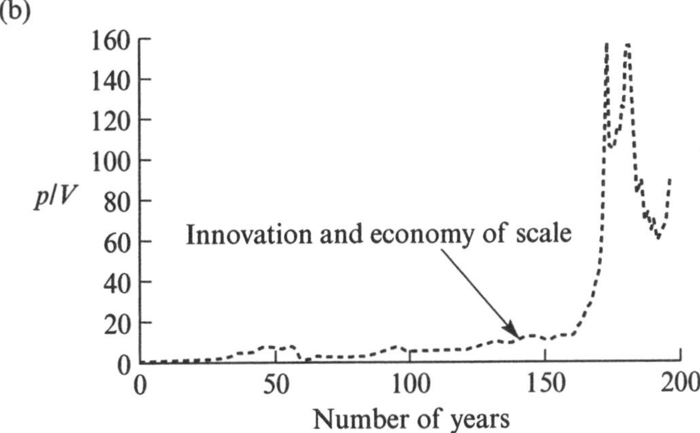

Figure 2.19 Price margin (average values)

greater than 2.3. A slightly greater price margin was observed in the presence of economy of scale (Figure 2.19a). The maximum value of this ratio was about five. Full exploitation of both innovation process and economy of scale caused very high price margin values (up to 160) and a high concentration of industry. But this does not mean that the largest firm in the third experiment (being in fact very close to a monopolist position) achieved enormous profits. The cost ratio AV at this stage of development was about 0.03 and, according to the findings presented in the previous section, this meant that in the last decades of the simulation the profit of this 'monopolist' was no larger than 0.7 per cent.

5 CONCLUSIONS

The results presented here are purely theoretical and should be the subject of further study and verification. In particular, a comparative study of real industrial processes ought to be made, focusing on the modes of development relating to different values of the cost ratio. But even at the current stage of research, after the preliminary simulation studies and a 'stylized' comparison of the models' behaviour with real industrial development, it transpires that for plausible values of the cost ratio those of our model strongly resemble those of real industries.

Values of the cost ratio are simple real numbers and do not depend on the units of measurement of capital and production. It seems that the cost ratio may be used as the practical characteristic for the classification of industries. Small values of the cost ratio indicate that in this type of industry a large amount of capital is required to manufacture products at a relatively low cost and, vice versa, a large cost ratio means that in this type of industry a relatively small amount of capital is needed to manufacture products at high cost. Industries with large productivity of capital and low unit cost of production may have the same cost ratio as industries with small productivity of capital and high unit cost of production; the result is that in all such cases the characteristics of the industry development, in the absence of innovations, are exactly the same.

It may be expected that labour cost is a major part of the unit cost of production, so the invariability of development of different types of industries for the same value of the cost ratio resembles the classical finding of substituting labour with capital.

On the basis of the results presented here, one may draw the conclusion that it is possible to imagine situations (industry regimes) in which highly concentrated industries behave similarly to industries where a large number of firms are in a state of pure competition (namely, for very small values of the cost ratio *AV*), and also industry regimes in which numerous competitors behave as oligopolists – or even as monopolists – (namely, for high values of the cost ratio *AV*). These findings are purely theoretical and ought to be verified using real data of industrial development. The open question is whether we observe such small and large values of the cost ratio in actual industrial processes?

As a hypothesis, it may be stated that in real industrial processes, because the introduction of innovations reduces the unit cost of production, the cost factor is reduced in successive stages of an industry development. Therefore, for matured innovative industries, in spite of their relatively high concentration, we can expect higher competitive conditions of development.

NOTE

1. In the model, the spaces of routines and of characteristics play a role analogous to the spaces of genotypes and of phenotypes in biology. The existence of these two types of space is a general property of evolutionary processes. Probably the search spaces (that is, spaces of routines and of genotypes) are discrete spaces in contrast to the evaluation spaces (that is, the spaces of characteristics and of phenotypes) which are continuous spaces. The dimension of the space of routines (space of genotypes) is much larger than the dimension of the space of characteristics (space of phenotypes).

REFERENCES

Kwasnicki, Witold (1996a), *Knowledge, Innovation, and Economy: An Evolutionary Exploration*, Cheltenham, UK, Brookfield, US: Edward Elgar (1st edn 1994, Oficyna Wydawnicza Politechniki Wroclawskiej; Wroclaw).

Kwasnicki, Witold (1996b), 'Innovation regimes, entry and market structure', *Journal of Evolutionary Economics*, **6**, 375–409.

Kwasnicki, W. and H. Kwasnicka (1992), 'Market, innovation, competition. An evolutionary model of industrial dynamics', *Journal of Economic Behavior and Organization*, **19**, 343–68.

Kwasnicki, W. and H. Kwasnicka (1994), 'Bounded rationality and fluctuations in industry development – an evolutionary model', in Robert Delorme and Kurt Dopfer (eds), *The Political Economy of Diversity: Evolutionary Perspectives on Economic Order and Disorder*, Cheltenham, UK, Brookfield, US: Edward Elgar, pp. 199–228.

Nelson, Richard R. and Sidney G. Winter (1982), *An Evolutionary Theory of Economic Change*, Cambridge, MA: Harvard University Press.

Simon, H.A. (1986), 'On behavioral and rational foundations of economic dynamics', in R.H. Day and G. Eliasson (eds), *The Dynamics of Market Economies*, Amsterdam: North-Holland.

Winter, Sidney (1984), 'Schumpeterian competition in alternative technological regimes', *Journal of Economic Behavior and Organization*, **5**, 287–320.

3. University knowledge and innovation activities: evidence from the German manufacturing industry

Wolfgang Becker and Jürgen Peters

1 INTRODUCTION

R&D activities determine the economic success and competitiveness of innovative firms on both the national as well as on the international level (Dosi 1988; Griliches 1995; Mairesse and Saasenou 1991). The basic reason for R&D activities is that expansion of technical and organizational know-how, together with the realization of dynamic (increasing) returns to scale, positively influence the rates of productivity, returns and profits.

As a consequence of the dynamics of economic and technological change, firms continuously have to improve the effectiveness of their R&D activities as well as expand their potential to develop new and improved products. Therefore, innovative firms constantly exercise considerable effort to optimize their technological capacities by drawing from external resources. This is closely related to the increasing importance of multi- and interdisciplinary R&D activities (for example, in the nanotechnology field as a new basis of technology) and the strengthened interrelation of basic research with industrial research (for example, science confinement to technological development).

The spectrum of possibilities for utilizing externally generated knowledge for innovation activities is manifold (Cohen 1995; Hagedoorn 1995; Harabi 1995). The adaptation of technological know-how can occur through acquisition, purchasing capital goods and advance commitments. Materialized R&D efforts embodied in goods and services are reflected in the prices on the goods market. Patenting, respectively licensing, imitation strategies (for example, by reverse engineering), mergers and takeovers are other possibilities. Furthermore, the adaptation of external knowledge can be performed by employment, that is, enticing personnel from the academic domain. Another possibility is the establishment of R&D cooperation external to the industrial sector; here in particular with universities (Faulkner and Senker 1994; Lee 1996; Sanchez and Tejedor 1995).

University knowledge[1] is a part of the total stock of externally available and relevant technological know-how for innovative firms (Becker 1996; Mansfield 1995). The importance of this kind of knowledge has increased continuously over time because the development of new and improved products depends increasingly on the findings of and results from academic research (David 1994; Narin et al. 1997; Rosenberg and Nelson 1994). In the fifth Kondratieff cycle, characterized primarily not by growing energy inputs but rather by the utilization of information as a resource, the development and production of high-tech goods in science-based firms has reached a technological level requiring the utilization and integration of academic expertise and capabilities to widen firms' in-house capacities efficiently.

For key technologies, such as micro-electronics, telecommunication technology and so on, the scientific foundation has become increasingly relevant for industrial innovation activities (Gibbons et al. 1994; Nelson and Wolff 1997). Important innovation impulses for modern computer technology and biotechnology draw from academic research and its adaptation and utilization in industrial research. But also technologies in mass production sectors, such as chemicals, machinery and electronics, have reached development levels requiring a specific degree of optimization of internal resources through knowledge from the academic sphere (Grupp 1996; Klevorick et al. 1995). As shown in Table 3.1, the relevance of university knowledge for technological advance in Germany can be derived from the fact that the share of universities in R&D expenditure has increased steadily since 1990 and reached about one-sixth of all registered R&D expenditures in 1993.

Empirical studies from the USA show that innovation effects induced by university (academic) knowledge occur on two levels (Becker 1996; Stephan 1996). On the one hand, internalizing scientific knowledge leads to an expansion of firms' *technological capacities* (*capabilities*) for developing new or improved products, which becomes evident in an increase of technological know-how and improved skills; on the other hand, knowledge generated in universities (scientific institutions) improves firms' *research efficiency*. They accelerate the lead time of innovations and increase the rates of return of R&D.

Jaffe (1989) delivers pathbreaking empirical proof that university (scientific) knowledge expands the technological capacities of innovative firms. His empirical investigations into US industry show a strong relation between university research and firm patenting. Knowledge from academic research significantly influences the number of applied patents

Table 3.1 R&D expenditures in Germany

Year	R&D expenditures				
	Total	Universities	Public sector without universities	Private sector	Foreign countries
	In millions of DM				
1990	67 370	9 850	9 030	48 490	2 545
1991	75 315	12 071	11 187	52 057	2 741
1992	77 113	12 710	11 759	52 644	3 097
1993	78 345	13 730	12 215	52 400	3 210
	Percentage of total R&D expenditures				
1990	100	14.6	13.4	72.0	3.8
1991	100	16.0	14.9	69.1	3.6
1992	100	16.5	15.2	68.3	4.0
1993	100	17.5	15.6	66.9	4.1
	Percentage of GNP				
1990	2.78	0.41	0.37	2.00	–
1991	2.64	0.42	0.39	1.82	–
1992	2.51	0.41	0.38	1.71	–
1993	2.48	0.43	0.39	1.66	–

Source: Bundesministerium für Bildung, Wissenschaft, Forschung und Technologie (1995, p. 334).

by in-state firms. This influence becomes even more evident when the number of firms' innovations are used as an explanatory variable rather than the frequency of patent applications (Acs et al. 1992). These findings can be interpreted that new advances in university research act not only at the basic research stage but affect the entire innovation chain and directly stimulate a market-oriented application of new knowledge.

Klevorick et al. (1995) come to a similar conclusion, investigating the relevance of academic research for technological progress in specific industries in the USA. The results of academic research are particularly relevant for high-tech firms in R&D-intensive industries, such as the computer industry, aircraft industry, astronautics, and the pharmaceutical industry. Firms in these industries directly and instantaneously utilize academic research findings from applied sciences (mechanical engineering, electrical engineering, chemical engineering) while new findings from basic research in physics and mathematics are of less relevance for industrial innovation. Mansfield (1991) finds that 11 per cent of all product innovations, and 9 per cent of all process innovations developed in research-intensive industries (drugs, metals, information processing and so on) in the USA between 1975 and 1985 could not have been realized without the respective results from university research.

The increasing use of external (university) knowledge enables even small and medium-sized firms to compensate for their competitive disadvantages conditioned by their firm size (Cohen and Klepper 1996a; Link and Rees 1991; Scherer 1991). The importance of externally generated knowledge for innovative firms in the USA, subject to the size of the firm, has been investigated by Acs et al. (1994). They show that for small firms, external R&D activities play a larger role in the innovation process than for larger enterprises which rather draw from in-house innovation inputs.

Comparable investigations, as conducted in the past in the USA, are not as yet available for Germany. The existing studies in Germany focus on *distinct* aspects of the science–technology–innovation interface, for example, the importance of academic research in science-based fields of technology (Grupp 1994, 1996), the role of universities in the technology transfer process especially for small and medium-sized firms (Beise et al. 1995; Harhoff and Licht 1996), or the relevance of the regional science and research infrastructure on the formation and location of firms (Harhoff 1995; Nerlinger 1996). Only Peters and Becker (1998) have analysed econometrically the interaction between university knowledge and innovation activities on both the input *and* the output sides for the German automobile supply industry. They found significant effects of academic research on the variation of supplier firms in R&D activities and innovation intensities as well as on the improvement of existing products rather than on the development of new ones. Because of the restrictions to firms in the German automobile industry, the results of their study cannot be generalized.

Against this background, the importance and effects of university knowledge on the innovation input and output activities of firms in the *whole* German manufacturing industry will be analysed in this chapter. In Germany, more than 90 per cent of the entire R&D invested by private enterprises is performed by firms from the manufacturing sector (Bundesministerium für Bildung, Wissenschaft, Forschung und Technologie 1995). In this context, we also want to investigate the impact of certain conditions on the probability of German firms cooperating with universities.

In Section 2, we discuss the connections between innovation activities, technological opportunities and university knowledge under more *theoretical* aspects. In Section 3, the importance and impacts of university knowledge on the innovation input and output of firms in the German manufacturing sector are investigated *empirically* using data from the Mannheim Innovation Panel (MIP) for the period 1990–92. In Section 4, factors determining the probability of R&D cooperation between universities and firms as the most systematic form of knowledge transfer are analysed. Section 5 summarizes the main findings.

2 INNOVATION ACTIVITIES, TECHNOLOGICAL OPPORTUNITIES AND UNIVERSITY KNOWLEDGE

In modern innovation research (Cohen 1995; Cohen and Levin 1989; Kline and Rosenberg 1986) it is generally agreed that targeted and systematic R&D activities are a main source of generating new products or improved production techniques. The basic reason for R&D is that the expansion of technical and organizational know-how, together with the realization of increasing returns to scale, positively influence the rates of productivity and profits.

A firm's research efficiency depends on the interaction of internal (in-house) R&D activities and the extent to which external knowledge sources can be adapted and implemented for own purposes. The adaptation of *external* knowledge changes the characteristics and influences the performance of product factor inputs required for innovations. For the recipients, targeted utilization of exogenously generated knowledge leads to an improved quality of the input factors. The importance of external knowledge is that firms can expand their innovation (technological) capacities to develop new or improved products (Cohen and Levinthal 1989; Klevorick et al. 1995). They can increase their innovative capabilities, which generally has a positive influence on the probability of being successful in R&D and on the quality of new technologies. The improvement of production performance resulting from the use of external knowledge leads – depending on the level of the firms' *absorptive capacities*[2] – to more efficient production processes and/or larger technological know-how and competences of the research personnel.

External knowledge, however, induces factor-embodied technological progress which does not effect all production technologies to the same extent, but unfolds its effects selectively and cumulatively within specific technological paths. In this context, economists agreed that the level of innovation activities of firms and the technical advance in industries depend on the possibilities of acquiring and internalizing usable technological knowledge (Geroski 1990; Harabi 1995; Klevorick et al. 1995). Thus, variances in R&D expenditures and innovation activities can better be explained by differences in the *level of technological opportunities*[3] each firm or industry is faced with than by factors such as firm size, appropriability conditions, market structure and so on.

Technological opportunities are related to the contribution of external (knowledge) sources to firms' innovation activities. The possibilities for innovators to obtain economic useful technological (technical) knowledge vary from industry to industry and, to a certain degree, from firm to firm 'Due to variations in the degree of availability of these technological opportunities, innovations are "cheaper" to realize. . . . This factor stands – in combination with others – behind the empirically observable interindustrial differences in the rates of technical progress, of total factor productivity and of economic growth' (Harabi 1995, p. 67). Strength and sources of technological opportunities are important factors for explaining firm-specific and cross-industry variations in R&D intensity, productivity of R&D and technological advance.

Numerous empirical studies underline the importance of different sources of technological opportunities in the innovative process (for an overview, see Cohen 1995). In general, one can distinguish *industrial* and *non-industrial* sources. Technological information stemming from suppliers, customers and competitors constitute the industrial knowledge pool. Non-industrial sources relate to knowledge stemming from institutions and organizations outside the business sector, here in particular the scientific sphere. Especially information generated in universities is an important knowledge source for firms with high-level R&D (innovation) activities due to the close interrelation of (basic) academic research and (applied) industrial research. Scherer (1992, p. 1424) pointed out that the concept of technological opportunities 'was originally constructed to reflect the richness of the scientific knowledge base tapped by firms'. Technological opportunities are 'mainly fostered by the advances of scientific knowledge and positively affect the productivity and thus the intensity of R&D' (Sterlacchini 1994, p. 124). It is generally acknowledged that a major factor in economic progress 'has been the growing dependence upon a quintessentially nonmarket activity – the organized pursuit of pure scientific knowledge' (David 1991, p. 1).

To analyse the impacts of technological opportunities empirically – especially related to university knowledge – on the innovative activities of firms, a simple theoretical framework can be used. The basic assumption is that the technological opportunities Ω_i result from the sum of the total stock of externally accumulated knowledge which firm *i* is faced with:

$$\Omega_i = \Omega(R_i^C, R_i^S, R_i^U), \tag{3.1}$$

where R_i^C represents the technological opportunities stemming from customers/competitors and R_i^S defines the knowledge pool generated by suppliers. R_i^U reflects the contributions of university (scientific) knowledge to the technological capacities of firm *i*. The contributions of these sources of technological opportunities 'serve to offset the diminishing returns to which a fixed pool would give rise' (Klevorick et al. 1995, pp. 188–9). Thus, the total stock of externally accumulated knowledge has to be replenished constantly.

Besides being influenced by the traditional product factors labour L_i and capital K_i as well as by conditions A regarding demand schedule, price elasticities of demand, market structure and so on, the innovation activities R_i are then affected by the level of technological opportunities Ω_i:

$$R_i = f(A, L_i, K_i, \Omega_i), \tag{3.2}$$

with $\partial R_i/\partial\Omega_i > 0$ and $\partial^2 R_i/(\partial\Omega_i)^2 \gtreqless 0$. Higher levels of technological opportunities increase the intensity of innovation activities with diminishing, constant, or increasing rates of return, depending on the initial level of the firms' in-house R&D. The impacts on the firms' innovation *input* depend on the relation of substitutive and complementive effects of using this kind of external knowledge (Becker and Peters 1998). If university knowledge is used as a substitute (complement) for in-house R&D, it is more likely that the adaptation of technological opportunities stemming from academic research will discourage (encourage) the firms' innovation/R&D activities.[4]

The level of technological opportunities not only affects the firms' innovation input (and therefore the performance of new technologies because of the higher efficiency of their in-house R&D), but also the innovation *output*. For example, using new materials or information technologies enables advances in product quality directly. As Becker and Peters (1998) show, it is also more likely that high (low) technological opportunities stemming from academic research will increase (decrease) the firms' innovation output y_i (or the quality of new technologies respectively) in the kind of

$$y_i = g(R_i, \Omega_i), \tag{3.3}$$

with $\partial y_i/\partial R_i > 0$ and $\partial^2 y_i/(\partial R_i)^2 \gtreqless 0$ and $\partial y_i/\partial\Omega_i > 0$ and $\partial^2 y_i/(\partial\Omega_i)^2 \gtreqless 0$.

We assume that the degree to which firms use technological information stemming from universities (and other scientific institutions) is closely correlated with their technological capacities to develop new or improved products (Arvanitis and Hollenstein 1994; Felder et al. 1996; Levin and Reiss 1988). In general, the higher the importance of the results of academic research for a firm's in-house innovation activities, the higher its capabilities in the innovation process. Along this line, the adaptation of university knowledge enlarges a firm's (in-house) innovation capacity with positive effects on the development of new or improved products. Further, we assume that the higher the level of technological opportunities, the larger a firm's incentive to invest in innovation (R&D) activities will be.

Against this background, the importance of universities as external knowledge sources for innovative firms in the German manufacturing industry will be investigated *empirically* in the next section. Universities are considered here as unities, without faculty-specific differences. It will be investigated whether the variation of innovation and R&D activities can be explained by differences in the technological opportunities caused by the adaptation of knowledge generated in academic research (and other external sources).

3 EMPIRICAL EVIDENCE OF UNIVERSITY KNOWLEDGE IN THE GERMAN MANUFACTURING INDUSTRY

The data for the empirical investigations originate from the first wave of the Mannheim Innovation Panel (MIP) conducted in Germany in 1993.[5] About 2900 firms participated in the survey and filled in a questionnaire about their innovation activities for the period 1990–92. They answered a broad range of questions related to input-, output- and market-oriented aspects of innovation activities.[6]

The original survey covers innovative as well as non-innovative firms. We restrict our analysis to *innovative* firms only. An innovative firm is defined as a company which had introduced new or improved products in the years 1990–92 or had intended to do so in the period 1993–95. After having excluded the non-innovative firms from the original data set, about 2300 firms were included in the empirical analysis.

To estimate the importance and effects of university knowledge on the innovation activities of firms in the German manufacturing industry, a set of input- and output-related innovation variables is used (see Table 3.2). Unless otherwise noted, all data relate to the year 1992. The descriptive

Table 3.2 List of variables, description and empirical measurement

Variable	Description	Empirical measurement	Value (range)
	Input-related innovation variables		
INNO_INT	Innovation intensity	Log of innovation expenditures to sales ratio	Metric
R&D_INT	R&D intensity	Log of R&D expenditures to sales ratio	Metric
R&D_PLAN	Mid-term planning of R&D	1 = Planning of R&D in 1993–95, 0 = otherwise	Nominal
	Output-related innovation variables		
INNO_NOV	Novelty of product innovations	Sales share of new products	Interval (0–9)
INNO_IMP	Novelty of product innovations	Sales share of improved products	Interval (0–9)
INNO_PA	Importance of patents	1 = Registration of patents, 0 = otherwise	Nominal
	Variables related to technological opportunities		
	Importance of external knowledge to firms' innovation activities		
TEC_CUCO		Customers and competitors as knowledge source: factor score	Metric
TEC_MARK		Suppliers as knowledge source: factor score	Metric
TEC_SCIE		Scientific institutions as knowledge source: factor score	Metric
TEC_UNIV		Universities as knowledge source: (1 = very low to 5 = very high)	Interval (1–5)
COOP_UNI	Joint R&D activities with universities	1 = R&D cooperation, 0 = otherwise	Nominal
	Barriers of innovation		
BAR_COST		Cost/riskness of innovation activities: factor score	Metric
BAR_FIN		Financial restrictions: factor score	Metric
BAR_GOV		Governmental intervention: factor score	Metric
BAR_MARK		Market impulse/demand condition: factor score	Metric
BAR_TEC		Internal technological resources: factor score	Metric

	Variables related to the firms' appropriability conditions		
	Extent to which technological knowledge can be protected from others regarding		
AP_FP_PR	– product innovations	Firm-specific mechanisms to protect product innovations: factor score	Metric
AP_LP_PR		Mechanisms to protect product innovations by law: factor score	Metric
AP_FP_PZ	– process innovations	Firm-specific mechanisms to protect process innovations: factor score	Metric
AP_LP_PZ		Mechanisms to protect process innovations by law: factor score	Metric
AP_FIRM	– product/process innovations	Firm-specific mechanism to protect innovations – all mechanism: factor score	Metric
AP_LAW		Mechanisms to protect innovations by law – all mechanism: factor score	Metric
	Market-related variables		
	Firm size (small firms with up to 50 employees as basic group)		
EMPL_MI		1 = 50 up to 249 employees, 0 = otherwise	Nominal
EMPL_G		1 = 250 and more employees, 0 = otherwise	Nominal
SALE_EXP	Sales expectations	Expected change of sales in 1993-1995 (1 = low to 5 = very high)	Interval (1–5)
INTERNAT	Intensity of international sales	Foreign sales/whole sales	Metric
	Aims of innovation		
AIM_COST		Reduction of innovation costs: factor score	Metric
AIM_DEMA		Expansion of demand abroad: factor score	Metric
AIM_DEMI		Expansion of demand at home: factor score	Metric
AIM_ENVI		Improvement of environment: factor score	Metric
AIM_PROD		Enlargement of production programme: factor score	Metric

statistics of the variables used in the model specifications are listed in Appendix Table 3A.1.

The *input-related* innovation variables reflect the engagement of firms to generate product and process innovations. This engagement includes expenditures for R&D, product design, trial production, market analysis, purchase of patents and licence, and training of employees related to innovation projects. We distinguish two special intensities (INNO_INT, R&D_INT). The logs of these intensities are computed because of the problems with non-normal distributions of these variables. The planning of R&D activities in the period 1993–95 (R&D_PLAN) is also defined as input-related variables. The *output-related* innovation variables are measured by the sales share of new and improved products (INNO_NOV, INNO_IMP), based on ordered information (ten-point scale). In addition, the registration of patents (INNO_PA) is used to identify the relevance of invention for firms. We argue that patents are a specific output factor of innovative activities without direct market references.[7] Unfortunately there was no information available on the number of patents.

Table 3.2 also summarizes information with respect to the further variables influencing the innovation process, especially the importance of external knowledge sources (*technological opportunities*) and the extent to which firms can protect their technological knowledge from other firms (*appropriability conditions*). These variables are closely associated with supply factors influencing the innovative activities of firms (Arvanitis and Hollenstein 1994; Rosenberg 1976). To measure the importance of university knowledge and other types of *technological opportunities* in the innovation process, the scores of factor analysis on external knowledge sources are employed (Appendix Table 3A.2). Following the considerations in the theoretical part of the chapter, we distinguish technological opportunities stemming from universities, among other scientific organizations (TEC_SCIE), suppliers (TEC_MARK) and competitors/ customers (TEC_CUCO). Further, in the estimations a variable reflecting *separately* the role of universities as knowledge sources (TEC_UNIV) is used. Finally, the empirical evidence of R&D cooperation as the most *systematic* form of knowledge transfer between universities will be checked. COOP_UNI defines firms having developed new or improved products jointly with universities formally or informally.

The degree to which external (technological) information is internalized depends on the extent to which firms can protect their knowledge from other companies (Cohen and Levinthal 1989; König and Licht 1995; Levin et al. 1987). *Appropriability conditions* define the ability of innovators to retain the returns of R&D.[8] The technological opportunities a single firm is faced with are not independent of these conditions, 'For example, one firm's feasible advances in technology may be blocked by the property rights of another' (Klevorick et al. 1995, p. 186). Therefore, several variables related to the firms' appropriability conditions are used in the regressions (AP_). We employ the scores of factor analysis (see Appendix Table 3A.3) on a specific mechanism of protecting technological knowledge from other firms regarding product and process innovations.

Market-related variables, well-known from other empirical studies,[9] are introduced in the estimations to reflect further factors determining the innovation process such as firm size (EMPL_MI, EMPL_G),[10] sales expectations in the medium term (SALE_EXP), and intensity of international sales (INTERNAT). These variables allow us to control the influence of size and demand factors on innovation activities, which may explain differences in innovation activities on the firm level. The role of firm size is a priori unknown, because these variables 'can be used as a proxy for various economic effects' (Arvanitis and Hollenstein 1996, p. 18). It can be expected that sales expectations (Schmookler 1966) and high shares of exports (Felder et al. 1996) will have positive signs in the estimations.

Restrictions of the data set make it impossible to include variables on the degree of competition in the firm's market (ten-firm concentration ratio or Herfindahl index) in the model specifications. But other studies have shown that the degree of competition has no significant effects or impacts of a comparable small order of magnitude on the innovation or R&D activities of firms, if the estimations are controlled by variables of technological opportunities (Arvanitis and Hollenstein 1996; Crépon et al. 1996; König and Licht 1995).

Depending on the kind of innovation variables, different methods are used to analyse the importance of university knowledge and to estimate the effects of technological opportunities stemming from the academic sphere (and other sources) to the innovation activities of firms in the German manufacturing industry. The estimation of our model specifications raises several statistical problems. First, some data were censored (Felder 1997). To avoid the identification of firms, the innovation and R&D intensities were censored in the upper tail of the distributions by 0.35 and 0.15 respectively. Second, some innovative firms did not perform any R&D and had no innovation expenditures. Accepting a misspecification of the model, the problem can be solved by using a Tobit model with censoring in both tails of the distributions. Possible misspecification may be attributed to the fact that independent parameters simultaneously determine the probability as well as the expenditures of innovation activities (Cohen et al. 1987). Therefore, we use the two-step version of the

Heckman model (Heckman 1979). This model specification allows us to identify the parameters affecting a firm's decision to participate in R&D (innovation activities) and the degree of its intensity. In the case of discrete variables, we employ the Probit methods for dichotomous response variables and ordered Probit models for the various multinomial variables. In all estimations, industry dummies were included to control industry effects.

In the following, the results of the investigations of the empirical evidence of university knowledge for *innovative* firms in the German manufacturing industry are presented. We concentrate on the question, if and to which extent

- universities are important knowledge sources for these firms;
- university knowledge (and other external sources) have significant effects on the innovation activities of such firms.

3.1 Importance of University Knowledge for Innovative Firms

On the first wave of the Mannheim Innovation Panel (1993), firms were asked to rate on a five-point scale the importance of several external knowledge sources of technological information for their innovation activities in the years 1990–92. As shown in Table 3.3, customers were rated as the most important sources for innovative firms in German manufacturing industry. Fairs and exhibitions, journals and conferences were also ranked as very important knowledge sources. Universities were ranked at a medium level, whereas the contribution of other scientific sources (for example, industry-financed research institutes and technical institutes) were rated as low.

Firms use technological information from customers, fairs and exhibitions as well as from journals and conferences to introduce new and improved products successfully by tracking down market needs. One important factor for success in competition is to evaluate future changes in demand and to address customers' needs (Christensen and Bower 1996). Thus, scientific information seems to be less important for industrial innovations, which apparently use more market-related information than new scientific findings. University knowledge seems to affect the development of innovations more *indirectly* by increasing a firm's R&D efficiency by enhancing in-house technological capacities rather than by generating technical advance *directly*.

Firms on the MIP 1993 were also asked whether they had formed R&D cooperations with other actors in 1992; 37.4 per cent of innovative firms had developed new or improved products jointly with other firms

Table 3.3 Importance of external knowledge sources

External knowledge sources	Mean	Standard deviation	Percentage of firms with valuation of no importance (1)	high importance (4 and 5)
Customers	4.2	1.04	2.6	81.5
Fairs and exhibitions	3.8	1.02	3.5	66.5
Journals and conferences	3.8	0.98	2.7	65.3
Competitors	3.7	1.14	6.5	63.8
Suppliers	3.5	1.22	8.7	55.7
Universities	2.6	1.33	31.8	29.2
Patent disclosures	2.5	1.35	33.2	28.4
Market research, advertising	2.2	1.12	37.0	13.8
Industry-financed research institutes	2.0	1.17	45.8	13.9
Technical institutes	2.0	1.15	48.2	13.2
Agencies of technology transfer	1.9	1.12	48.9	11.8

Source: Mannheim Innovation Panel 1993.

Table 3.4 Partners of R&D cooperations, 1992

Kind of partners*	Percentage of innovative firms
Universities	21.8
Customers	20.0
Suppliers	16.6
Other public-financed research institutes	12.8
Industry-financed research institutes	7.1
Competitors	7.0

Note: *Multiple answers possible.

Source Mannheim Innovation Panel 1993.

or public organizations. The various partners of these firms are listed in Table 3.4.

Although firms ranked the contribution of knowledge from universities as of moderate size, in 1992 most of the innovative firms (about 22 per cent) undertook R&D jointly with universities. Industry-financed research institutes as cooperation partners are less important for firms than universities or other public-financed research institutes. Not surprisingly, firms with academic R&D cooperations rated the importance of universities as external sources of technological information significantly higher than other firms.

3.2 Innovation Effects of University Knowledge

Innovation input effects

We investigate the impact of university (scientific) knowledge on innovation *input* activities using three models. In Model 1, we test the effects of the importance of universities among other scientific institutions (industry-financed research institutes, technical institutes, agencies of technology transfer and so on) on the level of the factor scores on external sources of information (TEC_SCIE). In Model 2, we check the contribution of universities alone as information sources (TEC_UNIV) with the two other non-scientific factors TEC_CUCO and TEC_MARK. In Model 3, we use the dummy variable COOP_UNI to identify firms which have developed new or improved products jointly with universities. The impacts of the science-related variables are estimated with regard to whether and how they can explain the probability and intensities of firms' innovation input activities in the German manufacturing industry (INNO_INT, R&D_INT). 'The relevance of planned R&D activities in the period 1993–95 (R&D_PLAN) is also taken into consideration. Table 3.5 summarizes the regression results.

Using the two-step version of the Heckman model, no significant effects of TEC_SCIE,TEC-UNIV and COOP_UNI on the probability of participating in the innovation process could be found. But TEC_SCIE and COOP_UNI have highly significant and positive effects on firms' innovation intensities. Further, the regression results indicate highly significant effects (at the 0.01 level) of the science-related variables on the participation in R&D. A high assessment for scientific/ university knowledge and for establishing R&D cooperation with universities increases the probability of R&D in the German manufacturing industry. With the exception of COOP_UNI, the estimations underline the importance of the stimulating effects of TEC_SCIE and TEC_UNIV on the intensity of R&D. Finally, the Probit models for planned R&D in the period 1993–95 point out significant and strong effects of the science-related variables.

The empirical investigation of the impacts of university (scientific) knowledge on the innovation input activities of firms in the German manufacturing industry underline that, in general, information stemming from these kinds of technological opportunities are used as *complements*. The adaptation of university (scientific) knowledge encourages the innovation and R&D intensities of German firms. In-house technological capacities can be expanded with positive effects on the engagement of developing new and improved products. In this context, Nelson and Wolff (1997) give empirical support on the level of certain lines of US business that the outcome of science can be regarded as pure opportunity enhancing. On the other hand, it has to be mentioned that 'by far the largest share of the work involved in creating and bringing to practice new industrial technology is carried out in industry, not in universities' (Rosenberg and Nelson 1994, p. 340).

However, as Harabi (1995) and Klevorick et al. (1995) remark, the impacts of academic research on R&D and innovation intensities may differ across industries (technologies). In some technology fields, university knowledge is used as a *substitute* for industrial research. Peters and Becker (1998), for example, find substitutive effects of academic research on in-house activities of firms in the German automobile supply industry. Some kinds of innovation activities, such as testing and prototype building, are outsourced by German automobile suppliers to university and scientific laboratories, which yields remarkable savings in innovation costs (see also Peters and Becker 1999). In this case, the extent of cost savings is larger than the stimulating (complementing) impact of academic research on in-house innovation activities.

In the estimations, no significant effects of TEC_MARK (information from suppliers) as another kind of external knowledge source on the innovation/R&D activities of firms in the German manufacturing industry could be found. The negative sign of the coefficients for firms' innovation intensities can partly be explained by the fact that suppliers' information as well as information from journals and conferences tend to be a *substitute* for the firms' in-house activities, as also found in studies for the USA (Cohen and Levinthal 1989; Levin and Reiss 1988; Nelson and Wolff 1997). Technological opportunities stemming from customers and competitors *together*, have a particular impact on the firms' R&D intensity. The coefficients for TEC_CUCO are positive and significant both for the probability as well as for the intensity of R&D.[11]

In general, the variables related to appropriability conditions have stimulating effects on innovation/R&D activities. Firm-specific strategies to protect knowledge from other companies (AP_FIRM) increase the

Table 3.5 Effects of university knowledge on the innovation input of firms in the German manufacturing industry

	INNO_INT					
	1		2		3	
Variables	Particip Coeff. (t-values)	Intensity Coeff. (t-values)	Particip. Coeff (t-values)	Intensity Coeff. (t-values)	Particip. Coeff. (t-values)	Intensity Coeff (t-values)
INTERCEPT	2.03***	−3.81***	2.09***	−3.88***	2.01***	−3.84***
	(3.80)	(−14.20)	(3.65)	(−13.64)	(3.74)	(−14.89)
EMPL_MI	0.75***	−0.05	0.75***	−0.04	0.75***	−0.08
	(2.64)	(−0.31)	(2.64)	(−0.21)	(2.64)	(−0.46)
EMPL_G	0.65**	−0.59***	0.65**	−0.55***	0.64**	−0.62***
	(2.38)	(−3.54)	(2.38)	(−3.22)	(2.25)	(−3.87)
SALE_EXP	0.03	0.22***	0.03	0.22***	0.03	0.22***
	(0.29)	(5.14)	(0.27)	(5.02)	(0.28)	(5.39)
AP_FIRM	0.07**	0.11*	0.23**	0.12**	0.23**	0.11**
	(2.39)	(1.88)	(2.38)	(2.13)	(2.34)	(2.04)
AP_LAW	0.67	−0.02	0.07	0.01	0.05	0.01
	(0.59)	(−0.43)	(0.59)	(0.15)	(0.45)	(0.11)
INTERNAT	0.77	0.28	0.77	0.29	0.75	0.24
	(1.13)	(1.29)	(1.13)	(1.32)	(1.09)	(1.16)
TEC_CUCO	0.08	0.09*	0.08	0.08	0.09	0.09*
	(0.81)	(1.71)	(0.82)	(1.56)	(0.88)	(1.79)
TEC_MARK	−0.19*	−0.01	−0.18	−0.02	−0.18	−0.01
	(−1.66)	(−0.24)	(−1.63)	(−0.28)	(−1.62)	(−0.05)
TEC_SCIE	−0.04	0.13**				
	(−0.32)	(2.44)				
TEC_UNIV			−0.03	0.04		
			(−0.34)	(0.97)		
COOP_UNI					0.02	0.31***
					(0.07)	(2.95)
χ^2 (d.o.f.)	2.27	19.45**	2.27	18.15*	2.15	20.73**
industry dummies	(10)	(10)	(10)	(10)	(10)	(10)
Number of observations	1309	1289	1309	1289	1284	1264
Log likelihood	−86.14		−86.13		−85.79	
McFaddens R^2	0.17		0.17		0.17	
Model F-statistic		12.03***		11.40***		12.00***

Table 3.5 Effects of university knowledge on the innovation input of firms in the German manufacturing industry (continued)

R&D_INT						R&D_PLAN		
1		2		3		1	2	3
Particip Coeff. (*t*-values)	Intensity Coeff. (*t*-values)	Particip. Coeff. (*t*-values)	Intensity Coeff. (*t*-values)	Particip. Coeff. (*t*-values)	Intensity Coeff (*t*-values)	Coeff. (*t*-values)	Coeff. (*t*-values)	Coeff (*t*-values)
–0.66***	–4.93***	–0.87***	–5.14***	–0.69***	–5.00***	–0.26	–0.63**	–0.31
(–3.69)	(–3.35)	(–4.43)	(–3.23)	(–3.69)	(–3.51)	(–1.16)	(–2.52)	(–1.30)
0.53***	–0.49	0.53***	–0.48	0.46***	–0.49	0.41***	0.41***	0.36***
(5.90)	(–1.24)	(5.88)	(–1.25)	(5.09)	(–1.41)	(3.49)	(3.50)	(3.05)
0.87***	–0.77	0.87***	–0.75	0.74***	–0.79*	0.45***	0.45***	0.37***
(8.76)	(–1.40)	(8.77)	(–1.37)	(7.24)	(–1.71)	(3.50)	(3.53)	(2.82)
0.03	0.17***	0.03	0.17***	0.03	0.18***	0.11**	0.11**	0.11**
(0.92)	(4.58)	(0.92)	(4.56)	(0.76)	(4.83)	(2.41)	(2.37)	(2.25)
0.16***	0.05	0.17***	0.07	0.15***	0.07	0.23***	0.24***	0.24***
(4.34)	(0.49)	(4.53)	(0.59)	(3.97)	(0.66)	(5.02)	(5.27)	(4.97)
0.16***	0.06	0.18***	0.09	0.19***	0.11	0.26***	0.28***	0.29***
(3.91)	(0.58)	(4.36)	(0.81)	(4.79)	(0.97)	(4.60)	(5.14)	(5.25)
1.00***	0.65	1.01***	0.66	0.92***	0.61	1.01***	1.01***	0.95***
(5.02)	(1.34)	(5.09)	(1.34)	(4.52)	(1.41)	(3.62)	(3.66)	(3.32)
0.07*	0.11**	0.06	0.10*	0.08*	0.11*	0.03	0.01	0.03
(1.69)	(1.96)	(1.49)	(1.84)	(1.94)	(1.88)	(0.55)	(0.22)	(0.60)
0.02	0.00	0.02	–0.01	0.02	0.01	0.06	0.05	0.06
(0.56)	(–0.01)	(0.47)	(–0.07)	(0.56)	(0.10)	(1.26)	(1.13)	(1.16)
0.13***	0.18**					0.22***		
(2.93)	(2.27)					(3.90)		
		0.09***	0.10*				0.16***	
		(2.88)	(1.72)				(3.95)	
				0.96***	0.39			1.52***
				(7.35)	(1.04)			(5.19)
77.86***	25.66***	75.92***	25.60***	66.24***	25.94***	67.98***	67.30***	58.28***
(10)	(10)	(10)	(10)	(10)	(10)	(10)	(10)	(10)
1610	1129	1610	1129	1580	1112	1454	1454	1434
–772.51		–772.65		–731.37		–451.38	–451.25	–425.41
0.21		0.21		0.24		0.26	0.26	0.29
	16.99***		16.50***		16.68***			

Notes: Industry dummies included. * significant at the 1.0 level. ** significant at the 0.05 level. *** significant at the 0.01 level.

probability of participating in the innovation process and the intensity of investment in the development of new and improved products significantly. Mechanisms of protecting innovations by law (AP_LAW) have strongly significant effects on firms' participation in R&D, not on the intensity of R&D investments. The firms' decisions to plan R&D activities in the future are positively influenced on the 0.01 significant level by AP_FIRM and AP_LAW.

Further, we found highly significant explanatory power of the market-related variables of firm-size classifications (EMPL_MI, EMPL_G) on the probability of engaging in innovation/R&D. Large and middle-sized firms in the German manufacturing industry have a higher probability of investing in their R&D and innovation process. The effects of the incurred firm-size variables on the intensity to undertake innovation/R&D expenditures are negative. These findings are in line with studies in other countries (Cohen and Klepper 1996b; Evangelista et al. 1997; Kleinknecht 1996). Large firms spend less money (compared to their sales) on their total innovation activities than small firms.

Finally, high shares of exports (SALE_EXP) motivate firms to invest in innovation/R&D activities (significant at the 0.01 level) which strengthens the hypothesis of Schmookler (1966). The intensity of international sales (INTERNAT) has positive and highly significant effects only on the decision of firms in the German manufacturing industry to engage in R&D. At least, the regression results underline strongly motivating influences of the market-related variables on the planned R&D in the period 1992–95.

Innovation output effects

We used the set of explanatory factors as on the input level but with slight modifications to estimate the effects of university (scientific) knowledge on the innovation *output* of firms in the German manufacturing industry. The impacts were tested for the sales share of new products (INNO_NOV), improved products (INNO_IMP) and the registration of patents (INNO_PA). Table 3.6 presents the results of these estimations.

Surprisingly, we found no statistically significant, positive effects of the proxy of technological opportunities stemming from scientific sources (TEC_SCIE). Empirical evidence for the importance of universities as well as other scientific organizations as external knowledge sources could not be found in the three models for the innovation output. Separated university knowledge (TEC_UNIV) has only strongly positive effects (at the 0.05 level) on the probability of registering patents in the German manufacturing industry, which reveals a pattern confirmed by the findings of Grupp (1996).

Similar to the input-related estimations, the regression results point out the positive impact of COOP_UNI. R&D cooperation with universities as the most systematic form of knowledge transfer increases the probability of higher sales shares of improved products (INNO_IMP) significantly but quite surprisingly, not on the probability of higher sales shares of new products (INNO_NOV) on a statistically significant level. This is contrary to our theoretical (a priori) considerations in Section 2, but in line with the findings of Arvanitis and Hollenstein (1996). They also found negative (but insignificant) effects of technological opportunities stemming from scientific knowledge sources on the sales share of new products in the case of Swiss manufacturing firms. The results of Peters and Becker (1999) for the German automobile supply industry are similar. The main motivation of automobile suppliers to cooperate (formally) with universities is to attain knowledge in order to improve their *specific* capacities rather than their *basic* R&D capabilities to develop automobile parts more successfully.

Comparing the regression results for the science-related variables, one main finding is that in general the explanation power of TEC_SCIE, TEC_UNIV and COOP_UNI is much stronger in the regressions of the innovation input than in the estimations of the innovation output. Especially the application of the ordered Probit models for INNO_NOV yield insufficient results, which can in general be seen by the level of the McKelvey and Zavoina R^2. Technological opportunities regarding university knowledge seem to improve the quality of products more *indirectly* by increasing firms' R&D efficiency by enhancing in-house technological capacities rather than by generating technical advance *directly*.

Nevertheless, the technological opportunity proxy TEC_MARK has positive and highly significant impacts on INNO_NOV. The higher firms rank the importance of external knowledge from suppliers, the higher the sales share of new products. The regressions underline stimulating and strong effects (significant at 0.01) of technological opportunities stemming from customers and competitors (TEC_CUCO) only on the sales share of improved products. Further, we found positive and mostly significant effects of appropriability conditions (AP_) on the innovation output of firms in the German manufacturing industry.

The estimations point out the empirical relevance of market-related factors on the innovation output side. The effects of the firm-size variables (EMPL_MI, EMPL_G) are ambiguous. On the one hand, larger firms have remarkably lower sales shares of new products. On the other hand, positive and significant effects (at the 0.05 level) on the sales share of improved products could be identified for EMPL_MI. Finally, larger firms have a higher probability of patenting their new findings. This finding strengthens the presumption that it is more likely for larger firms to invest in process innovation activities than for smaller firms, which per-

Table 3.6 Effects of university knowledge on the innovation output of firms in the German manufacturing industry

	INNO_NOV		
	1	2	3
Variables	Coeff. (*t*-values)	Coeff. (*t*-values)	Coeff. (*t*-values)
INTERCEPT	0.17 (1.06)	0.09 (0.58)	0.16 (0.98)
EMPL_MI	–0.04 (–0.51)	–0.04 (–0.52)	–0.04 (–0.57)
EMPL_G	–0.13 (–1.62)	–0.13* (–1.67)	–0.14* (–1.69)
SALE_EXP	0.11*** (3.82)	0.11*** (3.82)	0.11*** (3.82)
INTERNAT	–0.16 (–1.17)	–0.15 (–1.13)	–0.17 (–1.23)
AP_FP_PR	0.16*** (5.31)	0.16*** (5.36)	0.16*** (5.22)
AP_LP_PR	0.05* (1.61)	0.05* (1.72)	0.06* (1.94)
AP_FP_PZ			
AP_LP_PZ			
AP_FIRM			
AP_LAW			
TEC_CUCO	0.05* (1.67)	0.05 (1.58)	0.05 (1.59)
TECMARK	0.09*** (2.98)	0.09*** (2.94)	0.09*** (3.05)
TEC_SCIE	0.03 (1.01)		
TEC_UNIV		0.03 (1.33)	
COOP_UNI			0.09 (1.15)
χ^2 (d.o.f.) industry dummies	42.51*** (10)	41.92*** (10)	41.23*** (10)
Numbers of observations	1398	1398	1373
Log likelihood	–2867.57	–2867.27	–2816.80
McKelvey/Zavoina R^2	0.05	0.05	0.05

Table 3.6 Effects of university knowledge on the innovation output of firms in the German manufacturing industry (continued)

INNO_IMP			INNO_PA		
1	2	3	1	2	3
Coeff. (*t*-values)	Coeff. (*t*-values)	Coeff. (*t*-values)	Coeff. (*t*-values)	Coeff. (*t*-values)	Coeff. (*t*-values)
0.61***	0.62***	0.57***	−1.64***	−1.80***	−1.74***
(4.07)	(3.92)	(3.80)	(−7.88)	(−8.05)	(−8.01)
0.17**	0.17**	0.18**	0.52***	0.51***	0.47***
(2.38)	(2.43)	(2.55)	(4.45)	(4.38)	(3.93)
0.04	0.06	0.03	1.26***	1.25***	1.16***
(0.54)	(0.72)	(0.39)	(11.05)	(10.94)	(9.78)
0.06**	0.06**	0.06**	−0.02	−0.02	−0.02
(2.24)	(2.25)	(2.22)	(−0.56)	(−0.57)	(−0.49)
0.03	0.03	0.03	0.91***	0.92***	0.89***
(0.19)	(0.21)	(0.09)	(5.10)	(5.16)	(4.91)
0.07**	0.07**	0.06**			
(2.35)	(2.51)	(2.06)			
0.06*	0.06*	0.05			
(1.76)	(2.04)	(1.63)			
			0.11**	0.10**	0.09**
			(2.33)	(2.33)	(2.04)
			0.54***	0.54***	0.55***
			(11.67)	(12.01)	(12.19)
0.07***	0.07***	0.018***	0.06	0.06	0.06
(2.68)	(2.63)	(2.73)	(1.29)	(1.28)	(1.23)
0.03	0.03	0.03	−0.02	−0.02	−0.02
(0.96)	(0.98)	(0.90)	(−0.43)	(−0.50)	(0.35)
0.02			0.06		
(0.61)			(1.42)		
	−0.01			0.07**	
	(−0.24)			(2.06)	
		0.17**			0.52***
		(2.07)			(5.54)
23.39***	23.45**	22.86***	58.69***	55.89***	49.49***
(10)	(10)	(10)	(10)	(10)	(10)
1353	1353	1327	1592	1592	1563
−2741.85	−2742.02	−2687.67	−674.51	−673.39	−644.89
0.11	0.11	0.11	0.34	0.34	0.35

Notes: Industry dummies included. *significant at the 0. 1 level. **significant at the 0.05 level. ***significant at the 0.01 level.

form product R&D rather than process R&D. At least, high shares of exports (SALE_EXP) have significantly stimulating impacts on the sales share of new and improved products of firms in the German manufacturing industry. Firms' intensity of international trade (INTERNAT) shows empirical evidence only for patents.

4 DETERMINANTS OF R&D COOPERATIONS BETWEEN UNIVERSITIES AND FIRMS

The descriptive statistics in Section 3.1 as well as the regression results reported in Section 3.2 underline the empirical evidence of R&D cooperations as the most systematic form of knowledge transfer between universities and firms and their importance for industrial R&D. But what are the main factors determining the decision of firms in the German manufacturing industry to cooperate with universities?

In Section 2 we have discussed the motivations of firms for being interested in R&D cooperations with non-profit organizations. We have argued that a close relationship between academic research and industrial (application-oriented) research increases in a specific way the firms' possibilities for

- continuous anticipation of new research interests and technological development paths;
- approving the practical relevance of new scientific findings, new breakthroughs at an early stage;
- the task- and problem-oriented utilization and realization of new scientific knowledge and improved technological methods, instruments and skills.

In this context, variables reflecting the firms' restrictions in the innovation process ('barriers of innovation') are taken into consideration in investigating the conditions affecting the willingness to cooperate with universities (factor scores, see Appendix Table 3A.4). Further, we use specific market-oriented ambitions ('aims of innovative activities') generated by a factor analysis of 21 potential objectives on a five-point scale (see Appendix Table 3A.5). We also controlled for the effects of firm size and demand expectation. Especially, the significant impact of firm size on the willingness to cooperate with universities has been pointed out in other studies (Peters and Becker 1999; Veugelers 1997).

The regression results are listed in Table 3.7. The Wald test reveals that barriers of innovations (BAR_) have jointly positive and highly significant

Table 3.7 Determinants of R&D cooperations with universities

Variables	COOP_UNI 1	2	3
	Coeff. (*t*-values)	Coeff. (*t*-values)	Coeff. (*t*-values)
INTERCEPT	–1.56***	–1.18***	–0.13***
	(–8.44)	(–6.29)	(–6.65)
EMPL_MI	0.46***	0.38***	0.45***
	(4.08)	(3.48)	(3.87)
EMPL_G	1.14***	0.98***	1.07***
	(10.27)	(9.43)	(9.35)
SALE_EXP	0.05	0.01	0.03
	(1.31)	(0.30)	(0.81)
INTERNAT	0.45***	0.03	0.07
	2.73	(0.16)	(0.39)
BAR_COST	0.07*		0.06
	(1.91)		(1.63)
BAR_GOV	0.10***		0.10**
	(2.67)		(2.45)
BAR_MARK	0.16***		0.13***
	(4.24)		(3.33)
BAR_FIN	–0.02		0.02
	(–0.04)		(0.52)
BAR_TEC	–0.02		–0.04
	(–0.62)		(–1.04)
AIM_COST		–0.04	–0.07
		(–0.99)	(–1.73)
AIM_DEMA		0.23***	0.20***
		(5.38)	(4.63)
AIM_DEMI		–0.04	–0.06
		(–1.23)	(–1.56)
AIM_PROD		0.17***	0.16***
		(4.10)	(3.63)
AIM_ENVI		0.11***	0.08*
		(2.96)	(1.84)
χ^2 (d.o.f.) barriers	26.77***		20.44***
	(5)		(5)
χ^2 (d.o.f.) aims		38.05***	38.05***
		(5)	(5)
χ^2 (d.o.f.) industry dummies	43.90***	51.69***	44.29***
	(5)	(5)	(5)
Number of observations	1784	1844	1745
Log likelihood	–793.15	–802.80	–757.18
McFaddens R^2	0.17	0.18	0.19

Notes: Industry dummies included. *significant at the 0. 1 level. **significant at the 0.05 level. ***significant at the 0.01 level.

effects on the probability of cooperations with universities. In particular, firms in the German manufacturing industry are interested in such cooperations when the costs of innovations are high (BAR_COST). As shown in our study for the German automobile industry, suppliers cooperate with universities to save R&D costs (Peters and Becker 1999). But we have to bear in mind that university research may be used as a substitute for basic research done by industry which is on average very small compared to the whole industrial innovation expenditures (see also Nelson and Wolff 1997).[12]

The positive coefficients of BAR_MARK (significant at the 0.01 level) underline that insufficient technological opportunities stemming from the industrial sector (customers and suppliers) stimulate the firms' willingness to cooperate with universities. This can be interpreted as follows: firms are aware that they must establish R&D cooperations (networks) to obtain expertise which cannot be generated in-house. Realizing that they cannot learn from spillovers of firms in the value-added line because of low innovation activities of suppliers/customers or because firms try to develop technologies with high novelty for future markets, they organize and establish cooperations with scientific institutions with high technological opportunities. Within these cooperations, firms try to generate new *market potentials* in a direct way.

The effect of BAR_TEC is insignificant. But this variable is highly aggregated and inhibits different aspects of innovation barriers, making its interpretation very difficult. Regressions using only the firms' responses to questions of innovation barriers reveal, on the one hand, that the internalization of university knowledge expands the *generic* technological capacities of firms. But, on the other hand, firms have to build up absorptive capacities before they can cooperate with universities efficiently (Arora and Gambardella 1994; Malerba and Torrisi 1992; Peters and Becker 1999). The probability of cooperation increases if firms have sufficient knowledge about the state of the art in the research of new technologies and the capacities to adapt and implement knowledge generated in the academic sphere.

The variables defining the aims of innovation activities (AIM_) have jointly positive effects (at the 0.01 level) on the probability of R&D cooperations with universities. Our argument that the firms' probability of cooperating with universities is positively affected by the creation of new markets is still strengthened by the aims of innovation activities. AIM_DEMA and AIM_PROD, respectively, describe the objectives to open up new markets in foreign countries and to enlarge market shares (positions). The regression results indicate high significance of both factors for COOP_UNI. In special estimates, not illustrated here, we found empirical evidence for the assumption that firms without a lack of technological information assessed a higher importance of universities as partners in R&D.

Finally, the estimations for the three model specifications underline the great importance of the market-related variable firm size. Large and middle-sized firms in the German manufacturing industry (EMPL_G and EMPL_MI) have a higher probability of cooperating with universities than smaller ones. The significant positive impact of INTERNAT diminishes if we take into consideration the innovation aim of firms to expand or create new markets abroad (AIM_DEMA).

5 SUMMARY

The importance of university (scientific) knowledge in the innovation process as external knowledge sources for innovative firms has increased continuously over time. The development of new and improved products depends more and more on the adaptation of technological opportunities stemming from the academic sphere.

Against the background of theoretical considerations about the interrelation of innovation (R&D) activities and technological opportunities, the importance and effects of university (scientific) knowledge on the innovation activities of firms in the German manufacturing industry were investigated empirically. The regression results related to the science-related variables are summarized in Table 3.8. In general, knowledge generated in the academic sphere has significant effects on the innovation activities of German firms in the manufacturing sector. The impacts are stronger on variables relating to the innovation input than on variables relating to the innovation output.

Technological opportunities stemming from universities are used as *complements* on the innovation input side. The adaptation of university (scientific) knowledge encourages the innovation and R&D intensities of firms in the German manufacturing industry. Technological capacities can be expanded with positive effects on the development of new and improved products.

On the innovation output side, we found no empirical evidence for statistically significant, positive effects of university (scientific) knowledge on the sales share of new or improved products. These findings for the German manufacturing industry are contrary to comparable studies in other countries (Faulkner and Senker 1994; Mansfield 1995). We have to analyse the data set in more detail, especially to investigate sectoral peculiarities, to reveal the reasons for the specific constellations in Germany.

Table 3.8 Regression results of the innovation input and output effects of the science-related variables

	Scientific knowledge (TEC_SCIE)	University knowledge (TEC_UNIV)	R&D cooperations (COOP_UNI)
	Innovation input		
Innovation participation	–	–	+
Innovation intensity	+**	+	+***
R&D participation	+***	+***	+***
R&D intensity	+**	+*	+
Planning of R&D	+***	+***	+***
	Innovation output		
Sales share of new products	+	+	+
Sales share of improved products	+	–	+**
Patent registration	+	+**	+***

Notes: * significant at the 0.1 level. ** significant at the 0.05 level. *** significant at the 0.01 level.

Further, the regression results point out the relevance of R&D cooperations as the most systematic form of knowledge transfer between universities and firms and their importance for industrial R&D. In the German manufacturing industry, R&D cooperations with universities increase the probability of higher sales of improved products and patent registrations significantly.

On the basis of the regression results, further theoretical and empirical work has to be done to clarify the very complex relations between university (scientific) knowledge and innovation activities regarding industry-specific and product-specific aspects. Further investigations will be conducted to prove the relevance of faculty-specific differences of universities and the kind of academic research. In addition, the function of the firms' absorptive capacities to adapt external knowledge efficiently has to be analysed in detail. Finally, the impact of specific features of R&D cooperation on the industrial innovation process has to be examined. Our regression results underline in general the high empirical evidence for R&D cooperations. Obviously, what is exchanged is not only information. Thus, more investigation has to be done to reveal how firms cooperate with institutions in the academic sphere and how such arrangements are organized and managed efficiently.

APPENDIX 3A

Table 3A.1 Descriptive statistics of variables used in the model specifications

Variable	Description	Mean	Std. dev.
AIM_COST	Aim of innovation activities: reduction of innovaton costs (Factor score)	0	1
AIM_DEMA	Aim of innovation activities: expansion of demand abroad (Factor score)	0	1
AIM_DEMI	Aim of innovation activities: expansion of demand in the home country (Factor score)	0	1
AIM_ENVI	Aim of innovation activities: improvement of environmental issues (Factor score)	0	1
AIM_PROD	Aim of innovation activities: enlargement of production programme (Factor score)	0	1
AP_FIRM	Firm-specific mechanism to protect innovations – all mechanisms (Factor score)	0	1
AP_FP_PR	Firm-specific mechanisms to protect product innovations (Factor score)	0	1
AP_FP_PZ	Firm-specific mechanisms to protect product innovations (Factor score)	0	1
AP_LAW	Mechanisms to protect innovations by law – all mechanism (Factor score)	0	1
AP_LP_PR	Mechanisms to protect product innovations by law (Factor score)	0	1
AP_LP_PZ	Mechanisms to protect process innovations by law (Factor score)	0	1
BAR_COST	Barriers to innovation: costs and risks of innovation activities (Factor score)	0	1
BAR_FIN	Barriers to innovation: financial restrictions (Factor score)	0	1
BAR_GOV	Barriers to innovation: governmental intervention (Factor score)	0	1
BAR_MARK	Barriers to innovation: market impulse and demand condition (Factor score)	0	1
BAR_TEC	Barriers to innovation: internal technological resources (Factor score)	0	1

Variable	Description	Mean	Std. dev.
EMPL_MI	Firm size: 1 = 50 up to 249 employees, 0 = otherwise	0.33	0.47
EMPL_G	Firm size: 1 = 250 and more employees, 0 = otherwise	0.36	0.48
COOP_UNI	Joint R&D activities with universities: 1 = R&D cooperation, 0 = otherwise	0.22	0.41
INNO_IMP	Sales share of improved products (Interval 0-9)	4.33	2.56
INNO_INT	Log of innovation expenditures to sales ratio (Metric)	–3.09	1.65
INNO_NOV	Sales share of new products (Interval 0-9)	3.55	2.73
INNO_PA	Importance of patents: 1 = Registration of patents, 0 = otherwise	0.31	0.46
INTERNAT	Intensity in international sales (Metric)	0.19	0.23
R&D_INT	Log of R&D expenditures to sales ratio (Metric)	–5.87	2.77
R&D_PLAN	Mid-term planning of R&D: 1 = Planning of R&D in 1993-1995, 0 otherwise	1.07	0.77
SALE_EXP	Sales expectations (Interval 1-5)	3.28	1.08
TEC_SCIE	Scientific institutions as knowledge source (Factor score)	0	1
TEC_CUCO	Customers and competitors as information source (Factor source)	0	1
TEC_MARK	Suppliers as knowledge source (Factor source)	0	1
TEC_UNIV	Universities as information source (Interval 1-5)	2.56	1.33

Table 3A.2 External sources of technological knowledge: factor sources

	Factor TEC_SCIE	Factor TEC_MARK	Factor TEC_CUCO
TEC_TI	0.86	0.04	0.03
TEC_UNIV	0.81	0.04	0.05
TEC_AGEN	0.75	0.13	0.06
TEC_RI	0.73	0.07	0.09
TEC_PADI	0.58	0.10	0.27
TEC_JOUR	0.16	0.82	–0.03
TEC_FAIR	0.01	0.80	0.22
TEC_SUPP	0.06	0.50	0.09
TEC_CUST	0.10	0.13	0.82
TEC_COMP	0.14	0.13	0.80

Note: Kaiser–Meyer–Olkin Measure of Sampling Adequacy: 0.80; Bartlett Test of Sphericity: 4758.95.

Table 3A.3 Appropriability conditions of firms: factor scores

	Factor AP_LAW	Factor AP_FIRM
AP_PA_PR	0.83	0.04
AP_PA_PZ	0.82	0.14
AP_CO_PZ	0.79	0.16
AP_CO_PR	0.76	0.06
AP_DE_PZ	0.11	0.73
AP_LE_PZ	0.24	0.71
AP_LO_PZ	–0.05	0.70
AP_DE_PR	0.06	0.63
AP_LO_PR	–0.04	0.62
AP_SE_PZ	0.42	0.57
AP_LE_PR	0.32	0.56
AP_SE_PR	0.40	0.50

Note: Kaiser–Meyer–Olkin Measure of Sampling Adequacy: 0.68; Bartlett Test of Sphericity: 9709.27.

	Factor AP_FP_PR	Factor AP_LP_PR
AP_DE_PR	0.74	0.03
AP_LO_PR	0.69	–0.05
AP_LE_PR	0.68	0.32
AP_SE_PR	0.53	0.42
AP_PA_PR	0.06	0.87
AP_CO_PR	0.10	0.85

Note: Kaiser–Meyer–Olkin Measure of Sampling Adequacy: 0.71; Bartlett Test of Sphericity: 2007.44.

	Factor AP_FP_PZ	Factor AP_LP_PZ
AP_LE_PZ	0.80	0.22
AP_DE_PZ	0.77	0.12
AP_LO_PZ	0.77	0.01
AP_SE_PZ	0.61	0.41
AP_PA_PZ	0.14	0.90
AP_CO_PZ	0.14	0.90

Note: Kaiser–Meyer–Olkin Measure of Sampling Adequacy: 0.74; Bartlett Test of Sphericity: 3248.15.

Table 3A.4 Barriers of innovation: factor scores

	Factor BAR_COST	Factor BAR_GOV	Factor BAR_TEC	Factor BAR_MARK	Factor BAR_FIN
HEM_5	0.78	0.10	0.13	0.11	0.18
HEM_6	0.78	0.08	0.13	0.12	0.09
HEM_1	0.76	0.07	0.10	0.14	0.03
HEM_2	0.71	0.10	0.16	0.11	0.12
HEM_17	0.05	0.89	0.08	0.09	0.06
HEM_16	0.08	0.89	0.10	0.07	–0.05
HEM_18	0.21	0.63	0.04	0.24	0.16
HEM_11	0.07	0.08	0.75	0.14	0.14
HEM_10	0.11	0.08	0.71	0.11	–0.08
HEM_9	0.07	0.09	0.70	0.05	0.08
TEC_NOOP	0.23	–0.05	0.45	0.12	–0.02
HEM_14	0.10	0.10	0.40	0.26	0.32
HEM_19	0.15	0.09	0.11	0.83	0.05
HEM_20	0.09	0.21	0.19	0.76	0.10
DEMA_NO	0.20	0.06	0.17	0.69	0.02
HEM_3	0.17	0.07	0.05	0.01	0.91
HEM_4	0.14	0.04	0.07	0.06	0.90

Note: Kaiser–Meyer–Olkin Measure of Sampling Adequacy: 0.79; Bartlett Test of Sphericity: 11742.33.

Table 3A.5 Aims of innovation activities: factor scores

	Factor AIM_COST	Factor AIM_DEMA	Factor AIM_ENVI	Factor AIM_PROD	Factor AIM_DEMI
IN_AIM15	0.77	0.03	–0.13	0.11	0.04
IN_AIM16	0.73	0.06	0.10	0.09	0.04
IN_AIM18	0.71	0.07	0.21	0.10	0.07
IN_AIM19	0.62	0.07	0.34	0.12	0.02
IN_AIM17	0.60	–0.03	0.47	–0.09	0.11
IN_AIM14	0.44	–0.03	0.32	0.30	0.14
IN_AIM10	0.04	0.88	0.00	0.10	–0.05
IN_AIM9	0.01	0.82	0.09	0.08	–0.06
IN_AIM11	0.06	0.80	0.03	0.07	0.14
IN_AIM8	0.09	0.61	0.01	0.19	0.48
IN_AIM21	0.22	0.05	0.80	–0.05	0.00
IN_AIM13	–0.04	0.12	0.73	0.22	0.07
IN_AIM20	0.40	–0.03	0.63	0.10	0.10
IN_AIM12	0.27	–0.07	0.38	0.37	0.08
IN_AIM1	0.08	0.26	0.02	0.66	–0.07
IN_AIM2	0.10	0.02	0.01	0.64	0.12
IN_AIM3	0.04	0.01	0.08	0.60	0.17
IN_AIM4	0.05	0.07	0.06	0.42	0.05
IN_AIM5	0.09	0.13	0.04	0.17	0.81
IN_AIM6	0.09	–0.04	0.12	0.10	0.80

Note: Kaiser–Meyer–Olkin Measure of Sampling Adequacy: 0.83; Bartlett Test of Sphericity: 10914.748.

NOTES

We thank Jürgen Dietz for his help and support in data analysis.

1. We use 'university' knowledge (research) and 'academic' knowledge (research) synonymously.
2. Absorptive capacities can be defined as the ability 'to identify, assimilate, and exploit knowledge from the environment' (Cohen and Levinthal 1989, p. 569). Firms have to invest in complementary in-house R&D in order to understand and implement the results of externally performed R&D and to obtain full access to the research findings of other firms and institutions (Mowery and Rosenberg 1989; Veugelers 1997).
3. Different attempts have been undertaken to make the concept of technological opportunities precise and empirically operational (Cohen 1995; Dosi 1988; Harabi 1995; Klevorick et al. 1995). In this chapter, we follow the framework of the evolutionary theory of technical progress.
4. For more theoretical considerations, see Becker and Peters (1998).
5. The collection of the data was arranged by the Zentrum für Europäische Wirtschaftsforschung (ZEW) and the Institut für angewandte Sozialforschung (INFAS). We thank the ZEW for the permission to use the censored version of the survey data (Version 97- 1) for empirical investigations.
6. For more details, see Felder et al. (1995); Harhoff and Licht (1994).
7. For the general discussion of the status of patents in the innovation process, see Archibugi (1992); König and Licht (1995); OECD (1996).
8. The higher (lower) the appropriability conditions of firms, the fewer (more) R&D spillovers will occur. R&D spillovers are externalities of R&D activities beyond their primary definition, where not only the innovator has the benefit, but which can also be applied elsewhere (Eliasson 1996; Griliches 1992; Nelson 1992). In the case of intra-industry spillovers, the incentives of firms to invest in R&D will diminish (Harhoff 1996; Levin and Reiss 1988; Spence 1984). In the case of spillovers stemming from sources outside the market (for example from universities), positive effects of R&D spillovers on the efficiency and output of innovation activities can be expected (Bernstein 1989; Nadiri 1993; Peters 1998).
9. For an overview, see Cohen (1995).
10. Firm size is a categorial variable with three extensions. We define the category 'small firms' (up to 50 employees) as the basic group. EMPL_MI (50 up to 249 employess) and EMPL_G (250 and more employees) are used as dummies in the estimations.
11. In the regressions above we have not distinguished the effects of customer information and of competitor information on the firms' in-house R&D. In separate regressions not reported here, technological opportunities stemming from customers which inhibit technological-based as well as demand-related information have stimulating impacts on R&D activities. The effects of the contribution of competitors was positive on in-house R&D (but with lower significance) because firms can learn from competitors' R&D spillovers.
12. As basic research done by industry is very small in almost all cases, we could not find a negative sign of technological opportunities stemming from universities in our regressions. The pure enhancing effect of university knowledge on the firms' applied research (and other related innovation activities) is much larger than the substitution effect in basic research.

REFERENCES

Acs, Z.J., D.B. Audretsch and M.P. Feldman (1992), 'Real effects of academic research: comment', *American Economic Review*, **82**, 363–7.

Acs, Z.J., D.B. Audretsch and M.P. Feldman (1994), 'R&D spillovers and recipient firm size', *Review of Economics and Statistics*, **76**, 336–40.

Archibugi, D. (1992), 'Patenting as an indicator of technological innovation: a review', *Science and Public Policy*, **19**, 357–68.

Arora, A. and A. Gambardella (1994), 'Evaluating technological information and utilizing it: scientific knowledge, technological capability, and external linkages in biotechnology', *Journal of Economic Behavior and Organization*, **24**, 91–114.

Arvanitis, S. and H. Hollenstein (1994), 'Demand and supply factors in explaining the innovative activity of Swiss manufacturing firms', *Economics of Innovation and New Technology*, **3**, 15–30.

Arvanitis, S. and H. Hollenstein (1996), 'Industrial innovation in Switzerland: a model-based analysis with survey data', in A.K. Kleinknecht (ed.), *Determinants of Innovation. The Message of New Indicators*, London: Macmillan Press Ltd, pp. 13–62.

Becker, W. (1996), 'Ökonomische Bedeutung von Hochschulen als Produzenten von Humankapital' (Economic importance of universities as producers of human capital), Habilitationsschrift, Augsburg.

Becker, W. and J. Peters (1998), 'Innovation effects of academic research', Paper presented at the Workshop on 'Innovations Systems and Industrial Performance' at Wissenschaftszentrum Berlin für Sozialforschung (WZB), October, Berlin.

Beise, M., G. Licht and A. Spielkamp (1995), *Technologietransfer an kleine und mittlere Unternehmen* (Technology transfer to small and medium-sized firms), Baden-Baden: Nomos Verlag.

Bernstein, J.I. (1989), 'The structure of Canadian inter-industry R&D-spillovers, and the rate of return to R&D', *Journal of Industrial Economics*, **37**, 315–28.

Bundesministerium für Bildung, Wissenschaft, Forschung und Technologie (1995), *Grund- und Strukturdaten 1995/96* (Basic data 1995/96), Bonn.

Christensen, C. and J. Bower (1996), 'Customer power, strategic investment, and the failure of leading firms', *Strategic Management Journal*, **17**, 197–218.

Cohen, W. (1995), 'Empirical studies of innovative activity', in P. Stoneman (ed.), *Handbook of the Economics of Innovation and Technological Change*, Oxford, UK and Cambridge, MA: Blackwell, pp. 182–264.

Cohen, W.M. and S. Klepper (1996a), 'Firm size and the nature of innovation within industries: the case of process and product R&D', *Review of Economics and Statistics*, **78**, 232–43.

Cohen, W.M. and S. Klepper (1996b), 'A reprise of size and R&D', *Economic Journal*, **106**, 925–51.

Cohen, W. and R. Levin (1989), 'Empirical studies of innovation and market structures', in R. Schmalensee and R.D. Willig (eds), *Handbook of Industrial Organization*, Vol. 2, Amsterdam, New York, Oxford, Tokyo: Elsevier Science Publishers, pp. 1059–107.

Cohen, W., R. Levin and D. Mowery (1987), 'Firm size and R&D intensity: a re-examination', *Journal of Industrial Economics*, **35**, 543–65.

Cohen, W. and D. Levinthal (1989), 'Innovation and learning: the two faces of R&D', *Economic Journal*, **99**, 569–96.

Crépon, B., E. Duget and I. Kabla (1996), 'Schumpeterian conjectures: a moderate support from various innovation measures', in A.K. Kleinknecht (ed.), *Determinants of Innovation. The Message of New Indicators*, London: Macmillan Press Ltd, pp. 63–98.

David, P.A. (1991), 'Reputation and agency in the historical emergence of the institutions of "open science"', Discussion Paper series Center for Economic Policy Research No. 261, Stanford University.

David, P.A. (1994), 'Positive feedbacks and research productivity in science: reopening another black box', in O. Granstrand (ed.), *Economics of Technology*, Amsterdam, London, New York, Tokyo: Elsevier Science Publishers, pp. 5–89.

Dosi, G. (1988), 'Sources, procedures, and microeconomic effects of innovation', *Journal of Economic Literature*, **26**, 1120–71.

Eliasson, G. (1996), 'Spillovers, integrated production, and the theory of the firm', *Journal of Evolutionary Economics*, **6**, 122–40.

Evangelista, R., G. Perani, F. Rapiti and D. Archibugi (1997), 'Nature and impact of innovation in manufacturing industry: some evidence from the Italian innovation survey', *Research Policy*, **26**, 521–36.

Faulkner, W. and J. Senker (1994), 'Making sense of diversity. Public–private sector research linkage in three technologies', *Research Policy*, **23**, 673–95.

Felder, J. (1997), *Das Mannheimer Innovationspanel. Erläuterungen für externe Nutzer zum Datensatz der Erhebung 1993* (Mannheim Innovation Panel. Explanation for external users of the data set 1993), Version 97-1, Mannheim.

Felder, J., G. Licht, E. Nerlinger and H. Stahl (1995), 'Appropriability, opportunity, firm size and innovation activities', ZEW Discussion Paper No. 95–21, Mannheim.

Felder, J., G. Licht, E. Nerlinger and H. Stahl (1996), 'Factors determining R&D and innovation expenditure in German manufacturing industries', in A.K. Kleinknecht (ed.), *Determinants of Innovation. The Message of New Indicators*, London: Macmilllan Press Ltd, pp. 125–54.

Geroski, P.A. (1990), 'Innovation, technological opportunity, and market structure', *Oxford Economic Papers*, **42**, 586–602.

Gibbons, M., C. Limoges, H. Nowotny, S. Schwartzman, P. Scott and M. Trow (1994), *The New Production of Knowledge*. London: Sage.

Griliches, Z. (1992), 'The search for R&D spillovers', *Scandinavian Journal of Economics*, **94**, 29–47.

Griliches, Z. (1995), 'R&D and productivity. Econometric results and measurement issues', in P. Stoneman (ed.), *Handbook of the Economics of Innovation and Technological Change*, Oxford, UK and Cambridge, MA: Blackwell, pp. 52–89.

Grupp, H. (1994), 'The dynamics of science-based innovation reconsidered: cognitive models and statistical findings', in O. Granstrand (ed.), *Economics of Technology*, Amsterdam, London, New York, Tokyo: Elsevier Science Publishers, pp. 223–51.

Grupp, H. (1996), 'Spillover effects and the science base of innovations reconsidered: an empirical approach', *Journal of Evolutionary Economics*, **6**, 175–97.

Hagedoorn, J. (1995), 'Strategic technology partnering during the 1980s: trends, networks and corporate patterns in non-core technologies', *Research Policy*, **24**, 207–31.

Harabi, N. (1995), 'Sources of technical progress: empirical evidence from Swiss industry', *Economics of Innovation and New Technology*, **4**, 67–76.

Harhoff, D. (1995), 'Agglomerationen und regionale Spillovereffekte' (Agglomeration and regional spillover effects), in B. Gahlen, H. Hesse and H.J. Ramser (eds), *Standort und Region* (Location and region), Tübingen: Mohr Verlag (Paul Siebeck), 83–115.

Harhoff, D. (1996), 'Strategic spillovers and incentives for research and development', *Management Science*, **42**, 907–25.

Harhoff, D. and G. Licht (1994), 'Das Mannheimer Innovationspanel' (The Mannheim Innovation Panel), in U. Hochmuth and J. Wagner (eds), *Firmenpanelstudien in Deutschland* (Firm panel studies in Germany), Frankfurt am Main: Campus Verlag, pp. 255–84.

Harhoff, D. and G. Licht (1996), *Innovationsaktivitäten kleiner und mittlerer Unternemen* (Innovative activities of small and medium sized firms), Baden-Baden: Nomos Verlag.

Heckman, J. (1979), 'Sample selection bias as a specification error', *Econometrica*, **49**, 153–61.

Jaffe, A. (1989), 'Real effects of academic research', *American Economic Review*, **79**, 957–70.

Kleinknecht, A.K. (ed.) (1996), *Determinants of Innovation. The Message of New Indicators*, London: Macmillan Press Ltd.

Klevorick, A., R. Levin, R. Nelson and S. Winter (1995), 'On the sources and significance of inter-industry differences in technological opportunity', *Research Policy*, **24**, 185–205.

Kline, S. and N. Rosenberg (1986), 'An overview of innovation', in R. Landau and N. Rosenberg (eds), *The Positive Sum Strategy: Harnessing Technology for Economic Growth*, Washington DC: University Press, 275–306.

König, H. and G. Licht (1995), 'Patents, R&D and innovation. Evidence from the Mannheim Innovation Panel', *Ifo-Studien*, **33**, 521–43.

Lee, Y. (1996), 'Technology transfer and the research university: a search for the boundaries of university–industry collaboration', *Research Policy*, **25**, 843–63.

Levin, R.C., A.K. Klevorick, R.R. Nelson and S.G. Winter (1987), 'Appropriating the returns from industrial research and development', *Brookings Papers on Economic Activity*, **3**, 783–820.

Levin, R.C. and P.C. Reiss (1988), 'Cost-reducing and demand-creating R&D with spillovers', *Rand Journal of Economics*, **19**, 538–56.

Link, A.N. and J. Rees (1991), 'Firm size, university-based research and the returns to R&D', in Z.J. Acs and D.B. Audretsch (eds), *Innovation and Technological Change. An International Comparison*, New York: Harvester Wheatsheaf, pp. 60–70.

Mairesse, J. and M. Saasenou (1991), 'R&D and productivity. A survey of econometric studies at the firm level', *STI-Review*, **8**, 9–43.

Malerba, F. and S. Torrisi (1992), 'Internal capabilities and external networks in innovative activities. Evidence from the software industry', *Economics of Innovation and New Technology*, **2**, 49–71.

Mansfield, E. (1991), 'Academic research and industrial innovation', *Research Policy*, **20**, 1–12.

Mansfield, E. (1995), 'Academic research underlying industrial innovations: sources, characteristics, and financing', *Review of Economics and Statistics*, **77**, 55–65.

Mowery, D. and N. Rosenberg (1989), *Technology and the Pursuit of Economic Growth*, Cambridge, MA: Harvard University Press.

Nadiri, M.I. (1993), 'Innovations and technological spillovers', National Bureau of Economic Research Working Paper No. 4423, Cambridge, MA.

Narin, F., K. Hamilton and D. Olivastro (1997), 'The increasing linkage between U.S. technology and public science', *Research Policy*, **26**, 317–30.

Nelson, R.R. (1992), 'What is "commerical" and what is "public" about technology and what should be?', in N. Rosenberg, R. Landau and D.C. Mowery (eds), *Technology and the Wealth of Nations*, Stanford: University Press, 57–71.

Nelson, R.R. and E.N. Wolff (1997), 'Factors behind cross-industry differences in technical progress', *Structural Change and Economic Dynamics*, **8**, 205–20.

Nerlinger, E. (1996), 'Firm formation in high-tech industries: empirical results for Germany', ZEW Discussion Paper No. 96-07, Mannheim.

Organization for Economic Cooperation and Development (OECD) (1996), *Innovation, Patents and Technological Strategies*, Paris: OECD.

Peters, J. (1998), *Technologische Spillovers zwischen Zulieferer und Abnehmer. Ein spieltheoretische Analyse mit einer empirischen Studie für die deutsche Automobilindustrie* (Technological spillovers between suppliers and buyers. A game theoretic analysis with an empirical study for the German automobile industry), Heidelberg: Physica-Verlag.

Peters, J. and W. Becker (1998), 'Technological opportunities, academic research, and innovation activities in the German automobile supply industry', Working Paper Series of the Department of Economics No. 175, University of Augsburg, Augsburg.

Peters, J. and W. Becker (1999), 'Hochschulkooperationen und betriebliche Innovationsaktivitäten. Ergebisse aus der deutschen Automobilzulieferindustrie' (R&D cooperations with universities and the innovation activities of firms. Results from the German automobile supply industry), *Zeitschrift für Betriebswirtschaft*, **69** (11), 1293–1311.

Rosenberg, N. (1976), *Perspectives on Technology*, Cambridge, MA: Harvard University Press.

Rosenberg, N. and R.R. Nelson (1994), 'American universities and technical advance in industry', *Research Policy*, **23**, 323–48.

Sanchez, A.M. and A.C.P. Tejedor (1995), 'University–industry relationships in peripheral regions: the case of Aragon in Spain', *Technovation*, **15**, 613–25.

Scherer, F.M. (1991), 'Changing perspectives on the firm size problem', in J. Acs and D.B. Audretsch (eds), *Innovation and Technological Change. An International Comparison*, New York: Harvester Wheatsheaf, pp. 24–38.

Scherer, F.M. (1992), 'Schumpeter and plausible capitalism', *Journal of Economic Literature*, **30**, 1419–36.

Schmookler, J. (1966), *Invention and Economic Growth*, Cambridge, MA: Harvard University Press.

Spence, M. (1984), 'Cost reduction, competition and industry performance', *Econometrica*, **52**, 101–21.

Stephan, P. (1996), 'The economics of science', *Journal of Economic Literature*, **34**, 1199–235.

Sterlacchini, A. (1994), 'Technological opportunities, intra-industry spillovers and firm R&D intensity', *Economics of Innovation and New Technology*, **3**, 123–37.

Veugelers, R. (1997), 'Internal R&D expenditures and external technology sourcing', *Research Policy*, **26**, 303–15.

4. The empirical performance of a new inter-industry technology spillover measure

Bart Los*

1 INTRODUCTION

Technology has become an omnipresent topic in economic growth theory. At first this was true only for neo-Schumpeterian and evolutionary growth theories, but since the mid-1980s it also applies to mainstream growth theory. Even before technology became so prominent in mainstream theory, pioneers in the empirical field (such as Terleckyj 1974, Griliches 1979 and Scherer 1982) were aware of the influence that technology generation in an industry can exert on the productivity of other industries, through so-called technology spillovers. In the 1980s and 1990s, many similar studies confirmed their main finding that technology spillovers have significant positive productivity effects.

Although the significance of the spillover effects is beyond doubt, the estimated rates of return to research and development (R&D) expenditures, considered to be the major input in technology generation, vary over a large range. This variation does not seem to be caused only by differences in countries, industries and time periods, but also by the different spillover measurement methods that are applied. Los (1997) offers a survey of methods and argues that two broad classes of methods may be distinguished, based on the fundamental distinction between so-called 'rent spillovers' and 'pure knowledge spillovers', concepts originally introduced by Griliches (1979).[1] According to Griliches (1992) and Los (1997), the input–output-based measures of inter-industry technology spillovers as adopted by Terleckyj (1974), Sveikauskas (1981) and Wolff and Nadiri (1993), among others, refer to rent spillovers. In order to measure pure knowledge spillovers, reliance has always been placed on methods that utilize patent data (Jaffe 1986; Verspagen 1997a,b), R&D classified to product fields (Goto and Suzuki 1989) or data on the disciplinary composition of R&D staff (Adams 1990). Since classifications of these technological data have to be converted into classifications of economic data (for example, industry classifications) by means of particular concordances in order to carry out productivity studies, they are quite inflexible with regard to aggregation schemes. In this chapter, we introduce a more practical pure knowledge spillover measure, which is easily computable from input–output tables, and investigate its empirical performance *vis-à-vis* existing measures.

The chapter is organized as follows. The next section is devoted to a very short discussion of the basic distinction between rent and pure knowledge spillovers. Section 3 reviews technology spillover measures developed so far. In Section 4, we introduce the new pure knowledge spillover measure, show that two strands of economic theory can provide a justification for its use and conclude that the measure has some important practical advantages over more direct measures. Section 5 provides an empirical analysis of the most important technological inter-industry linkages within the American economy in 1987 as they are identified by the new method. In Section 6, the measurement method is applied to the US panel data on which Los and Verspagen (2000) based their study at the firm level, in order to see whether estimates of productivity effects differ from those obtained using alternative spillover measures. The final section contains a short summary of the results and some conclusions with regard to the usefulness of the proposed measure for future research.

2 TWO CATEGORIES OF TECHNOLOGY SPILLOVERS

In a widely cited paper, Griliches (1979) pointed out the basic difference between 'pure rent spillovers' and 'knowledge spillovers'. Pure rent spillovers relate to the fact that producers of product innovations are often unable to set a price for the improved product that reflects the quality increase relative to the old product, due to competition. This would not be problematic for productivity measurement, if both input and output deflators were so advanced that they could fully correct for quality differences. Although hedonic price indices are available for some products (for example, computers) in the US, researchers in the field of productivity have to rely on more conventional deflators if they are involved in economywide studies.[2] This implies that statistical deficiencies, together with competition, 'shift' rents connected to product innovations from the innovator to the user. Hence, productivity increases in the innovation-producing industry will be measured in the industries buying the product innovation.[3]

Knowledge spillovers have a completely different nature. They arise from the fact that knowledge has some public good characteristics: the use of a 'unit' of knowledge by one research employee does not prevent other researchers from using it (knowledge is non-rival) and knowledge can be appropriated only to a certain extent (knowledge is partly non-excludable).[4] As Los (1997) points out, a useful further distinction between two classes of knowledge spillovers can be made. The first class, having a strong intra-industry nature, enhances imitation of innovations. 'Reverse engineering' by competitors and imperfect patent protection are the most important sources of 'imitation-enhancing' knowledge spillovers, but they can also emerge from the mobility of R&D employees. Many R&D employees are not necessarily tied to one industry, however. In general, ideas from one industry may well induce new ideas in another industry, imitation playing no role at all. Los (1997) calls this second class of knowledge spillovers 'idea-creating' spillovers. They can also emerge from public information contained by patents and from scientific and professional journals and conferences. Other channels of 'idea-creating' knowledge spillovers are connected to supplier–buyer relationships: during trade negotiations or after-sales services, knowledge may be exchanged from supplier to buyer and vice versa.[5] It should be noted that the spill of knowledge does not affect the amount of knowledge present in the originating firm, contrary to the spill of rents of product innovations.

Although there is a basic difference between 'imitation-enhancing' and 'idea-creating' knowledge spillovers (at least at a theoretical level), the differences between pure rent spillovers and knowledge spillovers are more fundamental.[6] Therefore, it is important to determine which of these two kinds of spillovers are emphasized by a particular technology spillover measure, before the outcomes of empirical research with the measure are interpreted.

3 A BRIEF SURVEY OF EXISTING SPILLOVER MEASURES

In the last decades, the number of empirical studies concerning the productivity effects of technology spillovers has increased sharply. Mohnen (1990) and Nadiri (1993) offer good surveys of the results obtained for various countries, periods and aggregation levels. In general, spillover effects are found to be very significant and positive,[7] but the magnitude of the estimated effects appears to vary over a rather large range of values. One of the causes of this variation is the diversity of inter-industry spillover measurement methods used: in contrast to capital and labour inputs also present in the estimated productivity equations, technology cannot be measured very well because of its non-tangible nature. Often, the input of the technology production process (R&D expenditures) is used as a proxy for its output (technology). The intrinsic uncertainty about the outcome of the production process renders this proxy rather unreliable, but the problem is aggravated when we take into account that the effects of technology are not restricted to the producing industry itself, as was described in the previous section. How to measure the relevance of technology generated by industry i for industry j? Los (1997) provides an extensive survey of these methods, which have as a common feature that the amount of technology obtained through spillovers ('indirect R&D', *IRE*) is measured by a weighted sum of R&D expenditures by other industries (*RE*):

$$IRE_j = \sum_i \omega_{ij} RE_i \qquad \forall i \neq j, \tag{4.1}$$

in which i and j denote the spillover-producing and spillover-receiving industries, respectively. The weights ω_{ij} indicate to what extent the R&D undertaken by industry i may be considered to be part of the technology stock of industry j.

Many authors based the weights on input–output tables, following the pioneering contributions of Brown and Conrad (1967) and Terleckyj (1974).[8] These authors set their weights equal to the output coefficients, obtained by dividing the cell values by the corresponding row sums. The common idea behind this method is that the 'statistical benefits' obtained by industries through R&D embodied in intermediate goods are proportional to the parts of the output of the innovating industry they buy, through pure rent spillovers.[9] In some studies, capital flow matrix output coefficients are included in order to account for the fact that capital goods are carriers of pure rent spillovers, too. Wolff and Nadiri (1993) consider several input–output table-based sets of weights (such as input coefficients, coefficients of the Leontief inverse and backward linkage measures), but stress the knowledge spillovers related to supplier–buyer relationships. Van Meijl (1995) argues that pure rent spillovers and knowledge spillovers from supplier to buyer cannot be disentangled in economywide studies. Following him in that respect, and observing some confusion concerning the interpretation of some alternative measures not discussed here, Los (1997) proposes a classification of measurement methods according to the emphasis on either pure rent spillovers and supplier–buyer knowledge spillovers (together called 'rent spillovers'), or knowledge spillovers related to the relevance of produced knowledge to the R&D activities in other industries (called 'pure knowledge spillovers'). As may have become clear from the discussion above, the input–output table-based measures developed so far should be classified into the rent spillover measurement category.

In the empirical literature on *pure knowledge* spillover measurement, two classes of methods may be distinguished. The three most recent methods, introduced by Verspagen (1997a) and applied by Los and Verspagen (2000) and Verspagen (1997b) in firm-level and intercountry studies, respectively, are based on the information contained in patent documents. Verspagen's first two methods determine the industry in which the producer of each of the patented processes and products ('the spillover producer') is most likely to operate, as well as the industries most likely to use the information ('the spillover receiver').[10] This yields 'patent information input–output matrices', of which output coefficients can easily be computed. These coefficients are used as weights ω_{ij} in equation (4.1). The third method in Verspagen (1997a) utilizes a special feature of US Patent Office documents, the citation of related patents in each document. The holders of cited patents are considered to be spillover producers, the holder of the patent itself to be a spillover receiver. In the Verspagen papers the number of industries distinguished is 19 or 22, depending on the particular spillover matrix.

Another, somewhat older, class of measures follows Jaffe (1986). He proposed defining the weights ω_{ij} (in this case at the firm level) as the cosines of vectors $\mathbf{f}_i$ and $\mathbf{f}_j$ of elements representing the share of a particular patent class (k) in the total of patents granted to firms i and j in a certain period.

$$\omega_{ij} = \frac{\sum_{k=1}^{n} f_{ik} \cdot f_{jk}}{\sqrt{\left(\sum_{k=1}^{n} f_{ik}^2 \cdot \sum_{k=1}^{n} f_{jk}^2\right)}}, \tag{4.2}$$

If patent portfolios are perfectly identical, the knowledge resulting from R&D is assumed to be mutually relevant ($\omega_{ij} = 1$); the more different they are the lower the measure of relevance will be ($\omega_{ij} \rightarrow O$). Jaffe's (1986) approach has elicited a number of variants, the changes to the original being induced by data availability and/or research goals.[11]

With regard to the topic of this chapter (pure knowledge spillovers at the industry level), the study by Goto and Suzuki (1989) is the most interesting one. They used a product field classification (n = 30) of R&D investment (instead of patent classes) in order to determine the position of

Japanese industries in technological space and utilized equation (4.2) to assess the impact on other industries of R&D performed in the electronics industry. Although there are no major conceptual problems connected with this method, the empirical implementation seems to be restricted to countries in which a product field classification of R&D investments exists (at least if one is not willing to assume that inter-industry technological proximities are roughly equal across countries). Furthermore, an aggregation of R&D investment into 30 product classes may be far too rough to approximate 'technological proximity' well.[12] Another problem arises if one wants to repeat the productivity effects estimation for a new time interval. One has either to rely on the assumption that all R&D activity in all industries has been directed to the same product fields as before, or to perform a new survey in order to see how R&D has been redirected in the meantime. These practical disadvantages of the Goto and Suzuki (1989) method also hold concerning Jaffe's (1986) measure.

Measures constructed using a cosine measure also have a conceptual problem. As we focus on inter-industry idea-creating knowledge spillovers, the imposed symmetry of the proposed spillover matrix seems artificial: why should a dollar of R&D in industry *i* be as relevant to industry *j* as a dollar of *j*'s R&D is to *i*? Indeed, Los (1997) argues that Jaffe-like spillover measures originally have an emphasis on symmetrical 'imitation-enhancing' knowledge spillovers (which is understandable given Jaffe's, (1986) firm-level perspective partly inspired by Mansfield (1977), but he shows at the same time that Verspagen's (1997a) measures, which explicitly focus on idea-creating spillovers, have a rather high degree of symmetry, too. Hence, the theoretical differences between the two groups of pure knowledge spillover measures seem larger than their empirical differences.

To conclude this brief survey of technology spillover measurement, we may state that rent spillovers can be measured at a much lower degree of aggregation than pure knowledge spillovers. In the next section, we shall propose a pure knowledge spillover measure that could provide a theoretically justified solution to this problem of a practical nature.

4 A NEW MEASURE OF PURE KNOWLEDGE SPILLOVERS

The survey of the previous section showed that some theoretically sensible measures of rent spillovers almost automatically fulfil the wish of having industry classifications which are in line with the classification of productivity data, since they are based on input–output tables used in the compilation of national accounts. For measures of pure knowledge spillovers, though, such a match appeared to be much less natural, since these measures are all based on purely technological data which can only be linked to economic productivity data by means of specific concordances. This implies that existing measures of pure knowledge spillovers are not flexible with respect to changes in aggregation schemes (either imposed by changes in the compilation of productivity data or desired by the researcher linking productivity to R&D). Since the studies surveyed by Nadiri (1993) and the studies comparing the productivity effects of rent spillovers and pure knowledge spillovers by Verspagen (1997a) and Los and Verspagen (2000) indicate that pure knowledge spillovers are at least as important as rent spillovers from a productivity viewpoint, there seems to be a wish for a more flexible measure of pure knowledge spillovers.

In this section, a method which can fill the observed gap will be introduced. It takes advantage of the relatively low level of aggregation present in many present-day input–output tables. The new method utilizes the 'technology dimension' of input–output tables, instead of the 'trade dimension' on which the known input–output table-based methods for rent spillover measurement focus. First, we shall present this new, easy way of constructing a set of relevance weights ω (equation 4.1) and we shall show how elements of different strands of theory can support this method. Afterwards, the advantages and disadvantages of the new method compared to the existing ones will be set forth.

4.1 A New Measure of Knowledge Spillovers

The method proposed below is a member of the class of measures originating with Jaffe (1986), see equation (4.2). He introduced the concept of a 'potential spillover pool', which consists of all knowledge that cannot be appropriated by the firms that have generated this knowledge. The public knowledge in the pool is accessible to every firm in the economy. However, not all knowledge is relevant to every firm. For instance, recent insights concerning the air streams around aircraft wings is likely to have no impact on the R&D efforts of a wooden furniture manufacturing firm, while a producer of trucks may think of new opportunities to lower their drag. As said before, the more similar the technological activities of two firms are, the more they may benefit from each other's public knowledge.

Our measure is meant to capture *inter-industry* rather than of *inter-firm* knowledge spillovers.[13] As is well known (for example, Leontief

1989), a column of an input coefficient matrix derived from an input–output table is a strongly simplified representation of the technology of the corresponding industry, because it gives the amounts of the various inputs required to produce one value unit of the output of that industry. Therefore, similarity measures of production technologies of two industries can be derived from such columns.[14] Hence, we propose to apply the cosine measure introduced in equation (4.2) to input coefficient vectors, rather than to patent portfolio vectors as done by Jaffe (1986).[15]

$$\omega_{ij} = \frac{\sum_{k=1}^{n} a_{ik} \cdot a_{jk}}{\sqrt{\left(\sum_{k=1}^{n} a_{ik}^2 \cdot \sum_{k=1}^{n} a_{jk}^2\right)}}, \tag{4.3}$$

In which the *a*s stand for the input coefficients for the *n* intermediate inputs. If two industries have similar input compositions, ω_{ij} will be close to one. In this case, the indirect R&D stocks of both industries (equation (4.1)) include each other's R&D almost completely. Two industries with strongly different input structures (and therefore, by assumption, different technologies) barely contribute to each other's indirect R&D stock, as equation (4.3) yields a very small value in this case.

Having introduced the new measure, we should first judge its merits from a theoretical point of view: is there any argument for assuming that industries with similar input structures engage in similar R&D projects, thereby benefiting from one another's contributions to the 'potential spillover pool'? In order to answer this question, we refer briefly to two theories from completely different strands of economic theorizing.

First, some theories with a strong neoclassical flavour show how optimizing firms without uncertainty (reflected in a so-called 'innovation possibility frontier') will direct their R&D efforts at a lower use per unit of output of a particular input, the higher the cost share of this particular factor (the classic contribution is Kennedy 1964).[16] As a consequence, firms with roughly the same input structures (in value terms) will engage in roughly similar R&D projects, according to these so-called 'induced innovation' theories.[17] Extending the reasoning from firms to industries, similarity of industrial input structures would imply some similarity of technological activities, which would generate relatively large knowledge spillovers among those industries.

Second, an influential part of evolutionary growth theory deals explicitly with the directions in which firms search for technological progress. In their seminal contribution, Nelson and Winter (1982) replace the neoclassical assumptions of full rationality and optimizing behaviour with the concepts of bounded rationality and satisficing. Firms are assumed to decide on the basis of 'routines'. Regarding technological change, firms decide to search for alternative techniques only if a predetermined aspiration rate of return on capital is no longer attained. This search can focus either on imitation of more profitable technologies used by other firms or on 'world new' (but incremental) innovation. In the case of the search for innovation, which is most interesting regarding our industry-level point of view, new technologies which are 'close' to the one in use already are assumed to be discovered and implemented with a higher probability than technologies which are 'distant', even if the performance of the latter were superior. This is an immediate consequence of the bounded rationality hypothesis. Nelson and Winter (ibid.) define 'close' and 'distant', as 'having similar input structures' and 'having different input structures'.[18] Therefore, knowledge from firms (or industries) with similar input structures may be relatively productive for firms (or industries) searching for an innovation.

Both the induced innovation theories and the Nelson and Winter model consider only cost-reducing process innovations. This is not in line with empirical findings (for example, Scherer 1982) which show that R&D aiming at demand-creating product innovations accounts for more than half of total R&D in many countries. We are not aware of models of product innovation in which various inputs play an explicit role. Nevertheless, a non-formal argument based on bounded rationality can be put forward to support the measure for R&D aimed at product innovation. In order to attain product innovations, researchers require knowledge of properties of the materials that should be used for the product. In a world with rationality based on perfect information, firms would know of all knowledge available with respect to their (potential) inputs. In reality, however, firms do not have full information, but have to pick technological developments which can be monitored closely. In the case of the search for radical innovations, knowledge of properties of completely new inputs may be useful, but for incremental product innovation, knowledge of the materials already in use to produce the 'old' product is likely to suffice. Even this, though, is impossible for most firms, so they are likely to focus their monitoring on their most important inputs, that is, those with high input coefficients. Consequently, firms in an industry may obtain considerable (relatively) relevant knowledge of their most important inputs from innovative firms in industries mainly using the same inputs.

4.2 Comparison with Alternative Knowledge Spillover Measures

The practical inconveniences tied to the existing knowledge spillover methods are avoided by the method we propose. Columns of input coefficients are straightforwardly derivable from input–output tables. Nowadays, almost all countries of the world publish input–output tables on a regular basis, enabling researchers to make both international and intertemporal analyses. Moreover, statistical agencies tend to lower the level of aggregation of the tables published, thereby alleviating the heterogeneity problems paramount in traditional input–output analysis as well as in the aforementioned measures of technological proximity. Last but not least, the number of national statistical agencies collecting and publishing data on R&D expenditures is growing steadily. It seems plausible to assume that both the classification systems and the levels of aggregation used will become more and more comparable to those used with respect to traditional national accounts and input–output data.

Despite the theoretical considerations in favour of the proposed measure, it must be admitted that the input structure of an industry is a less 'direct' measure of the nature of its technological activity than the patent profiles, R&D expenditure profiles and patent information classifications used by Jaffe (1986), Goto and Suzuki (1989) and Verspagen (1997a), respectively. From a practical point of view, however, the input structures measure seems a worthwhile alternative to those methods. In order to assess the quality of the input coefficients-based measure, the next sections contain a report of two empirical exercises.

5 EMPIRICAL PERFORMANCE

We shall evaluate the quality of the proposed measure on the basis of two empirical studies. First, in this section, we shall look at the elements of a proximities matrix consisting of ω_{ij}s according to equation (4.3), in order to see whether the magnitudes of the elements are in line with common sense. In the next section, we compare estimation results of the productivity effects of technology spillovers based on the proposed measure with some other measures to investigate whether it can be considered as a reasonable alternative to some of them or not.

5.1 Inter-industry Technological Distances

Our empirical applications of the new measure are based on the commodity-by-commodity input–output table of the United States for 1987 derivable from the table of total (direct and indirect) requirements published on computer disk by the Bureau of Economic Analysis (BEA).[19] The choice for a US table is suggested by two considerations. First, as we argued above, we think one of the main advantages of the new measure is the possibility of obtaining indirect knowledge stocks at a relatively low level of aggregation. The US BEA tables distinguish 91 industries, which is a rather high number compared to most input–output tables. Second, the technological activity of an industry is not likely to depend on the geographic origin of its various inputs. Therefore, we should consider imported inputs as important as domestic inputs. Input–output tables, however, normally contain only one row or column for imports, without a classification of the industry of origin.[20] The USA is a country that does not depend heavily on imports, which implies that the omission of imported inputs will cause relatively small biases.[21]

We decided to calculate the cosines of 'truncated' input coefficients columns, which means that we did not include the coefficients for primary inputs. The reason for this decision is the enormous heterogeneity in the recorded values. Expenditures for engineers, secretaries, low-skilled manual workers and high-skilled managers are lumped together into one input category and the same happens to payments for the use of very different sorts of capital goods.[22] Although we immediately admit that these problems are also tied to intermediate input categories, we think that heterogeneity is far less a problem for these categories, at least with regard to our use of input–output tables. So we determined the cosines of each pair of industrial 91 elements input coefficients vectors. The resulting 'proximities' matrix is in Appendix Table 4A.2: a list of industries corresponding to the industry numbers can be found in Appendix Table 4A.1.

Examination of the proximities matrix shows that, although there are some counterintuitively high and low elements, the general pattern seems to be quite plausible: many high values are found in blocks in the neighbourhood of the main diagonal, which is a consequence of the SIC-related ordering of industries in the input–output table.[23] In order to provide a better overview of the technological clusters of manufacturing industries that are identified by the proposed measure, we provide a diagram generated by a so-called multidimensional scaling (MDS) algorithm.[24] This technique, mostly used by marketing researchers to reveal the position of products or product varieties in 'customer perception space', projects the $(n - 1)$ dimensional space containing n points, with interpoint distances given, on to a space of lower dimension, preserving the original interpoint distances as much as possible.[25] In our case, we decided to define the 'technological distance' between two industries i and j as $1 - \omega_{ij}$, one minus the cosine of the two input coefficients vectors.[26]

Figure 4.1 presents the two-dimensional 'map' of technological space, resulting from MDS applied to the inter-industry distance data.

It is hard, or even impossible, to give an interpretation of the two dimensions, since they contain information on 91 underlying dimensions (the number of intermediate inputs).[27] More important for our purposes, however, the diagram clearly identifies a number of technological clusters. Cluster 1 consists of food and textiles manufacturing industries (12–17) and footwear and leather products (32). Cluster II, denser than the first one, can be characterized as 'chemicals and plastics'. Cluster III is related to the manufacturing of paper products. Cluster IV, having a central position in the figure, consists mainly of electric products, while cluster V is strongly 'machines related'. Cluster VI is rather heterogeneous, containing both various metal producing industries as well as 'computers and office equipment' and 'audio, video and communication equipment'. Interesting solitary industries (as identified by the MDS plot) are 'petroleum refining' (30), 'lumber and wood products' (18), 'motor vehicles' (56) and 'aircraft and parts' (58).

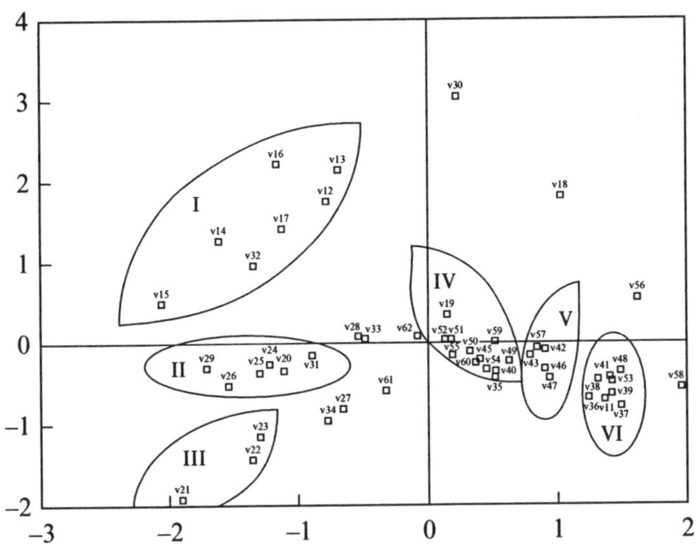

Note: See Appendix A for industry classification numbers.

Figure 4.1 Multidimensional scaling representation of 52 manufacturing industries in technological space

A closer look at Figure 4.1 reveals that some industries are close to each other, although no one would expect them to benefit from each other's R&D. For example, the distance between 'drugs' (27) and 'stone and clay products' (34) in the lower left quadrant is very small. In Appendix Table 4A.2, however, a proximity of only 0.165 is recorded for this pair of industries. Apparently, the 'true' distance between these industries has been distorted by the MDS algorithm, in reducing the 51-dimensional space with true distances to the two-dimensional space plotted in the figure.[28] The same applies for 'primary nonferrous metals manufacturing' (36) on the one hand, and 'computers and office equipment' and 'audio, video and communication equipment' (53) on the other, all present in cluster VI. Hence, these intuitively strange proximities should not be considered as indications of bad performance of the proposed measure itself, but as indications of bad performance of the MDS algorithm. In a few other cases, however, the measure itself seems to generate intuitively strange proximities, like in the case of cluster IV industry 'primary iron and steel manufacturing' (35), having ωs of more than 0.5 for many other industries in the same cluster, such as 'household appliances' (51) and 'electric lighting and wiring equipment' (52). It is impossible, however, to decide whether the measure or our intuition is wrong for such pairs of industries, since a firm like Philips, specialized in household appliances and lighting, holds some patents in metallurgy as well.[29]

In our view, the most important conclusion of this investigation into the technological linkages as identified by the proposed method should be that, despite a few counterintuitive results, the vast majority of elements in the proximities matrix seem to justify its use as a measure of pure knowledge spillovers.[30]

6 A COMPARISON OF SPILLOVER MEASURES IN ESTIMATIONS OF PRODUCTIVITY EFFECTS

Although the generation and diffusion of knowledge have entered mainstream economics only recently, their important productivity effects have long been recognized. Both the emergence of 'endogenous growth' theories and the strongly improved performance of various statistical agencies in gathering economic and technological data have caused an upsurge of empirical research in the relationship between innovative activity and economic performance. Following pioneering efforts by Terleckyj (1974) and Scherer (1982), the estimation of the influence of spillover effects has complemented the studies focusing on the productivity effects of technology generated by firms, industries or countries themselves. In this section, estimates of spillover effects obtained with the method proposed in the previous section will be compared with estimates based on existing methods.

6.1 Estimated Equations

Both the database and the method of analysis are identical to the firm-level study by Los and Verspagen (2000). We shall estimate the extended Cobb–Douglas production function

$$Q_{it} = A(IR)_{it}^{\eta} K_{it}^{\alpha} L_{it}^{\beta} R_{it}^{\gamma} e^{u_{it}}, \qquad (4.4)$$

in which Q is output, K the stock of physical capital, L the labour input, R the own R&D stock and IR the R&D stock obtained through technology spillovers. A is a constant and u a stochastic error term with zero mean. The indices i and t indicate the firm and the year under consideration, respectively. The elasticities α, β, γ and η, assumed to be constant across the sample, are the parameters to be estimated. To reduce heteroscedasticity and multicollinearity, equation (4.4) is usually written in labour-intensive form, after taking logarithms (indicated by lower-case letters):

$$(q_{it} - l_{it}) = a + \eta(ir)_{it} + \alpha(k_{it} - l_{it}) + \gamma(r_{it} - l_{it}) + (\mu - 1)l_{it} + \varepsilon_{it}, \qquad (4.5)$$

in which the dependent variable is labour productivity and μ is defined as $\alpha + \beta + \gamma$, the returns to scale with respect to all firm-specific inputs.[31] Los and Verspagen (1997) show that there is some preliminary evidence that the variables in their sample are not stationary (in the time-series dimension), but seem to be cointegrated. As is well known in time-series econometrics, coefficient estimates based on equation (4.5) are (super)consistent, but biased. The corresponding t-values and the adjusted R^2s have an (often strong) upward bias. To solve this problem, Engle and Granger (1987) proposed a two-step procedure (also known as the error correction model, ECM). The first step is simply to estimate the cointegration equation (4.5). In the second step, (4.5) is rewritten in first differences ($\Delta x_t = x_t - x_{t-1}$) and the residuals of the first step, u_{it} (lagged one period) are added to the set of independent variables:

$$\Delta(q-l)_{it} = \eta_s \Delta(ir)_{it} + \alpha_s \Delta(k-l)_{it} + \gamma_s \Delta(r-l)_{it} + (\mu_s - 1)\Delta l_{it} + \zeta u_{it-1} + \varepsilon_{it}. \qquad (4.6)$$

A significantly negative sign of the estimated coefficient for the lagged residual (ζ) is an indication for cointegration in the original level specification, equation (4.5). The subscripts s denote that equation (4.6) estimates short-run parameters. In order to find unbiased normally distributed estimators for the long-run parameters, which are more interesting from our point of view, we apply a third step introduced by Engle and Yoo (1991). This step estimates (in the time-series dimension):

$$\hat{\varepsilon}_{it} = \eta_3[-\hat{\zeta}(ir)_{it-1}] + \alpha_3[-\hat{\zeta}(k-l)_{it-1}] + \gamma_3[-\hat{\zeta}(r-l)_{it-1}] + (\mu_3 - 1)(-\hat{\zeta}l_{it-1}) + \upsilon_{it}, \quad (4.7)$$

in which the left-hand-side variable is the residual from the second step and hats denote estimated coefficients. Now, under the assumptions of a unique cointegration vector and weak exogeneity of the right-hand-side variables in the short-run ECM, the sums of the estimators in the first step (the level estimation, equation (4.5)) and the corresponding estimators of equation (4.7) are normally distributed unbiased estimators of the long-run relation. The standard deviations are estimated without bias by the standard error of the estimators in equation (4.7).

6.2 Construction of Variables

The databases and the construction of variables are discussed extensively in Los and Verspagen (2000), so only the most important topics will be repeated here. The first database used, made available by Bronwyn Hall through the NBER server, consists of annual data for nearly seven thousand US firms. For some of those firms data are available for the period 1974–91, whereas other firms are not covered for the full time interval. In this chapter, a non-random sample from the database will be used. It should be noted that we restrict the estimations to manufacturing firms, thus enabling us to make comparisons with results obtained by Los and Verspagen (2000). This decision reduces the size of the sample substantially. The sample consists of those firms that are covered for at least ten consecutive years and do not exhibit excessively large jumps which are probably due to mergers. The (unbalanced) sample consists of 680 firms, and 9223 observations. All variables, except IR, are constructed from this sample. Output Q is proxied by sales in millions of US dollars, because no better indicator for value added is available.[32] For the stock of physical capital K, we use 'net plant, property and equipment' (in millions of US dollars). Labour input L is measured by the number of employees (in thousands), because data on hours worked and/or skill levels are not available. R, the own R&D stock, is constructed from the annual data on R&D expenditures (in millions of US dollars) applying the well-known 'perpetual inventory method', with a depreciation rate of 15 per cent. To be sure that estimations will not be influenced by errors with respect to the initialization of the R&D stocks, the first two years for each firm are excluded from the sample. Furthermore, R is included in the regressions with a lag of one year to alleviate biases due to simultaneity. Q, K and R are expressed in constant prices.[33]

The various IR variables we want to compare with one another are constructed on the basis of industry-level, deflated R&D expenditures in the USA, available in the STAN database of the Organization for Economic Cooperation and Development (OECD). Defining RE_{it} as the R&D expenditures of industry i in year t, the indirect R&D expenditures of firm k operating in industry j are constructed according to a firm-level equivalent of equation (4.1):

$$IRE_{kjt} = \sum_{i} \omega_{ij} RE_{it}. \tag{4.8}$$

Note that when $i = j$, the R&D expenditures by firm k itself are excluded from RE_{it} to avoid double-counting. The indirect R&D stocks are then obtained according to the perpetual inventory model, again assuming a 15 per cent depreciation rate.

In this section, five spillover measures (corresponding to different sets of ωs) are considered. Two are taken from Los and Verspagen (2000). The first of them, IRT, is obtained by simply setting all ωs equal to one and may be regarded as a 'baseline measure', originally proposed by Bernstein (1988) in a study focusing on intra-industry technology spillovers. The second measure was introduced in Verspagen (1997a) and will be denoted by $IR2$. This measure relates explicitly to pure knowledge spillovers, being based on the assignment of probable producers and users of technical information present in patent documents of the European Patent Office.[34] The third measure (IRC) is the one proposed in this chapter. To make the measure both compatible with the STAN data and comparable with $IR2$, we had to recalculate our measure on the basis of a 21-industry input–output table of manufacturing industries only.[35] This table was obtained by aggregating the 91-industry US 1987 input–output table on the basis of the concordance between 4-digit SIC codes and input–output industries attached to the 1987 US tables. This caused trouble in two of the 52 manufacturing industries, represented by SIC codes belonging to two or three STAN classes. We decided to assign one-third of the inputs and outputs of 'ordnance and accessories' (11) to 'metal products', 'aerospace' and 'other transport' each, and half of the inputs and outputs of 'furniture and fixtures' (19) to 'metal products' and 'wood products' each.[36] The application of equation (4.3) on the 21-industry input–output table leads to the proximities matrix in Appendix Table 4A.3, which represents the ωs required to construct IRC. The fourth indirect R&D stock (*IRO*) is based on the same 21-industry input–output table for the US in 1987, but is the traditional measure on the basis of *output* coefficients. As we have argued in Section 3, such a measure captures rent spillovers. The cell values are divided by the total outputs of the corresponding row industries. An immediate consequence of this choice is that the ωs do not add to one for each *i*, representing the fact that the rents of some quality improvements are predominantly appropriated by service industries, consumers and foreign countries. Wolff and Nadiri (1993) and Wolff (1997), emphasizing the knowledge flows from suppliers to buyers, also propose using input coefficients as weights. Although we think knowledge spillovers and pure rent spillovers are mixed up in this approach, *IRI* is constructed along these lines.[37]

6.3 Estimation Results

Like Los and Verspagen (2000), we shall present the estimation results for the total sample as well as for three subsamples. The construction of these subsamples roughly corresponds to the division of industries into high-tech, medium-tech and low-tech as currently in use by the OECD. The high-tech subsample consists of those firms operating in 'electronics', 'drugs', 'aerospace', 'instruments' and 'computers'. The firms with SIC codes relating to 'electrical products', 'chemicals', 'motor vehicles', 'other transport', 'machines' and 'rubber and plastics' are included in the medium-tech subsample. Firms in other industries are assigned the low-tech status.[38] The 'fixed effects' estimates of the level equation (4.5) for the various samples, *IRC* being the specific indirect spillover measure, are in Table 4.1.

Table 4.1 'Fixed effects' estimates for level specification (unbalanced samples).

	NI	NOB	*k/l*	*r/l*	*l*	*IRC*
Total	680	9223	0.14	0.02	–0.11	0.60
			0.13	0.04		0.57
High-tech	245	3203	0.08	0.10	–0.04	0.44
			0.08	0.11		0.41
Med-tech	224	3110	0.16	0.00	–0.14	0.89
			0.15	0.02		0.87
Low-tech	211	2910	0.22	0.00	–0.11	0.39
			0.23	0.02		0.39

Note: NI: number of firms; NOB: number of observations.

Two sets of results are presented: with and without imposed constant returns to scale. We do not present *t*-values and R^2s because the estimators have non-normal distributions due to the non-stationarity of variables. If one were to insist on inference making on the basis of this fixed effects specification, all results would be significant at the usual levels, except those for own R&D in the med-tech and low-tech samples. The values are in line with earlier studies in this field (for example, Griliches and Mairesse 1984). Note the estimated decreasing returns to scale (the estimates for μ–1 are negative), common in time-series estimations. Another well-known phenomenon is the sensitivity of the results to the assumption of constant returns to scale. In this case, own R&D has more positive effects, even in medium-tech and low-tech industries. The estimated productivity effects of indirect R&D (*IRC*) are somewhat lower than in the unrestricted equation.

On the basis of the level estimates in Table 4.1, we performed the second and third step of the Engle–Granger–Yoo procedure. The corresponding ECM specification (equation (4.6)) yielded significantly negative estimates for ζ (the lagged residual coefficient), ranging from –0.311 to –0.421. This supports the assumption of cointegration and justifies the third step that yields the corrected long-run relationship estimates documented in Table 4.2.

Table 4.2 Engle–Granger–Yoo corrected estimates (unbalanced samples)

	k/l	*r/l*	*l*	*IRC*
Total	0.13	0.01	–0.03	0.68
	(7.34)	(0.61)	(1.54)	(19.4)
	0.12	0.04		0.72
	(7.20)	(4.33)		(21.0)
High-tech	0.06	0.07	0.05	0.47
	(2.56)	(2.59)	(1.39)	(7.27)
	0.06	0.12		0.48
	(2.42)	(5.32)		(8.17)
Med-tech	0.13	–0.01	–0.04	1.14
	(3.48)	(0.84)	(1.41)	(16.0)
	0.14	0.01		1.13
	(3.63)	(0.76)		(16.2)
Low-tech	0.21	0.00	–0.06	0.41
	(8.18)	(0.29)	(2.47)	(8.81)
	0.22	0.02		0.44
	(8.42)	(1.93)		(9.65)

Note: Number of observations: see Table 4.1; *t*-values in parentheses

The most important differences between the results in Table 4.1 and Table 4.2 are in the estimates with respect to the returns to scale parameter. For the total sample, the implausible decreasing returns to scale have disappeared. The same is true for the medium-tech sample. For the high-tech sample we find increasing returns to scale, partly as a result to the positive estimate of the elasticity for own R&D. Despite the small deviations from constant returns to scale, the results for the restricted estimations are different, in particular with respect to the productivity effects of own R&D. The physical capital estimates are not affected by a priori constant returns to scale, but are still rather low, which might be caused by mismeasurement: reliable indicators of capacity utilization are not available. The impacts of pure knowledge spillovers appear to be highly significant, although the variation across industry groups seems to be remarkable. Medium-tech firms seem to be the main beneficiaries of those spillovers, while high-tech firms tend to depend to a larger extent on their own technological activity, whereas the productivity of low-tech firms turns out to be simply less technology dependent.

From the point of view of this chapter, the most interesting issue to consider is how *IRC* performs compared to the alternative spillover measures. Ideally, we would like to estimate an equation simultaneously including a measure of knowledge spillovers (*IR*2 or *IRC*) and a measure of rent spillovers (*IRO* or *IRI*). In practice, this turns out to be impossible, because of the serious multicollinearity in the time-series dimension (even between first differences) apparent from Table 4.3.

Table 4.3 Correlation coefficients between first differences of five alternative measures of indirect R&D stocks (unbalanced samples)

	IRC	*IRO*	*IRI*	*IRT*
*IR*2	0.58	0.41	0.43	0.52
IRC		0.83	0.86	0.78
IRO			0.99	0.43
IRI				0.47

Given this impossibility, we estimated the equations with one *IR* variable at a time. In general, the estimates for the elasticities with respect to physical capital and own R&D, as well as the estimates for the returns to scale parameter and the lagged residual coefficient are rather stable, no matter which spillover variable is included. Therefore, we report in Table 4.4 only the estimated elasticities (obtained by the Engle–Granger–Yoo procedure) with respect to the various spillover variables.

Table 4.4 Engle–Granger–Yoo corrected indirect R&D elasticity estimates for various spillover measures (unbalanced samples)

	IRC	*IR2*	*IRO*	*IRI*	*IRT*
Total	0.68	0.68	0.47	0.54	0.62
	(19.4)	(19.0)	(12.2)	(14.1)	(20.5)
High-tech	0.47	0.36	0.45	0.45	0.44
	(7.27)	(4.76)	(8.15)	(7.93)	(6.39)
Med-tech	1.14	0.95	0.48	0.76	0.95
	(16.1)	(18.6)	(4.18)	(6.89)	(18.2)
Low-tech	0.41	0.33	0.42	0.43	0.34
	(8.81)	(6.16)	(7.64)	(8.19)	(8.95)

Note: *t*-values in parentheses; unrestricted estimates.

The differences between the estimates for the various indirect R&D stocks are not too large, except for the medium-tech sample. The rent spillover-oriented *IRO* and *IRI* yield remarkably low estimates, whereas *IRC* shows relatively high estimates. On the basis of these results, however, we cannot decide whether our spillover measures perform like other pure knowledge spillover measures but unlike input–output-based rent spillover measures: there is simply too little variation. The estimated short-run dynamics due to indirect R&D, obtained from the second step error correction specification (equation (4.6)), provides more clues (Table 4.5).

For the total sample, all *IR* measures yield significantly positive estimates again. Two remarkable results should be pointed out: first, the estimated elasticities for the pure knowledge spillover measures (*IRC* and *IR2*) and the unweighted indirect technology stock (*IRT*) are about twice as high as those for the rent spillover measures (*IRO* and *IRI*). This difference is even more apparent in the medium-tech subsample, in which *IRO* and *IRI* yield negative or insignificant estimates.

The most important conclusion of the estimates presented in this section is that the proposed measure seems to provide a good alternative to *IR2*, one of the pure knowledge spillover measures introduced by Verspagen (1997a). Furthermore, the results obtained with *IRC* are generally different from those obtained with *IRO* and *IRI*, the existing input–output-based measures which were argued to measure a different aspect of inter-industry technology effects.

Table 4.5 ECM estimates for alternative measures of indirect R&D (unbalanced samples)

	IRC	*IR2*	*IRO*	*IRI*	*IRT*
Total	0.55	0.55	0.25	0.31	0.54
	(14.3)	(14.7)	(6.45)	(8.20)	(15.7)
High-tech	0.56	0.58	0.49	0.50	0.59
	(8.59)	(8.03)	(8.87)	(8.68)	(8.54)
Med-tech	0.65	0.64	–0.42	–0.11	0.66
	(8.56)	(10.7)	(4.55)	(1.26)	(10.7)
Low-tech	0.32	0.18	0.09	0.15	0.29
	(4.84)	(2.66)	(1.30)	(2.15)	(5.42)

Note: *t*-values in parentheses; unrestricted estimation.

7 CONCLUSIONS

In the first part of the chapter, we introduced a new measure of pure knowledge spillovers between industries. It was argued that existing measures of pure knowledge spillovers are rather inflexible concerning the number of industries and the time periods that are considered, due to the nature of the data needed to construct them. The new measure does not have this disadvantage, because its basic data (input–output tables) are widely available. Although some theoretical arguments that support the measure were adduced, existing methods are clearly more straightforward measures of knowledge flows from industry to industry. It was argued that existing input–output based measures focus on rent spillovers, a conceptually different notion of technology spillovers. Therefore, the new measure could fill a gap in the case where it would yield results comparable to the existing, more sophisticated knowledge spillover measures.

The second part of the chapter was devoted to empirical analyses to evaluate whether the measure generated plausible outcomes. First, a rather disaggregated US input–output table was used to compute an actual spillover matrix. This matrix enabled us to distinguish a number of technological clusters which, in general, appeal to intuition. The second, and most elaborate, empirical exercise compared estimation results obtained with several technology spillover measures. A large cross-section time-series sample of US manufacturing firms was used to estimate the impact of indirect R&D. Application of an estimation framework to correct for potential cointegration of variables yielded evidence that the new measure is likely to be a good alternative to existing pure knowledge spillover measures: the estimated elasticities are of the same order of magnitude as those obtained with the alternatives and sometimes differ from input–output-based rent spillover measures, in particular with respect to short-run dynamics. The most important conclusion of this chapter is that the proposed measure seems to be a flexible, easy-to-construct-and-adapt measure of pure knowledge spillovers, which may prove to be a worthwhile alternative to current measures.

In our opinion, the proposed measure opens up some interesting directions for future research. First, until now, most productivity studies either are restricted to the impacts of technology spillovers on manufacturing industries or evaluate the effects of rent spillovers on service industries only. The new measure enables researchers to study whether knowledge about how to run a bank in a more efficient way influences the productivity in the real estate and insurance industries, or not. Of course, a host of problems should be solved (like how to measure inputs or outputs of this knowledge-creating process, how to measure productivity in services industries in an adequate way, and so on), but the increasing importance of services and government industries in developed countries seems to justify more intense empirical work on the subject. Second, a number of studies have appeared recently that investigate whether international technology spillovers have significant effects or not. In an influential study in this field, Coe and Helpman (1995) used a spillover measure comparable to *IRI* in this chapter to weight R&D performed by different countries. As Verspagen (1997b) has already observed, they hereby seem to confuse the two notions of technology spillovers. Verspagen himself used a spillover measure with a strong resemblance to the pure knowledge spillover measure *IR*2, but had to assume that the weights are equal within and across the countries studied. This seems to be a reasonable assumption as long as developed (OECD) countries are studied, but must be questioned when other countries are included in the analysis. In principle, the proposed measure is able to correct for differences in technology within industries across countries, due to the fact that different technologies are likely to be reflected in differing input coefficients derivable from input–output tables. Again, data availability is likely to be the limiting factor, as the reliability of the proposed measure probably increases as the level of aggregation of input–output tables decreases.

APPENDIX 4A

Table 4A.1 Industry classification US IO tables

1 Livestock and livestock products
2 Other agricultural products
3 Forestry and fishery products
4 Agricultural, forestry, and fishery services
5 Metallic ores mining
6 Coal mining
7 Crude petroleum and natural gas
8 Nonmetallic minerals mining
9 New construction
10 Maintenance and repair construction
11 Ordnance and accessories
12 Food and kindred products
13 Tobacco products
14 Broad and narrow fabrics, yam and thread mills
15 Miscellaneous textile goods and floor coverings
16 Apparel
17 Miscellaneous fabricated textile products
18 Lumber and wood products
19 Furniture and fixtures
20 Paper and allied products, except containers
21 Paperboard containers and boxes
22 Newspapers and periodicals
23 Other printing and publishing
24 Industrial and other chemicals
25 Agricultural fertilisers and chemicals
26 Plastics and synthetic materials
27 Drugs
28 Cleaning and toilet preparations
29 Paints and allied products
30 Petroleum refining and related products
31 Rubber and miscellaneous plastics products
32 Footwear, leather, and leather products
33 Glass and glass products
34 Stone and clay products
35 Primary iron and steel manufacturing
36 Primary nonferrous metals manufacturing
37 Metal containers
38 Heating, plumbing, and fabricated structural metal products
39 Screw machine products and stampings
40 Other fabricated metal products
41 Engines and turbines
42 Farm, construction, and mining machinery
43 Materials handling machinery and equipment
44 Metalworking machinery and equipment
45 Special industry machinery and equipment
46 General industrial machinery and equipment
47 Miscellaneous machinery, except electrical
48 Computer and office equipment
49 Service industry machinery
50 Electrical industrial equipment and apparatus
51 Household appliances

52 Electric lighting and wiring equipment
53 Audio, video, and communication equipment
54 Electronic components and accessories
55 Miscellaneous electrical machinery and supplies
56 Motor vehicles (passenger cars and trucks)
57 Truck and bus bodies, trailers, and motor vehicles parts
58 Aircraft and parts
59 Other transportation equipment
60 Scientific and controlling instruments
61 Ophthalmic and photographic equipment
62 Miscellaneous manufacturing
63 Railroads and related services; passenger ground transportation
64 Motor freight transportation and warehousing
65 Water transportation
66 Air transportation
67 Pipelines, freight forwarders, and related services
68 Communications, except radio and TV
69 Radio and TV broadcasting
70 Electric services (utilities)
71 Gas production and distribution (utilities)
72 Water and sanitary services
73 Wholesale trade
74 Retail trade
75 Finance
76 Insurance
77 Owner-occupied dwellings
78 Real estate and royalties
79 Hotels and lodging places
80 Personal and repair services (except auto)
81 Computer and data processing services
82 Legal, engineering, accounting, and related services
83 Other business and professional services. except medical
84 Advertising
85 Eating and drinking places
86 Automotive repair and services
87 Amusements
88 Health services
89 Educational and social services, and membership organizations
90 Federal Government enterprises
91 State and local government enterprises

Table 4A.2 91-industry US 1987 proximities matrix (shaded cells in upper right part indicate values > 0.4)

	1	2	3	4	5	6	7	8	9	10	11	12	13	14	15	16	17	18	19	20	21	22	23	24	25	26	27	28	29	30	31	32	33	34	35	36	37	38	39	40	41	42	43	44	45	46	47
1	1.00	0.47	0.28	0.61	0.06	0.06	0.10	0.09	0.06	0.06	0.04	0.82	0.31	0.23	0.03	0.02	0.05	0.03	0.07	0.06	0.02	0.03	0.04	0.06	0.06	0.04	0.08	0.11	0.05	0.02	0.05	0.22	0.06	0.06	0.06	0.04	0.03	0.04	0.03	0.06	0.04	0.06	0.07	0.06	0.07	0.06	0.06
2	0.47	1.00	0.62	0.63	0.17	0.23	0.58	0.28	0.21	0.21	0.12	0.25	0.19	0.15	0.05	0.06	0.11	0.10	0.21	0.14	0.06	0.14	0.13	0.12	0.40	0.08	0.17	0.18	0.07	0.05	0.13	0.09	0.18	0.15	0.15	0.09	0.08	0.12	0.09	0.16	0.12	0.18	0.19	0.19	0.21	0.17	0.18
3	0.28	0.62	1.00	0.19	0.12	0.10	0.07	0.16	0.23	0.23	0.08	0.23	0.05	0.05	0.03	0.03	0.06	0.08	0.13	0.08	0.02	0.09	0.07	0.09	0.11	0.06	0.15	0.14	0.06	0.05	0.07	0.16	0.08	0.07	0.09	0.05	0.04	0.07	0.06	0.10	0.07	0.10	0.11	0.10	0.11	0.09	0.10
4	0.61	0.63	0.19	1.00	0.13	0.13	0.09	0.17	0.18	0.18	0.10	0.44	0.27	0.20	0.05	0.06	0.11	0.08	0.19	0.13	0.05	0.10	0.10	0.11	0.61	0.08	0.18	0.17	0.06	0.04	0.11	0.11	0.16	0.11	0.15	0.08	0.08	0.11	0.09	0.15	0.12	0.18	0.18	0.17	0.20	0.16	0.14
5	0.06	0.17	0.12	0.13	1.00	0.32	0.19	0.66	0.25	0.25	0.17	0.08	0.05	0.11	0.09	0.05	0.08	0.13	0.26	0.25	0.05	0.10	0.13	0.39	0.25	0.32	0.22	0.30	0.24	0.06	0.22	0.11	0.37	0.27	0.49	0.29	0.17	0.24	0.24	0.32	0.25	0.32	0.30	0.34	0.32	0.32	0.30
6	0.06	0.23	0.10	0.13	0.32	1.00	0.23	0.42	0.19	0.19	0.12	0.07	0.05	0.07	0.05	0.05	0.08	0.09	0.19	0.16	0.04	0.08	0.09	0.15	0.13	0.12	0.15	0.18	0.09	0.05	0.13	0.08	0.19	0.20	0.28	0.10	0.07	0.12	0.09	0.17	0.14	0.26	0.22	0.19	0.23	0.22	0.17
7	0.10	0.58	0.07	0.09	0.19	0.23	1.00	0.25	0.09	0.09	0.12	0.03	0.04	0.03	0.03	0.03	0.04	0.04	0.11	0.06	0.02	0.13	0.09	0.16	0.11	0.09	0.11	0.11	0.08	0.24	0.08	0.05	0.13	0.10	0.10	0.04	0.04	0.07	0.06	0.10	0.06	0.08	0.10	0.14	0.14	0.10	0.14
8	0.09	0.28	0.16	0.17	0.66	0.42	0.25	1.00	0.33	0.33	0.20	0.11	0.08	0.13	0.10	0.07	0.12	0.12	0.28	0.32	0.10	0.18	0.20	0.42	0.44	0.33	0.26	0.38	0.26	0.09	0.26	0.15	0.47	0.59	0.39	0.24	0.16	0.20	0.17	0.30	0.23	0.34	0.36	0.33	0.37	0.34	0.30
9	0.06	0.21	0.23	0.18	0.25	0.19	0.09	0.33	1.00	1.00	0.20	0.09	0.07	0.07	0.05	0.06	0.11	0.49	0.60	0.28	0.07	0.17	0.13	0.16	0.17	0.12	0.32	0.23	0.11	0.06	0.17	0.11	0.29	0.45	0.32	0.18	0.23	0.33	0.27	0.35	0.28	0.40	0.41	0.41	0.40	0.34	0.37
10	0.06	0.21	0.23	0.18	0.25	0.19	0.09	0.33	1.00	1.00	0.20	0.09	0.07	0.07	0.05	0.06	0.11	0.49	0.59	0.28	0.06	0.17	0.13	0.16	0.17	0.12	0.32	0.23	0.11	0.06	0.17	0.11	0.29	0.45	0.30	0.18	0.21	0.31	0.25	0.34	0.27	0.39	0.40	0.39	0.39	0.32	0.36
11	0.04	0.12	0.08	0.10	0.17	0.12	0.12	0.20	0.20	0.20	1.00	0.07	0.08	0.05	0.05	0.04	0.07	0.08	0.27	0.11	0.03	0.11	0.09	0.13	0.11	0.11	0.16	0.20	0.09	0.03	0.13	0.08	0.18	0.14	0.31	0.25	0.30	0.31	0.26	0.36	0.31	0.30	0.32	0.33	0.34	0.33	0.36
12	0.82	0.25	0.23	0.44	0.08	0.07	0.03	0.11	0.09	0.09	0.07	1.00	0.17	0.13	0.04	0.04	0.07	0.06	0.13	0.13	0.06	0.09	0.10	0.10	0.11	0.07	0.16	0.24	0.11	0.02	0.10	0.37	0.17	0.09	0.11	0.06	0.06	0.07	0.06	0.11	0.07	0.12	0.12	0.11	0.13	0.10	0.09
13	0.31	0.19	0.05	0.27	0.05	0.05	0.04	0.08	0.07	0.07	0.08	0.17	1.00	0.13	0.04	0.03	0.06	0.04	0.13	0.10	0.05	0.13	0.12	0.07	0.07	0.06	0.10	0.17	0.04	0.02	0.07	0.06	0.17	0.07	0.09	0.04	0.05	0.06	0.05	0.10	0.06	0.10	0.11	0.10	0.11	0.08	0.09
14	0.23	0.15	0.05	0.20	0.11	0.07	0.03	0.13	0.07	0.07	0.05	0.13	0.13	1.00	0.91	0.67	0.88	0.05	0.31	0.17	0.05	0.04	0.06	0.14	0.09	0.19	0.09	0.18	0.30	0.02	0.50	0.19	0.13	0.10	0.10	0.07	0.04	0.05	0.05	0.10	0.05	0.08	0.09	0.10	0.11	0.08	0.08
15	0.03	0.05	0.03	0.05	0.09	0.05	0.03	0.10	0.05	0.05	0.05	0.04	0.04	0.91	1.00	0.57	0.78	0.05	0.28	0.21	0.08	0.05	0.08	0.19	0.11	0.27	0.08	0.24	0.48	0.02	0.73	0.19	0.14	0.11	0.08	0.07	0.03	0.04	0.04	0.09	0.04	0.06	0.06	0.10	0.09	0.07	0.07
16	0.02	0.06	0.03	0.06	0.05	0.05	0.03	0.07	0.06	0.06	0.04	0.04	0.03	0.67	0.57	1.00	0.74	0.03	0.26	0.09	0.03	0.04	0.05	0.04	0.04	0.06	0.06	0.09	0.07	0.01	0.16	0.15	0.07	0.06	0.06	0.04	0.03	0.04	0.03	0.06	0.04	0.07	0.07	0.07	0.08	0.06	0.06
17	0.05	0.11	0.06	0.11	0.08	0.08	0.04	0.12	0.11	0.11	0.07	0.07	0.06	0.88	0.78	0.74	1.00	0.06	0.43	0.16	0.04	0.06	0.09	0.10	0.09	0.12	0.12	0.18	0.12	0.02	0.27	0.26	0.14	0.09	0.10	0.07	0.05	0.07	0.06	0.11	0.08	0.12	0.12	0.11	0.14	0.12	0.10
18	0.03	0.10	0.08	0.08	0.13	0.09	0.04	0.12	0.49	0.49	0.08	0.06	0.04	0.05	0.05	0.03	0.06	1.00	0.73	0.38	0.03	0.05	0.06	0.10	0.09	0.08	0.09	0.14	0.08	0.02	0.11	0.07	0.24	0.11	0.12	0.07	0.05	0.10	0.07	0.15	0.08	0.13	0.15	0.12	0.15	0.11	0.10
19	0.07	0.21	0.13	0.19	0.26	0.19	0.11	0.28	0.60	0.59	0.27	0.13	0.13	0.31	0.28	0.26	0.43	0.73	1.00	0.43	0.07	0.15	0.16	0.20	0.18	0.16	0.28	0.39	0.13	0.05	0.28	0.23	0.39	0.22	0.30	0.23	0.38	0.51	0.45	0.61	0.43	0.52	0.54	0.51	0.52	0.49	0.48
20	0.06	0.14	0.08	0.13	0.25	0.16	0.06	0.32	0.28	0.28	0.11	0.13	0.10	0.17	0.21	0.09	0.16	0.38	0.43	1.00	0.88	0.67	0.83	0.35	0.25	0.34	0.22	0.39	0.31	0.04	0.35	0.15	0.34	0.26	0.21	0.12	0.08	0.12	0.09	0.20	0.11	0.17	0.17	0.19	0.21	0.16	0.14
21	0.02	0.06	0.02	0.05	0.05	0.04	0.02	0.10	0.07	0.06	0.03	0.06	0.05	0.05	0.08	0.03	0.04	0.03	0.07	0.88	1.00	0.72	0.88	0.11	0.08	0.11	0.09	0.11	0.09	0.01	0.15	0.04	0.08	0.13	0.06	0.04	0.04	0.05	0.04	0.07	0.04	0.05	0.06	0.07	0.07	0.06	0.05
22	0.03	0.14	0.09	0.10	0.10	0.08	0.13	0.18	0.17	0.17	0.11	0.09	0.13	0.04	0.05	0.04	0.06	0.05	0.15	0.67	0.72	1.00	0.93	0.14	0.14	0.12	0.22	0.21	0.08	0.02	0.12	0.07	0.14	0.15	0.11	0.05	0.08	0.07	0.06	0.11	0.07	0.10	0.13	0.13	0.14	0.11	0.15
23	0.04	0.13	0.07	0.10	0.13	0.09	0.09	0.20	0.13	0.13	0.09	0.10	0.12	0.06	0.08	0.05	0.09	0.06	0.16	0.83	0.88	0.93	1.00	0.22	0.17	0.21	0.20	0.28	0.16	0.03	0.19	0.09	0.19	0.17	0.13	0.08	0.09	0.09	0.07	0.13	0.08	0.12	0.14	0.14	0.17	0.12	0.12
24	0.06	0.12	0.09	0.11	0.39	0.15	0.16	0.42	0.16	0.16	0.13	0.10	0.07	0.14	0.19	0.04	0.10	0.10	0.20	0.35	0.11	0.14	0.22	1.00	0.58	0.96	0.32	0.77	0.83	0.27	0.37	0.21	0.59	0.31	0.29	0.13	0.08	0.10	0.11	0.24	0.09	0.14	0.15	0.20	0.25	0.13	0.13
25	0.06	0.40	0.11	0.61	0.25	0.13	0.11	0.44	0.17	0.17	0.11	0.11	0.07	0.09	0.11	0.04	0.09	0.09	0.18	0.25	0.08	0.14	0.17	0.58	1.00	0.51	0.24	0.46	0.44	0.21	0.23	0.15	0.42	0.39	0.24	0.11	0.08	0.10	0.09	0.19	0.09	0.15	0.16	0.17	0.20	0.13	0.13
26	0.04	0.08	0.06	0.08	0.32	0.12	0.09	0.33	0.12	0.12	0.11	0.07	0.06	0.19	0.27	0.06	0.12	0.08	0.16	0.34	0.11	0.12	0.21	0.96	0.51	1.00	0.29	0.78	0.89	0.05	0.44	0.20	0.55	0.26	0.23	0.10	0.06	0.07	0.08	0.20	0.06	0.11	0.11	0.17	0.21	0.10	0.09
27	0.08	0.17	0.15	0.18	0.22	0.15	0.11	0.26	0.32	0.32	0.16	0.16	0.10	0.09	0.08	0.06	0.12	0.09	0.28	0.22	0.09	0.22	0.20	0.32	0.24	0.29	1.00	0.42	0.23	0.05	0.22	0.15	0.33	0.17	0.21	0.11	0.09	0.13	0.13	0.22	0.13	0.22	0.22	0.22	0.27	0.20	0.21
28	0.11	0.18	0.14	0.17	0.30	0.18	0.11	0.38	0.23	0.23	0.20	0.24	0.17	0.18	0.24	0.09	0.18	0.14	0.39	0.39	0.11	0.21	0.28	0.77	0.46	0.78	0.42	1.00	0.69	0.07	0.48	0.32	0.64	0.29	0.29	0.15	0.12	0.17	0.14	0.34	0.16	0.29	0.29	0.29	0.36	0.24	0.22
29	0.05	0.07	0.06	0.06	0.24	0.09	0.08	0.26	0.11	0.11	0.09	0.11	0.04	0.30	0.48	0.07	0.12	0.08	0.13	0.31	0.09	0.08	0.16	0.83	0.44	0.89	0.23	0.69	1.00	0.04	0.69	0.21	0.46	0.27	0.20	0.10	0.07	0.07	0.08	0.18	0.06	0.08	0.09	0.16	0.17	0.08	0.08
30	0.02	0.05	0.05	0.04	0.06	0.05	0.24	0.09	0.06	0.06	0.03	0.02	0.02	0.02	0.02	0.01	0.02	0.02	0.05	0.04	0.01	0.02	0.03	0.27	0.21	0.05	0.05	0.07	0.04	1.00	0.04	0.03	0.06	0.04	0.05	0.03	0.02	0.03	0.02	0.04	0.03	0.04	0.05	0.05	0.05	0.04	0.04
31	0.05	0.13	0.07	0.11	0.22	0.13	0.08	0.26	0.17	0.17	0.13	0.10	0.07	0.50	0.73	0.16	0.27	0.11	0.28	0.35	0.15	0.12	0.19	0.37	0.23	0.44	0.22	0.48	0.69	0.04	1.00	0.19	0.31	0.22	0.20	0.15	0.10	0.15	0.12	0.25	0.13	0.20	0.20	0.23	0.24	0.19	0.18
32	0.22	0.09	0.16	0.11	0.11	0.08	0.05	0.15	0.11	0.11	0.08	0.37	0.06	0.19	0.19	0.15	0.26	0.07	0.23	0.15	0.04	0.07	0.09	0.21	0.15	0.20	0.15	0.32	0.21	0.03	0.19	1.00	0.19	0.11	0.12	0.07	0.05	0.08	0.06	0.13	0.08	0.13	0.13	0.13	0.15	0.11	0.10
33	0.06	0.18	0.08	0.16	0.37	0.19	0.13	0.47	0.29	0.29	0.18	0.17	0.17	0.13	0.14	0.07	0.14	0.24	0.39	0.34	0.08	0.14	0.19	0.59	0.42	0.55	0.33	0.64	0.46	0.06	0.31	0.19	1.00	0.44	0.35	0.17	0.12	0.18	0.15	0.30	0.16	0.24	0.25	0.30	0.33	0.23	0.23
34	0.06	0.15	0.07	0.11	0.27	0.20	0.10	0.59	0.45	0.45	0.14	0.09	0.07	0.10	0.11	0.06	0.09	0.11	0.22	0.26	0.13	0.15	0.17	0.31	0.39	0.26	0.17	0.29	0.27	0.04	0.22	0.11	0.44	1.00	0.32	0.13	0.11	0.16	0.14	0.23	0.14	0.19	0.19	0.27	0.22	0.19	0.22
35	0.06	0.15	0.09	0.15	0.49	0.28	0.10	0.39	0.32	0.30	0.31	0.11	0.09	0.10	0.08	0.06	0.10	0.12	0.50	0.21	0.06	0.11	0.13	0.29	0.24	0.23	0.21	0.29	0.20	0.05	0.20	0.12	0.35	0.32	1.00	0.34	0.68	0.86	0.88	0.88	0.75	0.79	0.74	0.82	0.72	0.79	0.75
36	0.04	0.09	0.05	0.08	0.29	0.10	0.04	0.24	0.18	0.18	0.25	0.06	0.04	0.07	0.07	0.04	0.07	0.07	0.23	0.12	0.04	0.05	0.08	0.13	0.11	0.10	0.11	0.15	0.10	0.03	0.15	0.07	0.17	0.13	0.34	1.00	0.76	0.42	0.21	0.41	0.35	0.20	0.25	0.34	0.34	0.34	0.49
37	0.03	0.08	0.04	0.08	0.17	0.07	0.04	0.16	0.23	0.21	0.30	0.06	0.05	0.04	0.03	0.03	0.05	0.05	0.38	0.08	0.04	0.08	0.09	0.08	0.08	0.06	0.09	0.12	0.07	0.02	0.10	0.05	0.12	0.11	0.68	0.76	1.00	0.86	0.74	0.80	0.71	0.59	0.57	0.68	0.59	0.66	0.76
38	0.04	0.12	0.07	0.11	0.24	0.12	0.07	0.20	0.33	0.31	0.31	0.07	0.06	0.05	0.04	0.04	0.07	0.10	0.51	0.12	0.05	0.07	0.09	0.10	0.10	0.07	0.13	0.17	0.07	0.03	0.15	0.08	0.18	0.16	0.86	0.42	0.86	1.00	0.96	0.96	0.84	0.82	0.78	0.85	0.72	0.81	0.82
39	0.03	0.09	0.06	0.09	0.24	0.09	0.06	0.17	0.27	0.25	0.26	0.06	0.05	0.05	0.04	0.03	0.06	0.07	0.45	0.09	0.04	0.06	0.07	0.11	0.09	0.08	0.13	0.14	0.08	0.02	0.12	0.06	0.15	0.14	0.88	0.21	0.74	0.96	1.00	0.91	0.81	0.81	0.74	0.85	0.68	0.78	0.77
40	0.06	0.16	0.10	0.15	0.32	0.17	0.10	0.30	0.35	0.34	0.36	0.11	0.10	0.10	0.09	0.06	0.11	0.15	0.61	0.20	0.07	0.11	0.13	0.24	0.19	0.20	0.22	0.34	0.18	0.04	0.25	0.13	0.30	0.23	0.88	0.41	0.80	0.96	0.91	1.00	0.81	0.82	0.81	0.86	0.77	0.81	0.82
41	0.04	0.12	0.07	0.12	0.25	0.14	0.06	0.23	0.28	0.27	0.31	0.07	0.06	0.05	0.04	0.04	0.08	0.08	0.43	0.11	0.04	0.07	0.08	0.09	0.09	0.06	0.13	0.16	0.06	0.03	0.13	0.08	0.16	0.14	0.75	0.35	0.71	0.84	0.81	0.81	1.00	0.89	0.78	0.81	0.76	0.80	0.75
42	0.06	0.18	0.10	0.18	0.32	0.26	0.08	0.34	0.40	0.39	0.30	0.12	0.10	0.08	0.06	0.07	0.12	0.13	0.52	0.17	0.05	0.10	0.12	0.14	0.15	0.11	0.22	0.29	0.08	0.04	0.20	0.13	0.24	0.19	0.79	0.20	0.59	0.82	0.81	0.82	0.89	1.00	0.84	0.82	0.77	0.83	0.77
43	0.07	0.19	0.11	0.18	0.30	0.22	0.10	0.36	0.41	0.40	0.32	0.12	0.11	0.09	0.06	0.07	0.12	0.15	0.54	0.17	0.06	0.13	0.14	0.15	0.16	0.11	0.22	0.29	0.09	0.05	0.20	0.13	0.25	0.19	0.74	0.25	0.57	0.78	0.74	0.81	0.78	0.84	1.00	0.83	0.87	0.87	0.76
44	0.06	0.19	0.10	0.17	0.34	0.19	0.14	0.33	0.41	0.39	0.33	0.11	0.10	0.10	0.10	0.07	0.11	0.12	0.51	0.19	0.07	0.13	0.14	0.20	0.17	0.17	0.22	0.29	0.16	0.05	0.23	0.13	0.30	0.27	0.82	0.34	0.68	0.85	0.85	0.86	0.81	0.82	0.83	1.00	0.87	0.85	0.85
45	0.07	0.21	0.11	0.20	0.32	0.23	0.14	0.37	0.40	0.39	0.34	0.13	0.11	0.11	0.09	0.08	0.14	0.15	0.52	0.21	0.07	0.14	0.17	0.25	0.20	0.21	0.27	0.36	0.17	0.05	0.24	0.15	0.33	0.22	0.72	0.34	0.59	0.72	0.68	0.77	0.76	0.77	0.87	0.87	1.00	0.90	0.78
46	0.06	0.17	0.09	0.16	0.32	0.22	0.10	0.34	0.34	0.32	0.33	0.10	0.08	0.08	0.07	0.06	0.12	0.11	0.49	0.16	0.06	0.11	0.12	0.13	0.13	0.10	0.20	0.24	0.08	0.04	0.19	0.11	0.23	0.19	0.79	0.34	0.66	0.81	0.78	0.81	0.80	0.83	0.87	0.85	0.90	1.00	0.88

Table 4A.2 (continued)

	48	49	50	51	52	53	54	55	56	57	58	59	60	61	62	63	64	65	66	67	68	69
1	0.05	0.07	0.08	0.08	0.08	0.04	0.06	0.08	0.04	0.06	0.02	0.09	0.06	0.05	0.09	0.06	0.08	0.05	0.05	0.08	0.02	0.02
2	0.13	0.18	0.22	0.21	0.23	0.11	0.18	0.23	0.11	0.16	0.06	0.26	0.18	0.16	0.22	0.20	0.23	0.20	0.20	0.34	0.09	0.11
3	0.07	0.10	0.13	0.11	0.13	0.05	0.10	0.12	0.06	0.10	0.04	0.15	0.10	0.09	0.13	0.21	0.14	0.14	0.21	0.15	0.04	0.03
4	0.13	0.19	0.23	0.22	0.23	0.10	0.15	0.22	0.11	0.18	0.06	0.22	0.16	0.15	0.22	0.18	0.15	0.11	0.17	0.15	0.04	0.09
5	0.11	0.25	0.31	0.28	0.29	0.10	0.24	0.29	0.10	0.27	0.09	0.31	0.21	0.19	0.26	0.27	0.20	0.22	0.20	0.46	0.07	0.03
6	0.10	0.20	0.22	0.21	0.22	0.09	0.17	0.21	0.09	0.18	0.06	0.26	0.16	0.15	0.20	0.25	0.15	0.16	0.17	0.24	0.05	0.04
7	0.07	0.08	0.12	0.09	0.11	0.06	0.12	0.12	0.02	0.07	0.05	0.19	0.13	0.09	0.14	0.17	0.13	0.23	0.14	0.46	0.12	0.17
8	0.15	0.31	0.39	0.33	0.36	0.14	0.32	0.36	0.13	0.27	0.11	0.37	0.25	0.27	0.35	0.35	0.33	0.29	0.31	0.52	0.10	0.05
9	0.16	0.40	0.41	0.41	0.40	0.15	0.27	0.35	0.14	0.33	0.11	0.52	0.28	0.22	0.41	0.20	0.26	0.18	0.21	0.22	0.07	0.04
10	0.16	0.39	0.40	0.40	0.39	0.14	0.27	0.34	0.14	0.32	0.11	0.51	0.27	0.21	0.40	0.20	0.25	0.17	0.21	0.22	0.07	0.04
11	0.20	0.32	0.45	0.38	0.36	0.38	0.43	0.44	0.13	0.31	0.91	0.31	0.47	0.35	0.37	0.13	0.09	0.13	0.32	0.20	0.10	0.05
12	0.08	0.13	0.16	0.16	0.17	0.07	0.12	0.16	0.07	0.11	0.04	0.13	0.12	0.14	0.20	0.06	0.08	0.10	0.09	0.08	0.02	0.01
13	0.05	0.10	0.15	0.15	0.14	0.06	0.09	0.14	0.05	0.10	0.04	0.10	0.11	0.13	0.19	0.05	0.05	0.06	0.07	0.11	0.03	0.01
14	0.06	0.10	0.12	0.19	0.21	0.05	0.10	0.13	0.05	0.09	0.03	0.13	0.13	0.11	0.27	0.06	0.05	0.05	0.05	0.08	0.02	0.01
15	0.04	0.09	0.11	0.21	0.23	0.05	0.09	0.13	0.04	0.09	0.03	0.12	0.14	0.14	0.31	0.04	0.06	0.06	0.05	0.07	0.02	0.01
16	0.05	0.08	0.10	0.11	0.11	0.04	0.07	0.09	0.05	0.07	0.03	0.09	0.10	0.07	0.19	0.04	0.04	0.05	0.05	0.07	0.02	0.01
17	0.09	0.14	0.16	0.20	0.20	0.08	0.14	0.17	0.09	0.12	0.04	0.16	0.16	0.13	0.29	0.06	0.06	0.08	0.07	0.08	0.03	0.01
18	0.07	0.16	0.15	0.20	0.17	0.07	0.12	0.14	0.08	0.14	0.04	0.31	0.12	0.10	0.30	0.08	0.08	0.06	0.07	0.09	0.02	0.01
19	0.20	0.50	0.36	0.65	0.57	0.20	0.40	0.48	0.20	0.46	0.16	0.59	0.38	0.31	0.62	0.19	0.16	0.19	0.16	0.22	0.06	0.03
20	0.11	0.20	0.26	0.28	0.26	0.12	0.22	0.28	0.10	0.18	0.06	0.25	0.20	0.62	0.39	0.13	0.16	0.09	0.11	0.17	0.04	0.02
21	0.04	0.07	0.11	0.09	0.08	0.04	0.06	0.10	0.03	0.05	0.02	0.07	0.08	0.56	0.16	0.04	0.07	0.03	0.04	0.05	0.01	0.01
22	0.08	0.11	0.19	0.13	0.14	0.09	0.12	0.18	0.05	0.10	0.07	0.12	0.17	0.50	0.25	0.11	0.10	0.16	0.16	0.26	0.07	0.05
23	0.09	0.14	0.21	0.18	0.18	0.10	0.16	0.22	0.08	0.13	0.05	0.14	0.17	0.58	0.28	0.10	0.11	0.09	0.11	0.16	0.05	0.03
24	0.09	0.17	0.20	0.22	0.23	0.09	0.25	0.31	0.08	0.14	0.05	0.17	0.16	0.32	0.28	0.15	0.14	0.12	0.14	0.17	0.04	0.02
25	0.09	0.16	0.20	0.20	0.21	0.09	0.19	0.25	0.09	0.15	0.05	0.17	0.15	0.23	0.25	0.13	0.25	0.13	0.11	0.15	0.04	0.02
26	0.07	0.13	0.15	0.20	0.21	0.08	0.24	0.29	0.08	0.12	0.03	0.13	0.14	0.31	0.26	0.08	0.08	0.07	0.06	0.10	0.03	0.01
27	0.16	0.23	0.29	0.29	0.31	0.15	0.27	0.33	0.13	0.21	0.08	0.24	0.24	0.29	0.36	0.15	0.11	0.25	0.19	0.22	0.08	0.05
28	0.18	0.32	0.38	0.47	0.45	0.20	0.42	0.50	0.20	0.30	0.09	0.31	0.31	0.43	0.50	0.16	0.17	0.17	0.19	0.17	0.06	0.03
29	0.05	0.12	0.13	0.22	0.24	0.06	0.18	0.23	0.06	0.11	0.03	0.14	0.13	0.26	0.28	0.09	0.12	0.06	0.07	0.06	0.03	0.01
30	0.03	0.05	0.06	0.06	0.06	0.03	0.05	0.06	0.03	0.04	0.01	0.06	0.04	0.04	0.06	0.10	0.07	0.08	0.14	0.11	0.01	0.01
31	0.13	0.24	0.28	0.43	0.44	0.14	0.28	0.34	0.13	0.24	0.07	0.28	0.25	0.30	0.46	0.12	0.18	0.13	0.11	0.14	0.05	0.02
32	0.08	0.14	0.17	0.20	0.19	0.08	0.16	0.20	0.09	0.13	0.04	0.15	0.14	0.15	0.25	0.07	0.08	0.07	0.08	0.09	0.03	0.01
33	0.14	0.28	0.35	0.40	0.53	0.14	0.32	0.38	0.15	0.25	0.09	0.33	0.26	0.34	0.43	0.19	0.17	0.15	0.16	0.25	0.10	0.02
34	0.08	0.20	0.28	0.24	0.23	0.08	0.17	0.22	0.09	0.20	0.07	0.20	0.17	0.22	0.30	0.16	0.37	0.12	0.12	0.21	0.08	0.02
35	0.16	0.61	0.70	0.72	0.62	0.14	0.28	0.43	0.13	0.58	0.17	0.56	0.34	0.24	0.47	0.24	0.14	0.17	0.14	0.24	0.06	0.02
36	0.14	0.48	0.59	0.32	0.54	0.12	0.41	0.56	0.07	0.38	0.23	0.31	0.25	0.16	0.59	0.09	0.12	0.13	0.08	0.11	0.03	0.01
37	0.13	0.62	0.73	0.59	0.65	0.11	0.34	0.51	0.08	0.34	0.24	0.46	0.29	0.14	0.56	0.07	0.07	0.08	0.06	0.07	0.02	0.01
38	0.15	0.67	0.74	0.76	0.69	0.13	0.31	0.44	0.14	0.61	0.20	0.60	0.33	0.18	0.47	0.14	0.12	0.10	0.09	0.10	0.04	0.02
39	0.11	0.57	0.62	0.70	0.57	0.10	0.19	0.31	0.12	0.54	0.16	0.51	0.27	0.14	0.34	0.11	0.09	0.08	0.07	0.10	0.03	0.02
40	0.19	0.70	0.79	0.84	0.75	0.19	0.42	0.55	0.19	0.65	0.22	0.64	0.41	0.28	0.57	0.19	0.14	0.17	0.14	0.19	0.07	0.02
41	0.17	0.67	0.72	0.70	0.63	0.14	0.28	0.43	0.20	0.60	0.22	0.75	0.32	0.18	0.42	0.14	0.09	0.13	0.10	0.12	0.05	0.02
42	0.20	0.64	0.71	0.76	0.63	0.18	0.32	0.46	0.23	0.64	0.17	0.78	0.37	0.25	0.47	0.17	0.13	0.17	0.15	0.17	0.06	0.02
43	0.23	0.75	0.77	0.81	0.69	0.20	0.37	0.51	0.21	0.63	0.18	0.72	0.42	0.28	0.49	0.22	0.13	0.18	0.16	0.19	0.08	0.03
44	0.21	0.74	0.80	0.78	0.71	0.18	0.34	0.50	0.18	0.64	0.22	0.64	0.41	0.27	0.53	0.21	0.16	0.19	0.17	0.24	0.09	0.04
45	0.28	0.85	0.83	0.84	0.75	0.23	0.42	0.58	0.20	0.61	0.21	0.70	0.47	0.33	0.56	0.25	0.13	0.21	0.19	0.24	0.10	0.04
46	0.22	0.78	0.78	0.79	0.68	0.18	0.33	0.48	0.17	0.62	0.19	0.68	0.39	0.25	0.48	0.20	0.11	0.20	0.14	0.21	0.07	0.03

	70	71	72	73	74	75	76	77	78	79	80	81	82	83	84	85	86	87	88	89	90	91
1	0.03	0.01	0.05	0.09	0.09	0.05	0.03	0.09	0.10	0.10	0.10	0.05	0.05	0.07	0.05	0.42	0.09	0.05	0.15	0.16	0.10	0.03
2	0.13	0.07	0.19	0.37	0.42	0.21	0.21	0.49	0.54	0.39	0.48	0.22	0.28	0.33	0.21	0.19	0.36	0.19	0.53	0.53	0.27	0.12
3	0.13	0.02	0.15	0.22	0.17	0.13	0.11	0.22	0.16	0.23	0.20	0.09	0.21	0.20	0.10	0.29	0.19	0.11	0.17	0.20	0.12	0.11
4	0.11	0.02	0.14	0.26	0.15	0.09	0.05	0.11	0.13	0.20	0.21	0.12	0.14	0.19	0.15	0.13	0.20	0.12	0.18	0.17	0.18	0.10
5	0.17	0.10	0.36	0.31	0.36	0.16	0.06	0.14	0.18	0.40	0.35	0.15	0.20	0.25	0.14	0.13	0.25	0.12	0.30	0.24	0.25	0.34
6	0.73	0.03	0.17	0.23	0.24	0.12	0.04	0.16	0.21	0.24	0.26	0.11	0.15	0.19	0.12	0.11	0.22	0.09	0.26	0.23	0.18	0.15
7	0.11	0.25	0.20	0.37	0.68	0.21	0.10	0.61	0.81	0.42	0.60	0.26	0.37	0.37	0.26	0.20	0.32	0.26	0.73	0.77	0.27	0.19
8	0.28	0.18	0.41	0.39	0.39	0.30	0.08	0.19	0.26	0.50	0.42	0.26	0.26	0.36	0.21	0.16	0.33	0.15	0.39	0.33	0.34	0.35
9	0.12	0.03	0.18	0.40	0.29	0.21	0.08	0.09	0.13	0.31	0.41	0.15	0.43	0.36	0.18	0.12	0.38	0.14	0.25	0.20	0.18	0.11
10	0.12	0.03	0.18	0.39	0.28	0.21	0.08	0.09	0.12	0.31	0.40	0.15	0.43	0.36	0.18	0.12	0.38	0.13	0.24	0.19	0.18	0.11
11	0.08	0.04	0.15	0.25	0.24	0.10	0.04	0.10	0.15	0.21	0.29	0.17	0.17	0.26	0.13	0.09	0.21	0.10	0.21	0.18	0.13	0.12
12	0.04	0.02	0.06	0.14	0.11	0.04	0.02	0.02	0.04	0.12	0.11	0.06	0.04	0.10	0.08	0.66	0.10	0.06	0.16	0.17	0.11	0.04
13	0.03	0.01	0.04	0.17	0.14	0.07	0.03	0.02	0.06	0.10	0.12	0.07	0.05	0.10	0.09	0.07	0.10	0.05	0.08	0.12	0.06	0.02
14	0.04	0.02	0.07	0.10	0.06	0.03	0.01	0.02	0.03	0.10	0.12	0.04	0.04	0.08	0.05	0.05	0.08	0.03	0.09	0.06	0.05	0.06
15	0.03	0.02	0.06	0.08	0.07	0.04	0.02	0.02	0.04	0.08	0.10	0.04	0.03	0.06	0.05	0.04	0.06	0.03	0.08	0.05	0.07	0.04
16	0.02	0.01	0.03	0.09	0.07	0.03	0.01	0.02	0.04	0.10	0.14	0.04	0.04	0.06	0.04	0.04	0.07	0.04	0.07	0.07	0.04	0.03
17	0.04	0.02	0.06	0.14	0.08	0.06	0.02	0.03	0.05	0.14	0.17	0.07	0.06	0.10	0.07	0.06	0.11	0.04	0.12	0.08	0.07	0.04
18	0.05	0.02	0.10	0.15	0.08	0.04	0.02	0.03	0.05	0.09	0.10	0.05	0.04	0.08	0.06	0.06	0.13	0.04	0.09	0.07	0.08	0.05
19	0.11	0.03	0.19	0.39	0.24	0.16	0.05	0.09	0.16	0.32	0.33	0.16	0.20	0.29	0.16	0.14	0.34	0.12	0.28	0.21	0.16	0.12
20	0.11	0.06	0.18	0.25	0.20	0.08	0.03	0.04	0.07	0.23	0.22	0.13	0.09	0.18	0.58	0.11	0.16	0.06	0.24	0.17	0.18	0.11
21	0.03	0.01	0.05	0.11	0.11	0.04	0.01	0.01	0.03	0.11	0.10	0.07	0.05	0.09	0.61	0.04	0.06	0.02	0.11	0.10	0.08	0.03
22	0.06	0.02	0.11	0.45	0.32	0.24	0.08	0.10	0.20	0.36	0.37	0.36	0.38	0.47	0.74	0.10	0.19	0.15	0.33	0.53	0.28	0.07
23	0.06	0.02	0.11	0.29	0.21	0.13	0.05	0.06	0.12	0.24	0.26	0.24	0.17	0.27	0.73	0.09	0.15	0.08	0.27	0.37	0.20	0.06
24	0.14	0.25	0.27	0.19	0.16	0.08	0.03	0.05	0.09	0.19	0.21	0.08	0.12	0.19	0.16	0.09	0.15	0.07	0.33	0.13	0.14	0.18
25	0.13	0.23	0.20	0.21	0.13	0.12	0.03	0.04	0.07	0.23	0.19	0.09	0.10	0.19	0.14	0.10	0.15	0.06	0.23	0.13	0.25	0.13
26	0.08	0.06	0.22	0.12	0.10	0.05	0.02	0.03	0.05	0.12	0.17	0.06	0.09	0.14	0.14	0.06	0.09	0.05	0.30	0.08	0.08	0.14
27	0.09	0.05	0.18	0.41	0.30	0.18	0.06	0.09	0.15	0.34	0.40	0.18	0.39	0.39	0.21	0.15	0.24	0.15	0.48	0.24	0.18	0.14
28	0.11	0.05	0.22	0.34	0.22	0.10	0.04	0.05	0.11	0.28	0.36	0.17	0.14	0.28	0.21	0.24	0.27	0.10	0.43	0.22	0.17	0.12
29	0.07	0.03	0.17	0.09	0.06	0.04	0.01	0.03	0.05	0.07	0.12	0.04	0.05	0.11	0.10	0.09	0.08	0.03	0.24	0.07	0.12	0.10
30	0.06	0.63	0.05	0.07	0.04	0.03	0.01	0.02	0.03	0.06	0.06	0.03	0.03	0.05	0.03	0.03	0.07	0.02	0.06	0.04	0.03	0.03
31	0.08	0.04	0.15	0.19	0.13	0.08	0.03	0.04	0.08	0.18	0.21	0.11	0.09	0.16	0.14	0.09	0.17	0.06	0.23	0.12	0.18	0.10
32	0.04	0.02	0.08	0.15	0.11	0.06	0.02	0.03	0.06	0.14	0.23	0.08	0.07	0.13	0.08	0.50	0.12	0.07	0.20	0.15	0.09	0.05
33	0.19	0.23	0.29	0.33	0.25	0.10	0.04	0.08	0.13	0.33	0.28	0.13	0.10	0.22	0.15	0.13	0.24	0.09	0.31	0.20	0.21	0.21
34	0.17	0.11	0.21	0.22	0.20	0.13	0.04	0.06	0.11	0.22	0.23	0.10	0.10	0.17	0.15	0.10	0.17	0.07	0.19	0.15	0.35	0.15
35	0.26	0.14	0.25	0.28	0.22	0.09	0.03	0.09	0.13	0.27	0.24	0.11	0.11	0.21	0.11	0.12	0.21	0.09	0.20	0.17	0.16	0.21
36	0.07	0.04	0.10	0.13	0.09	0.05	0.02	0.02	0.04	0.12	0.12	0.06	0.05	0.09	0.07	0.06	0.11	0.04	0.10	0.07	0.13	0.07
37	0.04	0.02	0.05	0.11	0.07	0.03	0.02	0.02	0.04	0.07	0.09	0.06	0.04	0.08	0.08	0.05	0.10	0.03	0.08	0.08	0.08	0.03
38	0.08	0.03	0.12	0.17	0.10	0.06	0.02	0.06	0.08	0.14	0.14	0.08	0.08	0.13	0.08	0.07	0.20	0.04	0.13	0.11	0.12	0.09
39	0.06	0.03	0.11	0.14	0.11	0.06	0.02	0.05	0.06	0.12	0.14	0.06	0.10	0.11	0.07	0.06	0.15	0.04	0.10	0.09	0.09	0.08
40	0.12	0.05	0.20	0.27	0.19	0.11	0.04	0.09	0.13	0.23	0.24	0.13	0.12	0.20	0.12	0.11	0.31	0.08	0.22	0.16	0.15	0.14
41	0.08	0.03	0.11	0.17	0.10	0.06	0.03	0.05	0.07	0.13	0.15	0.09	0.07	0.14	0.07	0.08	0.21	0.04	0.13	0.10	0.10	0.08
42	0.10	0.04	0.15	0.26	0.17	0.09	0.04	0.06	0.10	0.20	0.23	0.13	0.10	0.20	0.11	0.12	0.27	0.07	0.20	0.15	0.14	0.10
43	0.11	0.04	0.17	0.30	0.19	0.10	0.04	0.08	0.12	0.22	0.25	0.15	0.12	0.23	0.13	0.13	0.33	0.08	0.21	0.18	0.13	0.12
44	0.11	0.05	0.18	0.33	0.25	0.13	0.06	0.10	0.16	0.26	0.29	0.16	0.17	0.27	0.14	0.13	0.29	0.10	0.25	0.21	0.18	0.14
45	0.12	0.04	0.19	0.34	0.23	0.14	0.05	0.10	0.16	0.26	0.31	0.17	0.16	0.28	0.16	0.14	0.31	0.10	0.27	0.21	0.15	0.14
46	0.10	0.04	0.15	0.27	0.18	0.10	0.04	0.08	0.12	0.21	0.24	0.14	0.13	0.23	0.12	0.11	0.24	0.08	0.20	0.16	0.13	0.11

Table 4A.2 (continued)

	48	49	50	51	52	53	54	55	56	57	58	59	60	61	62	63	64	65	66	67	68	69	70	71	72	73	74	75	76	77	78	79	80	81	82	83	84	85	86	87	88	89	90	91
47	0.18	0.67	0.77	0.68	0.67	0.18	0.38	0.53	0.16	0.66	0.26	0.59	0.40	0.26	0.56	0.20	0.13	0.23	0.16	0.26	0.09	0.04	0.11	0.04	0.19	0.39	0.28	0.18	0.05	0.13	0.20	0.33	0.31	0.18	0.25	0.34	0.15	0.12	0.32	0.13	0.25	0.24	0.17	0.16
48	1.00	0.27	0.37	0.31	0.31	0.44	0.48	0.42	0.12	0.20	0.10	0.24	0.49	0.38	0.28	0.11	0.06	0.12	0.13	0.10	0.07	0.05	0.06	0.02	0.09	0.19	0.09	0.07	0.02	0.05	0.08	0.13	0.29	0.30	0.09	0.21	0.11	0.08	0.15	0.04	0.15	0.12	0.10	0.06
49	0.27	1.00	0.84	0.87	0.78	0.22	0.44	0.60	0.24	0.58	0.22	0.66	0.47	0.30	0.59	0.20	0.12	0.15	0.14	0.16	0.06	0.02	0.10	0.03	0.14	0.27	0.16	0.07	0.03	0.05	0.09	0.19	0.25	0.13	0.10	0.21	0.12	0.13	0.29	0.07	0.24	0.15	0.12	0.09
50	0.37	0.84	1.00	0.86	0.87	0.45	0.66	0.81	0.25	0.66	0.30	0.67	0.68	0.52	0.78	0.25	0.16	0.23	0.26	0.29	0.12	0.05	0.15	0.05	0.19	0.41	0.31	0.18	0.06	0.09	0.17	0.32	0.40	0.24	0.17	0.34	0.20	0.17	0.35	0.12	0.28	0.24	0.19	0.14
51	0.31	0.87	0.86	1.00	0.87	0.35	0.57	0.68	0.27	0.62	0.22	0.69	0.59	0.45	0.67	0.20	0.16	0.18	0.17	0.18	0.06	0.03	0.11	0.04	0.17	0.33	0.18	0.11	0.03	0.05	0.10	0.26	0.33	0.19	0.11	0.26	0.15	0.15	0.36	0.08	0.30	0.18	0.16	0.09
52	0.31	0.78	0.87	0.87	1.00	0.32	0.58	0.73	0.28	0.62	0.24	0.65	0.54	0.42	0.72	0.22	0.16	0.19	0.20	0.24	0.09	0.04	0.13	0.04	0.18	0.38	0.23	0.15	0.06	0.09	0.14	0.32	0.34	0.20	0.16	0.29	0.16	0.16	0.39	0.10	0.30	0.22	0.19	0.12
53	0.44	0.22	0.45	0.35	0.32	1.00	0.87	0.68	0.14	0.22	0.16	0.21	0.91	0.77	0.35	0.08	0.06	0.10	0.10	0.11	0.10	0.09	0.05	0.02	0.08	0.18	0.13	0.07	0.02	0.04	0.08	0.15	0.38	0.31	0.09	0.30	0.15	0.08	0.16	0.06	0.16	0.12	0.08	0.05
54	0.48	0.44	0.66	0.57	0.58	0.87	1.00	0.87	0.21	0.41	0.22	0.40	0.90	0.78	0.58	0.18	0.11	0.19	0.15	0.22	0.11	0.09	0.10	0.04	0.20	0.29	0.20	0.14	0.04	0.10	0.15	0.30	0.47	0.34	0.16	0.37	0.19	0.12	0.31	0.09	0.32	0.21	0.14	0.15
55	0.42	0.60	0.81	0.68	0.73	0.68	0.87	1.00	0.28	0.56	0.27	0.52	0.79	0.69	0.73	0.19	0.16	0.20	0.21	0.26	0.10	0.08	0.11	0.04	0.19	0.39	0.27	0.16	0.06	0.08	0.15	0.31	0.46	0.31	0.18	0.38	0.23	0.16	0.36	0.12	0.34	0.23	0.19	0.12
56	0.12	0.24	0.25	0.27	0.28	0.14	0.21	0.28	1.00	0.79	0.07	0.34	0.19	0.16	0.22	0.11	0.10	0.06	0.07	0.07	0.02	0.01	0.05	0.01	0.17	0.14	0.07	0.03	0.02	0.01	0.03	0.11	0.13	0.08	0.04	0.11	0.06	0.06	0.72	0.03	0.13	0.07	0.19	0.03
57	0.20	0.58	0.66	0.62	0.62	0.22	0.41	0.56	0.79	1.00	0.19	0.61	0.39	0.29	0.55	0.19	0.19	0.18	0.15	0.17	0.07	0.02	0.09	0.04	0.22	0.30	0.20	0.08	0.04	0.05	0.09	0.21	0.24	0.14	0.10	0.21	0.11	0.11	0.67	0.07	0.21	0.15	0.27	0.09
58	0.10	0.22	0.30	0.22	0.24	0.16	0.22	0.27	0.07	0.19	1.00	0.19	0.26	0.17	0.24	0.07	0.04	0.09	0.32	0.13	0.05	0.02	0.04	0.02	0.08	0.15	0.13	0.09	0.03	0.05	0.08	0.13	0.15	0.09	0.09	0.14	0.07	0.05	0.12	0.05	0.12	0.10	0.08	0.06
59	0.24	0.65	0.67	0.69	0.65	0.21	0.40	0.52	0.34	0.61	0.19	1.00	0.40	0.28	0.52	0.33	0.16	0.22	0.17	0.23	0.11	0.04	0.17	0.04	0.28	0.33	0.23	0.10	0.04	0.20	0.24	0.29	0.31	0.16	0.16	0.26	0.15	0.15	0.40	0.10	0.31	0.26	0.19	0.22
60	0.49	0.47	0.68	0.59	0.54	0.91	0.90	0.79	0.19	0.39	0.26	0.40	1.00	0.81	0.54	0.16	0.12	0.17	0.17	0.24	0.13	0.10	0.09	0.04	0.16	0.35	0.29	0.14	0.06	0.10	0.17	0.28	0.50	0.35	0.19	0.41	0.22	0.14	0.28	0.12	0.31	0.23	0.15	0.11
61	0.38	0.30	0.52	0.45	0.42	0.77	0.78	0.69	0.16	0.29	0.17	0.28	0.81	1.00	0.53	0.15	0.12	0.22	0.20	0.22	0.10	0.09	0.10	0.03	0.17	0.36	0.27	0.18	0.07	0.07	0.14	0.30	0.46	0.32	0.18	0.40	0.49	0.12	0.25	0.12	0.30	0.22	0.17	0.10
62	0.28	0.59	0.78	0.67	0.72	0.35	0.58	0.73	0.22	0.55	0.24	0.52	0.54	0.53	1.00	0.20	0.22	0.35	0.30	0.31	0.11	0.05	0.12	0.04	0.19	0.47	0.40	0.20	0.07	0.09	0.19	0.35	0.49	0.22	0.20	0.35	0.25	0.19	0.32	0.15	0.32	0.29	0.28	0.12
63	0.11	0.20	0.25	0.20	0.22	0.08	0.18	0.19	0.11	0.19	0.07	0.33	0.16	0.15	0.20	1.00	0.24	0.26	0.48	0.36	0.18	0.05	0.61	0.08	0.78	0.32	0.27	0.15	0.07	0.58	0.51	0.44	0.26	0.16	0.17	0.27	0.14	0.10	0.33	0.11	0.24	0.39	0.24	0.76
64	0.06	0.12	0.16	0.16	0.16	0.06	0.11	0.16	0.10	0.19	0.04	0.16	0.12	0.12	0.22	0.24	1.00	0.18	0.32	0.26	0.09	0.03	0.14	0.03	0.19	0.26	0.21	0.18	0.08	0.10	0.15	0.21	0.23	0.13	0.15	0.21	0.13	0.08	0.29	0.07	0.21	0.19	0.77	0.07
65	0.12	0.15	0.23	0.18	0.19	0.10	0.19	0.20	0.06	0.18	0.09	0.22	0.17	0.22	0.35	0.26	0.18	1.00	0.55	0.45	0.15	0.06	0.12	0.04	0.14	0.46	0.29	0.42	0.08	0.15	0.26	0.44	0.32	0.23	0.32	0.42	0.16	0.13	0.32	0.16	0.30	0.31	0.31	0.11
66	0.13	0.14	0.26	0.17	0.20	0.10	0.15	0.21	0.07	0.15	0.32	0.17	0.17	0.20	0.30	0.48	0.32	0.55	1.00	0.49	0.16	0.05	0.28	0.05	0.19	0.41	0.28	0.28	0.10	0.13	0.20	0.29	0.27	0.20	0.20	0.32	0.15	0.12	0.35	0.11	0.22	0.27	0.33	0.13
67	0.10	0.16	0.29	0.18	0.24	0.11	0.22	0.26	0.07	0.17	0.13	0.23	0.24	0.22	0.31	0.36	0.26	0.45	0.49	1.00	0.31	0.11	0.21	0.12	0.43	0.60	0.72	0.62	0.35	0.48	0.60	0.68	0.62	0.48	0.47	0.57	0.29	0.20	0.42	0.26	0.51	0.60	0.36	0.38
68	0.07	0.06	0.12	0.06	0.09	0.10	0.11	0.10	0.02	0.07	0.05	0.11	0.13	0.10	0.11	0.18	0.09	0.15	0.16	0.31	1.00	0.07	0.12	0.03	0.22	0.28	0.24	0.20	0.07	0.19	0.23	0.27	0.26	0.28	0.17	0.28	0.14	0.06	0.14	0.11	0.20	0.24	0.12	0.21
69	0.05	0.02	0.05	0.03	0.04	0.09	0.09	0.08	0.01	0.02	0.02	0.04	0.10	0.09	0.05	0.05	0.03	0.06	0.05	0.11	0.07	1.00	0.03	0.02	0.05	0.15	0.15	0.09	0.03	0.12	0.16	0.12	0.17	0.10	0.13	0.15	0.68	0.06	0.08	0.96	0.15	0.20	0.07	0.05
70	0.06	0.10	0.15	0.11	0.13	0.05	0.10	0.11	0.05	0.09	0.04	0.17	0.09	0.10	0.12	0.61	0.14	0.12	0.28	0.21	0.12	0.03	1.00	0.28	0.55	0.19	0.16	0.13	0.07	0.38	0.33	0.30	0.17	0.10	0.09	0.15	0.08	0.06	0.19	0.07	0.15	0.25	0.14	0.53
71	0.02	0.03	0.05	0.04	0.04	0.02	0.04	0.04	0.01	0.04	0.02	0.04	0.04	0.03	0.04	0.08	0.03	0.04	0.05	0.12	0.03	0.02	0.28	1.00	0.23	0.10	0.11	0.04	0.02	0.10	0.12	0.18	0.12	0.05	0.05	0.07	0.04	0.03	0.07	0.04	0.11	0.13	0.06	0.19
72	0.09	0.14	0.19	0.17	0.18	0.08	0.20	0.19	0.17	0.22	0.08	0.28	0.16	0.17	0.19	0.78	0.19	0.14	0.19	0.43	0.22	0.05	0.55	0.23	1.00	0.27	0.33	0.16	0.23	0.74	0.60	0.51	0.30	0.12	0.18	0.25	0.14	0.10	0.31	0.13	0.28	0.42	0.22	0.96
73	0.19	0.27	0.41	0.33	0.38	0.18	0.29	0.39	0.14	0.30	0.15	0.33	0.35	0.36	0.47	0.32	0.26	0.46	0.41	0.60	0.28	0.15	0.19	0.10	0.27	1.00	0.70	0.50	0.15	0.30	0.53	0.81	0.74	0.42	0.73	0.87	0.45	0.24	0.55	0.36	0.63	0.65	0.44	0.22
74	0.09	0.16	0.31	0.18	0.23	0.13	0.20	0.27	0.07	0.20	0.13	0.23	0.29	0.27	0.40	0.27	0.21	0.29	0.28	0.72	0.24	0.15	0.16	0.11	0.33	0.70	1.00	0.39	0.15	0.49	0.73	0.66	0.81	0.36	0.63	0.63	0.38	0.26	0.43	0.36	0.65	0.74	0.33	0.31
75	0.07	0.07	0.18	0.11	0.15	0.07	0.14	0.16	0.03	0.08	0.09	0.10	0.14	0.18	0.20	0.15	0.18	0.42	0.28	0.62	0.20	0.09	0.13	0.04	0.16	0.50	0.39	1.00	0.23	0.23	0.38	0.65	0.42	0.43	0.45	0.32	0.24	0.12	0.39	0.18	0.33	0.39	0.23	0.11
76	0.02	0.03	0.06	0.03	0.06	0.02	0.04	0.06	0.02	0.04	0.03	0.04	0.06	0.07	0.07	0.07	0.08	0.08	0.10	0.35	0.07	0.03	0.07	0.02	0.23	0.15	0.15	0.23	1.00	0.49	0.27	0.15	0.14	0.10	0.13	0.17	0.08	0.04	0.22	0.06	0.15	0.16	0.07	0.06
77	0.05	0.05	0.09	0.05	0.09	0.04	0.10	0.08	0.01	0.05	0.05	0.20	0.10	0.07	0.09	0.58	0.10	0.15	0.13	0.48	0.19	0.12	0.38	0.10	0.74	0.30	0.49	0.23	0.49	1.00	0.90	0.48	0.42	0.17	0.28	0.32	0.19	0.14	0.29	0.21	0.49	0.66	0.19	0.70
78	0.08	0.09	0.17	0.10	0.14	0.08	0.15	0.15	0.03	0.09	0.08	0.24	0.17	0.14	0.19	0.51	0.15	0.26	0.20	0.60	0.23	0.16	0.33	0.12	0.60	0.53	0.73	0.38	0.27	0.90	1.00	0.65	0.63	0.29	0.44	0.30	0.30	0.20	0.39	0.30	0.69	0.85	0.29	0.58
79	0.13	0.19	0.32	0.26	0.32	0.15	0.30	0.31	0.11	0.21	0.13	0.29	0.28	0.30	0.35	0.44	0.21	0.44	0.29	0.68	0.27	0.12	0.30	0.18	0.51	0.81	0.66	0.65	0.15	0.48	0.65	1.00	0.74	0.36	0.63	0.76	0.38	0.25	0.52	0.33	0.63	0.67	0.35	0.47
80	0.29	0.25	0.40	0.33	0.34	0.38	0.47	0.46	0.13	0.24	0.15	0.31	0.50	0.46	0.49	0.26	0.23	0.32	0.27	0.62	0.26	0.17	0.17	0.12	0.30	0.74	0.81	0.42	0.14	0.42	0.63	0.74	1.00	0.67	0.72	0.77	0.43	0.25	0.48	0.34	0.72	0.73	0.35	0.26
81	0.30	0.13	0.24	0.19	0.20	0.31	0.34	0.31	0.08	0.14	0.09	0.16	0.35	0.32	0.22	0.16	0.13	0.23	0.20	0.45	0.28	0.10	0.10	0.05	0.12	0.42	0.36	0.43	0.10	0.17	0.29	0.36	0.47	1.00	0.57	0.69	0.31	0.12	0.24	0.17	0.49	0.47	0.23	0.08
82	0.09	0.10	0.17	0.11	0.16	0.09	0.16	0.18	0.04	0.10	0.09	0.16	0.19	0.18	0.20	0.17	0.15	0.32	0.20	0.47	0.17	0.13	0.09	0.05	0.18	0.73	0.63	0.45	0.13	0.28	0.44	0.63	0.72	0.57	1.00	0.88	0.37	0.16	0.35	0.33	0.58	0.56	0.29	0.15
83	0.21	0.21	0.34	0.26	0.29	0.30	0.37	0.38	0.11	0.21	0.14	0.26	0.41	0.40	0.35	0.27	0.21	0.42	0.32	0.57	0.28	0.15	0.15	0.07	0.25	0.87	0.63	0.52	0.17	0.32	0.50	0.76	0.77	0.69	0.88	1.00	0.45	0.21	0.46	0.35	0.67	0.65	0.41	0.20
84	0.11	0.12	0.20	0.15	0.16	0.15	0.19	0.23	0.06	0.11	0.07	0.15	0.22	0.49	0.25	0.14	0.13	0.16	0.15	0.29	0.14	0.68	0.08	0.04	0.14	0.45	0.38	0.24	0.08	0.19	0.30	0.38	0.43	0.31	0.37	0.45	1.00	0.13	0.23	0.73	0.40	0.51	0.25	0.11
85	0.08	0.13	0.17	0.15	0.16	0.08	0.12	0.16	0.06	0.11	0.05	0.15	0.14	0.12	0.19	0.10	0.08	0.13	0.12	0.20	0.06	0.06	0.06	0.03	0.10	0.24	0.26	0.12	0.04	0.14	0.20	0.25	0.25	0.12	0.16	0.21	0.13	1.00	0.16	0.15	0.32	0.35	0.14	0.09
86	0.15	0.29	0.35	0.36	0.39	0.16	0.31	0.36	0.72	0.67	0.12	0.40	0.28	0.25	0.32	0.33	0.29	0.32	0.35	0.42	0.14	0.08	0.19	0.07	0.31	0.55	0.43	0.39	0.22	0.29	0.39	0.52	0.48	0.24	0.35	0.46	0.23	0.16	1.00	0.19	0.44	0.42	0.32	0.14
87	0.04	0.07	0.12	0.08	0.10	0.06	0.09	0.12	0.03	0.07	0.05	0.10	0.12	0.12	0.15	0.11	0.07	0.16	0.11	0.26	0.11	0.96	0.07	0.04	0.13	0.36	0.36	0.18	0.06	0.21	0.30	0.33	0.34	0.17	0.33	0.35	0.73	0.15	0.19	1.00	0.30	0.36	0.14	0.12
88	0.15	0.24	0.28	0.30	0.30	0.16	0.32	0.34	0.13	0.21	0.12	0.31	0.31	0.30	0.32	0.24	0.21	0.30	0.22	0.51	0.20	0.15	0.15	0.11	0.28	0.63	0.65	0.33	0.15	0.49	0.69	0.63	0.72	0.49	0.58	0.67	0.40	0.32	0.44	0.30	1.00	0.78	0.35	0.20
89	0.12	0.15	0.24	0.18	0.22	0.12	0.21	0.23	0.07	0.15	0.10	0.26	0.23	0.22	0.29	0.39	0.19	0.31	0.27	0.60	0.24	0.20	0.25	0.13	0.42	0.65	0.74	0.39	0.16	0.66	0.85	0.67	0.73	0.47	0.56	0.65	0.51	0.35	0.42	0.36	0.78	1.00	0.43	0.39
90	0.10	0.12	0.19	0.16	0.19	0.08	0.14	0.19	0.19	0.27	0.08	0.19	0.15	0.17	0.28	0.24	0.77	0.31	0.33	0.36	0.12	0.07	0.14	0.06	0.22	0.44	0.33	0.23	0.07	0.19	0.29	0.35	0.35	0.23	0.29	0.41	0.25	0.14	0.32	0.14	0.35	0.43	1.00	0.14
91	0.06	0.09	0.14	0.09	0.12	0.05	0.15	0.12	0.03	0.09	0.06	0.22	0.11	0.10	0.12	0.76	0.07	0.11	0.13	0.38	0.21	0.05	0.53	0.19	0.96	0.22	0.31	0.11	0.06	0.70	0.58	0.47	0.26	0.08	0.15	0.20	0.11	0.09	0.14	0.12	0.20	0.39	0.14	1.00

Table 4A.3 22-industry US 1987 proximities matrix

		1	2	3	4	5	6	7	8	9	10	11	12	13	14	15	16	17	18	19	20	21	22
1	Electrical prod.	1.000	0.569	0.238	0.165	0.102	0.000	0.378	0.274	0.679	0.559	0.507	0.693	0.667	0.287	0.699	0.141	0.114	0.433	0.265	0.213	0.169	0.696
2	Electronics	0.569	1.000	0.137	0.093	0.050	0.000	0.207	0.234	0.365	0.092	0.200	0.203	0.944	0.465	0.192	0.075	0.068	0.270	0.093	0.087	0.101	0.420
3	Chemicals	0.238	0.137	1.000	0.314	0.302	0.000	0.121	0.038	0.154	0.187	0.071	0.146	0.171	0.032	0.104	0.127	0.096	0.750	0.384	0.258	0.112	0.300
4	Drugs	0.165	0.093	0.314	1.000	0.087	0.000	0.079	0.025	0.090	0.059	0.033	0.075	0.125	0.029	0.061	0.121	0.063	0.326	0.187	0.232	0.054	0.203
5	Refined oil	0.102	0.050	0.302	0.087	1.000	0.000	0.049	0.019	0.070	0.081	0.038	0.057	0.062	0.015	0.045	0.034	0.030	0.204	0.167	0.074	0.053	0.110
6	Ships	0.000	0.000	0.000	0.000	0.000	1.000	0.000	0.000	0.000	0.000	0.000	0.000	0.000	0.000	0.000	0.000	0.000	0.000	0.000	0.000	0.000	0.000
7	Automotive	0.378	0.207	0.121	0.079	0.049	0.000	1.000	0.118	0.614	0.183	0.113	0.273	0.246	0.084	0.369	0.080	0.130	0.221	0.099	0.053	0.137	0.261
8	Aerospace	0.274	0.234	0.038	0.025	0.019	0.000	0.118	1.000	0.530	0.149	0.218	0.248	0.286	0.101	0.235	0.031	0.031	0.059	0.042	0.022	0.042	0.208
9	Other transport	0.679	0.365	0.154	0.090	0.070	0.000	0.614	0.530	1.000	0.465	0.255	0.597	0.457	0.158	0.799	0.115	0.118	0.211	0.184	0.071	0.289	0.420
10	Ferrous metals	0.559	0.092	0.187	0.059	0.081	0.000	0.183	0.149	0.465	1.000	0.210	0.940	0.202	0.052	0.730	0.035	0.014	0.139	0.205	0.047	0.090	0.318
11	Non-fer. metals	0.507	0.200	0.071	0.033	0.038	0.000	0.113	0.218	0.255	0.210	1.000	0.393	0.184	0.080	0.297	0.019	0.027	0.110	0.059	0.035	0.045	0.630
12	Metal products	0.693	0.203	0.146	0.075	0.057	0.000	0.273	0.248	0.597	0.940	0.393	1.000	0.313	0.087	0.758	0.081	0.072	0.167	0.148	0.096	0.187	0.490
13	Instruments	0.667	0.944	0.171	0.125	0.062	0.000	0.246	0.286	0.457	0.202	0.184	0.313	1.000	0.475	0.307	0.121	0.126	0.278	0.141	0.245	0.118	0.466
14	Computers	0.287	0.465	0.032	0.029	0.015	0.000	0.084	0.101	0.158	0.052	0.080	0.087	0.475	1.000	0.107	0.024	0.024	0.089	0.020	0.037	0.024	0.167
15	Machines	0.699	0.192	0.104	0.061	0.045	0.000	0.369	0.235	0.799	0.730	0.297	0.758	0.307	0.107	1.000	0.075	0.044	0.162	0.131	0.076	0.126	0.396
16	Food etc.	0.141	0.075	0.127	0.121	0.034	0.000	0.080	0.031	0.115	0.035	0.019	0.081	0.121	0.024	0.075	1.000	0.045	0.124	0.134	0.227	0.048	0.155
17	Textiles	0.114	0.068	0.096	0.063	0.030	0.000	0.130	0.031	0.118	0.014	0.027	0.072	0.126	0.024	0.044	0.045	1.000	0.283	0.059	0.082	0.081	0.347
18	Rubber, plastics	0.433	0.270	0.750	0.326	0.204	0.000	0.221	0.059	0.211	0.139	0.110	0.167	0.278	0.089	0.162	0.124	0.283	1.000	0.303	0.279	0.129	0.563
19	Glass, stone, etc.	0.265	0.093	0.384	0.187	0.167	0.000	0.099	0.042	0.184	0.205	0.059	0.148	0.141	0.020	0.131	0.134	0.059	0.303	1.000	0.256	0.129	0.232
20	Paper, printing	0.213	0.087	0.258	0.232	0.074	0.000	0.053	0.022	0.071	0.047	0.035	0.096	0.245	0.037	0.076	0.227	0.082	0.279	0.256	1.000	0.157	0.390
21	Wooden prod.	0.169	0.101	0.112	0.054	0.053	0.000	0.137	0.042	0.289	0.090	0.045	0.187	0.118	0.024	0.126	0.048	0.081	0.129	0.129	0.157	1.000	0.322
22	Other manuf.	0.696	0.420	0.300	0.203	0.110	0.000	0.261	0.208	0.420	0.318	0.630	0.490	0.466	0.167	0.396	0.155	0.347	0.563	0.232	0.390	0.322	1.000

Note: Industry classification similar to Verspagen (1997a). Zeros in row and column 6 due to the aggregation of the 'ships' industry into a broader industry in US input–output tables. This does not cause problems because the firm database does not distinguish 'ship manufacturing' firms.

NOTES

* The first versions of this chapter were written when the author was a PhD student at the University of Twente, The Netherlands. The chapter has benefited greatly from my previous work on technology spillovers with Bart Verspagen. 1 would like to thank him as well as Erik Dietzenbacher, Richard Nahuis, Paolo Saviotti, Albert Steenge, Katy Wakelin and participants of the Twelfth Annual Congress of the European Economic Association (Toulouse, 1997), the European Association for Evolutionary Political Economy (Athens, 1997) and the Twelfth International Conference on Input–Output Techniques (New York, 1998) for their useful comments on earlier versions of the chapter. Of course, the usual disclaimer applies.

1. The *labels* 'rent spillovers' and 'pure knowledge spillovers' (as well as 'pure rent spillovers' and 'knowledge spillovers', which will be referred to below) were introduced by Van Meijl (1995), and the *concepts* by Griliches (1979).
2. Although they surely outperform conventional deflators, even hedonic deflators are not able to measure quality improvements perfectly, see Trajtenberg (1990).
3. See Griliches (1979), Van Meijl (1995) and Los (1997) for more comprehensive treatments of the pure rent spillover concept. Triplett (1996) provides an excellent empirical example of the disturbing effects connected to the use of conventional price deflators on productivity estimates.
4. Verspagen (1997b) argues that both pure rent spillovers and knowledge spillovers are important elements in the advanced endogenous growth theories (for example, Romer 1990). The switch from the assumption of perfect competition in the older models (for example, Romer 1986) to monopolistic competition and product varieties has been an immediate consequence of the notion that no firm would undertake any R&D if the full rents were to spill over to buyers (as is the case if markets are perfectly competitive). Knowledge spillovers were already at the heart of the older generation of models, being the sole cause of the increasing returns to scale at the macro-level that enable the economy to stick to a positive growth rate of output per worker.
5. For a more elaborate discussion of spillover channels, see, for example, Los (1997).
6. Griliches (1992) even argues that pure rent spillovers are not 'real' spillovers, as they occur as a part of a transaction between the firms or industries involved.
7. Schumpeterian 'creative destruction' might also yield negative spillovers. See the theoretical model by Aghion and Howitt (1992).
8. See, for instance, Terleckyj (1980), Sveikaukas (1981), Odagiri (1985), Goto and Suzuki (1989), Wolff and Nadiri (1993) and Wolff (1997).
9. Griliches and Lichtenberg (1984) show that, under the (strong) hypothesis that R&D improves an industry's outputs sold to all buying industries to the same extent, quality mismeasurement leads to overestimations of total factor productivity in the user industries proportional to the output coefficients.
10. Note that these methods determine the users of the *information* provided in the patent document, not the users of the patented innovation itself. Scherer (1982), Englander et al. (1988) and Mohnen and Lépine (1991) used methods based on the latter notion, thereby (sometimes unconsciously) stressing rent spillovers (see Los 1997, and Los and Verspagen 2000).
11. See Jaffe (1988, 1989), Goto and Suzuki (1989), Adams (1990) and Park (1995).
12. Jaffe (1986) distinguished 49 patent classes.
13. Note that a firm can obtain both 'idea-creating' as well as 'imitation-enhancing' spillovers from the potential spillover pool. If the concept of a potential spillover pool is applied at the industry level, the latter class of knowledge spillovers no longer plays a role.
14. Blin and Cohen (1977) used this feature to construct measures of technological similarity, computing Euclidean distances between pairs of input coefficients vectors. Their ultimate aim was to find a good algorithm to aggregate n-dimensional input–output tables into m-dimensional tables, independent of the particular values of m and n.

15. See Oksanen and Williams (1992) for a recommendation of the cosine measure as a measure of similarity.
16. In these contributions, only two factors of production (physical capital and labour) are distinguished. The analysis, however, can easily be extended to any number of (intermediate) inputs. In an international endogenous growth theory context, Basu and Weil (1998) recently proposed using the similarity of capital requirements per unit of labour between countries as a measure of the mutual relevance of international technology spillovers.
17. For a comprehensive survey of these theories, see Thirtle and Ruttan (1987).
18. Like Kennedy (1964), Nelson and Winter (1982) describe technologies by their physical capital and labour requirements. Again, nothing seems to preclude extensions to more than two inputs in our case.
19. See also Lawson and Teske (1994).
20. For some countries, additional tables are available in which imports are assigned to a number of industries or commodities. One of those commodities is 'non-comparable imports' which can include very different inputs.
21. According to the Penn World Tables (Summers and Heston, 1991, series 'OPEN') the USA was second least open developed market economy in 1987, after Japan.
22. Admittedly the composition of the labour force admittedly is a major determinant of the human capital of industries, which could provide more information on technologies used. We lack data on these compositions, however. The disturbing effects of inclusion of one aggregate labour input and one aggregate capital input (the only alternative to the chosen procedure given the data limitations) can be illustrated by a hypothetical example. Assume that the ratio of labour costs to gross output is 0.3 for two industries and that capital costs per unit of gross output are 0.2 (these are reasonable values, for many industries in the USA the value added to gross output ratios exceeds 0.6). Ten intermediate inputs are distinguished. Five of them are used by the first industry (the input coefficients are assumed to be 0.1), the other five by the second industry (input coefficients also 0.1). Now, given the fact that the industries have no intermediate inputs in common, the cosine would be 0. If the extremely heterogeneous labour and capital inputs were included, the proximity would equal 0.96!
23. At high levels of aggregation the SIC ordering is relatively arbitrary, although the 'agriculture–mining–manufacturing–construction–services' order is always preserved. At lower aggregation levels, however, industries constituting a 'superindustry' at higher levels of aggregation levels are put next to each other (see, for example, Appendix Table 4.A.1).
24. Schmoch et al. (1996) present a similar diagram for their application of the Jaffe measure on German industries.
25. For details on MDS, see Green et al. (1989).
26. MDS has several variants, one of them treating the distances as measurable on the ordinal level. We used this variant, so the exact construction of the distance measure does not influence the configuration as long as the distance measure is a monotone decreasing function of ω.
27. Closer examination of the spread of industries over the MDS plot indicates that the position of industries along the horizontal axis might be determined to some extent by the 'metal content' of the intermediate inputs: industries using relatively more metal inputs tend to be located to the right. We are not able to draw such 'impressionistic' conclusions for the vertical dimension.
28. These distortions are the most important reason for presenting an MDS plot for manufacturing industries only. The MDS output reports an '*RSQ*' of 0.75 in this case, indicating that 75 per cent of the variation in the 'true' distances is reflected in the two-dimensional representation. Applying MDS on all 91 industries reduces *RSQ* to 0.64. The most important (but also expected) conclusion of the latter exercise should be that two 'super-clusters' are identified: one corresponding to manufacturing, the other to services. Primary industries are scattered over the plot, which (after examination of the proximities matrix) turns out to be the consequence of reducing the number of dimensions.

29. Using more direct data on technological activity (information contained in patent documents), Verspagen (1997a) finds that 'electrical products' ranks about fifth (out of 21) as receiver of knowledge spillovers from 'ferrous metal manufacturing', while the rank of the latter industry as receiver of knowledge spillovers from 'electrical products' varies (across his three measures) from five to a joint 21. This variation, as well as the higher level of aggregation, renders inference on the correctness of the high proximity value impossible.
30. We also investigated the sensitivity over time. Comparisons on the basis of US input–output tables for 1963 and 1967 showed that for all industries Pearson's correlation coefficients between rows of proximity values were in excess of 0.8, and 80 per cent of the industries showed coefficients over 0.95. Comparisons for a longer interval (1963–87) showed more instability: 70 per cent of the industries had correlation coefficients over 0.8 and only 20 per cent coefficients between 0.95 and one. The latter results may (partly) be a reflection of technological change.
31. To analyse the sensitivity of the obtained estimation results, constant returns to scale will sometimes be imposed by setting μ equal to one.
32. Cuneo and Mairesse (1984) find that that estimates using value added instead of sales do not differ significantly.
33. See Los and Verspagen (2000) for the deflators used.
34. The choice for Verspagen's (1997a) *IR*2 measure instead of his *IR*1 is rather arbitrary. Generally, *IR*1 and *IR*2 yield comparable results in estimations like the ones presented here (see Los and Verspagen, 2000).
35. Note that the inputs of primary and tertiary industries are 'moved' to the primary input block. This implies that they play no role at all in the determination of the proximities ω and, consequently, the *IRC*s of the firms. An alternative course would be to leave these input coefficients unchanged in the intermediate block, but then discrepancies across industries concerning the levels of aggregation would be unavoidable.
36. This is an admittedly crude solution. Examination of other trade or production statistics would have yielded a 'better' aggregated table, but we do not think this effort would serve the goals of this chapter.
37. In Verspagen's (1997a) *IR*2 measure some R&D benefits a 22nd industry, 'ships and boats manufacturing'. The STAN database on R&D investments and the US input–output tables do not distinguish such an industry. Therefore, firms in this industry are added to 'other transport' and no attention should be paid to the row and column corresponding to 'ships' in the matrix presented in Appendix Table 4A.3.
    Note that we implicitly assume homogeneity of firms within an industry, for all spillover measures.
38. The OECD classification is based on average R&D to value-added ratios. This is a rather crude criterion which, for example, causes the 'other transports' industry to be low-tech. Bearing in mind the huge R&D efforts of consortia in the field of rail transportation (Verspagen 1997a), we think firms in this industry would be better included in the medium-tech category. Of course, this adjustment does not remove all undesired heterogeneity in the categories. A firm specialized in the field of high-tech superconducting ceramics will still be in the low-tech 'glass, stone, etc.' industry.

REFERENCES

Adams, J.D. (1990), 'Fundamental stocks of knowledge and productivity growth', *Journal of Political Economy*, **98**, 673–702.

Aghion, P. and P. Howitt (1992), 'A model of growth through creative destruction', *Econometrica*, **60**, 323–51.

Basu, S. and D.N. Weil (1998), 'Appropriate technology and growth', *Quarterly Journal of Economics*, **114**, 1027–54.

Bernstein, J.I. (1988), 'Costs of production, intra- and interindustry R&D spillovers: Canadian evidence', *Canadian Journal of Economics*, **21**, 324–47.

Blin, J. and C. Cohen (1977), 'Technological similarity and aggregation in input–output systems: a cluster-analytic approach', *Review of Economics and Statistics*, **59**, 82–91.

Brown, M. and A. Conrad (1967), 'The influence of research on CES production relations', in M. Brown (ed.), *The Theory and Empirical Analysis of Production*, New York: Columbia University Press for NBER, pp. 341–72.

Coe, D.T. and E. Helpman (1995), 'International R&D spillovers', *European Economic Review*, **39**, 859–87.

Cuneo, P. and J. Mairesse (1984), 'Productivity and R&D at the firm level in French manufacturing', in Z. Griliches (ed.), *R&D, Patents and Productivity*, Chicago: University of Chicago Press, pp. 375–92.

Englander, A.S., R. Evenson and M. Hanazaki (1988), 'R&D, innovation and the total factor productivity slowdown', *OECD Economic Studies*, **11**, pp. 7–42.

Engle, R.F. and C.W.J. Granger (1987), 'Co-integration and error correction: representation, estimation and testing', *Econometrica*, **55**, 251–76.

Engle, R.F. and B.S. Yoo (1991), 'Cointegrated economic time series: an overview with new results', in R.F Engle and C.W.J. Granger (eds), *Long-run Economic Relationships*, Oxford: Oxford University Press, pp. 237–66.

Goto, A. and K. Suzuki (1989), 'R&D capital, rate of return on R&D investment and spillover of R&D in Japanese manufacturing industries', *Review of Economics and Statistics*, **71**, 555–64.

Green, P.E., F.J. Carmone and S.M. Smith (1989), *Multidimensional Scaling: Concepts and Applications*, Boston: Allyn & Bacon.

Griliches, Z. (1979), 'Issues in assessing the contribution of research and development to productivity growth', *Bell Journal of Economics*, **10**, 92–116.

Griliches, Z. (1992), 'The search for R&D spillovers', *Scandinavian Journal of Economics*, **94**, S29–S47.

Griliches, Z. and F.R. Lichtenberg (1984), 'Interindustry technology flows and productivity growth: a reexamination', *Review of Economics and Statistics*, **66**, 324–9.

Griliches, Z. and J. Mairesse (1984), 'Productivity and R&D at the firm level', in Z. Griliches (ed.), *R&D, Patents and Productivity*, Chicago: University of Chicago Press, pp. 339–74.

Jaffe, A.B. (1986), 'Technological opportunity and spillovers of R&D: evidence from firms' patents, profits, and market value', *American Economic Review*, **76**, 984–1001.

Jaffe, A.B. (1988), 'Demand and supply influences in R&D intensity and productivity growth', *Review of Economics and Statistics*, **70**, 431–7.

Jaffe, A.B. (1989), 'Real effects of academic research', *American Economic Review*, **79**, 957–70.

Kennedy, C. (1964), 'Induced bias in innovation and the theory of distribution', *Economic Journal*, **74**, 541–7.

Lawson, A.M. and D.A. Teske (1994), 'Benchmark input–output accounts for the U.S. economy, 1987', *Survey of Current Business*, **74**, 73–115.

Leontief, W. (1989), 'Input–output data base for analysis of technological change', *Economic Systems Research*, **1**, 287–95.

Los, B. (1997), 'A review of interindustry technology spillover measurement methods', Working Paper, University of Twente, Enschede, The Netherlands.

Los, B. and B. Verspagen (2000), 'R&D spillovers and productivity: evidence from U.S. manufacturing microdata', *Empirical Economics*, **25** (forthcoming).

Mansfield, E. (1977), *The Production and Application of New Industrial Technology*, New York: W.W. Norton.

Mohnen, P. (1990), 'New technology and interindustry spillovers', *Science/Technology/Industry Review*, **7**, 131–47.

Mohnen, P. and N. Lépine (1991), 'R&D, R&D spillovers and payments for technology: Canadian evidence', *Structural Change and Economic Dynamics*, **2**, 213–28.

Nadiri, M.I. (1993), 'Innovations and technological spillovers', National Bureau of Economic Research Working Paper no. 4423.

Nelson, R.R. and S.G. Winter (1982), *An Evolutionary Theory of Economic Change*, Cambridge, MA: Harvard University Press.

Odagiri, H. (1985), 'Research activity, output growth, and productivity increase in Japanese manufacturing industries', *Research Policy*, **14**, 117–30.

Oksanen, E.H. and J.R. Williams (1992), 'An alternative factor-analytic approach to aggregation of input–output tables', *Economic Systems Research*, **4**, 245–56.

Park, W.G. (1995), 'International R&D spillovers and OECD economic growth', *Economic Inquiry*, **33**, 571–91.

Romer, P.M. (1986), 'Increasing returns and long-run growth', *Journal of Political Economy*, **94**, 1002–37.

Romer, P.M. (1990), Endogenous technological change', *Journal of Political Economy*, **98**, S71–S102.

Scherer, F.M. (1982), 'Inter-industry technology flows and productivity measurement', *Review of Economics and Statistics*, **64**, 627–34.

Schmoch, U., G. Münt and H. Grupp (1996), 'New patent indicators for the knowledge-based economy', Paper prepared for the OECD Conference on 'New Science and Technology Indicators for the Knowledge-Based Economy', June, Paris, France.

Summers, R. and A. Heston (1991), 'The Penn World Table (Mark 5): an expanded set of international comparisons (1950–1988)', *Quarterly Journal of Economics*, **106**, 1–41.

Sveikauskas, L. (1981), 'Technological inputs and multifactor productivity growth', *Review of Economics and Statistics*, **63**, 275–82.

Terleckyj, N.E. (1974), *Effects of R&D on the Productivity Growth of Industries: An Exploratory Study*, Washington, DC: National Planning Association.

Terleckyj, N. (1980), 'Direct and indirect effects of industrial research and development on the productivity growth of industries', in J. Kendrick and B. Vaccara (eds), *New Developments in Productivity Measurement and Analysis*, Chicago: University of Chicago Press, pp. 359–77.

Thirtle, C.G. and V.W. Ruttan (1987), *The Role of Demand and Supply in the Generation and Diffusion of Technical Change*, New York: Harwood Academic Publishers.

Trajtenberg, M. (1990), *Economic Analysis of Product Innovations*, Cambridge, MA: Harvard University Press.

Triplett, J.E. (1996), 'High tech industry productivity and hedonic price indexes', Paper presented at the 24th General Conference of the International Association for Research in Income and Wealth, 19–23 August, Lillehammer, Norway.

Van Meijl, H. (1995), 'Endogenous technological change: the case of information technology', PhD thesis, University of Limburg, Maastricht.

Verspagen, B. (1997a), 'Measuring inter-sectoral technology spillovers: estimates from the European and US patent office databases', *Economic Systems Research*, **9**, 47–64.

Verspagen, B. (1997b), 'Estimating international technology spillovers using technology flow matrices', *Weltwirtschaftliches Archiv*, **133**, 226–48.

Wolff, E.N. (1997), 'Spillovers, linkages, and technical change', *Economic Systems Research*, **9**, 9–23.

Wolff, E.N. and M.I. Nadiri (1993), 'Spillover effects, linkage structure and research and development', *Structural Change and Economic Dynamics*, **4**, 315–31.

5. Classifying technological systems: an empirical application to eight OECD countries*

Riccardo Leoncini and Sandro Montresor

1 INTRODUCTION

The analysis of technological change has recently been revived by the contributions of the neo-Schumpeterian (evolutionary) research programme (Freeman 1994; Nelson 1995), paying particular attention to themes only implicitly or superficially treated by the previous literature. In particular, the role of institutions and of systemic relationships among innovative agents has been emphasized. New interpretative tools have been developed, such as, for example, the application of Kuhnian paradigms to the process of technological change (Freeman 1979; Dosi 1982), and the identification of sectoral patterns of technological change (Pavitt 1984; Malerba and Orsenigo 1995). Particularly fruitful have been concepts such as that of national system of innovation (Freeman 1988; Lundvall 1988 and 1992; Nelson 1993; Edquist 1997) and of technological system (Carlsson and Stankiewicz 1991; De Liso and Metcalfe 1996). These concepts proved to be useful for a wider understanding of the innovative processes in which complex elements, even not strictly pertaining to the sphere of economics as such, are drawn into the analysis: tacit knowledge, competencies, human capital and institutions are the most relevant examples. However, despite their originality and variety, these approaches are hardly consistent and far from being homogeneous, and moreover, their empirical applications are not assisted by a coherent unit of analysis.

The aim of this chapter is therefore to define a classifying method for these systemic concepts, based on a particular notion of technological system and on an applied kind of analysis. The notion of technological system to which we refer (Leoncini 1998) is quite different from those already proposed. In fact, on the one hand, it has a macroeconomic nature which differs radically from the 'sectoral' approach proposed in Carlsson and Stankiewicz (1991), and, on the other hand, unlike that of De Liso and Metcalfe (1996) it is more suitable for empirical applications. Moreover, as far as the various concepts of the national system of innovation are concerned (McKelvey 1991), the concept of technological system has been used instead for the following reasons. The notion to which we refer is defined in a quite precise and operational way. In fact, it is composed of four interacting building blocks (hard core of techno-scientific knowledge, industrial subsystem, market subsystem and the institutional interface) for which it is possible to identify proper proxies. Thus a technological system explicitly encompasses several dimensions which are either ignored or treated only in an indirect and heterogeneous way in the case studies on national systems of innovation (this is particularly true for the industrial structure and the market).[1]

Such a notion of technological system is analysed here in an applied, taxonomic way. That is, the analysis focuses on the configurations it assumes by combining the structure of the internal relationships among its building blocks (subsystems), and that of its external relationships with the outer environment. We argue that structural indicators of intersectoral innovation flows and foreign trade are, respectively, the most suitable proxies.

Combining these two dimensions, a technological system can take one of the following configurations: a pervasive technological system (that is, one that could be defined as a system of innovation), either outward or inward oriented; or a segmented (or trajectories-based) technological system which again, can be either outward or inward oriented.

This approach is then applied to a group of eight OECD countries during three temporal spans (early 1980s, middle 1980s and early 1990s). It is thus shown that Japan and Canada are, respectively, the most pervasive and the most segmented inward-oriented technological systems. The rest of the countries have different outward-oriented configurations: pervasive, in the cases of Germany and France, segmented, in the cases of the UK, Canada, Australia and the Netherlands. All these configurations are quite stable, except for Denmark which experiences a structural change moving from one cluster (that of the inward-oriented segmented technological systems) to another (that of the outward-oriented segmented technological systems).

The chapter is organized as follows. In Section 2 the methodological background of our approach is briefly outlined. In Section 3 the results of the empirical application are presented and discussed. Conclusions follow.

2 METHODOLOGICAL BACKGROUND

As already said in the introduction, by mapping different technological systems with respect to an indicator of internal and one of external connection, we shall be able to provide a taxonomy of different configurations in terms of structure and international specialization. In particular, the former is derived by calculating the dispersion of the elements of a matrix of intersectoral innovation flows, while the latter is derived by analysing the variance of the well-known Balassa index of revealed comparative advantages.

2.1 The Internal Connection Indicator

By 'intersectoral innovation flows matrix' we mean a **D** $(n \times n)$ matrix[2] whose generic element d_{ij} represents the percentage of the whole intersectoral innovative acquisitions that sector j gets from sector i, embodied in the intermediate inputs they exchange.

The intersectoral focus of these matrices makes them suitable for mapping the structure of the internal relationships of a technological system as follows: First, if each sector j 'acquires' innovation from a unique sector i_{j^*}, we define this case *maximum polarization*. The **D** matrix is then characterized as follows:

$$d_{ij} = 1 \text{ if } i = i_{j^*};\ d_{ij} = 0 \text{ else.} \tag{5.1}$$

Second, if the intersectoral acquisitions of each sector are equally distributed among the others, we define this case *maximum pervasiveness*, and the **D** matrix is characterized as follows:

$$d_{ij} = \frac{1}{n-1} \text{ if } i \neq j;\ d_{ij} = 0 \text{ else.} \tag{5.2}$$

Since (5.1) and (5.2) are the two extremes comprising all the other feasible maps, it is possible to consider the distribution of the elements of the maximum pervasiveness matrix as a reference term and to characterize all the others as deviations from it. An internal connection indicator can thus be obtained for a technological system as follows:

$$TS_{IN} = \frac{\sum_i \sum_j (d_{ij} - d^{+}_{ij})^2}{n} \tag{5.3}$$

where d^{+}_{ij} is the general element of **D** in case of maximum pervasiveness.

The domain of (5.3) thus falls between the two extreme cases of maximum pervasiveness and of maximum polarization:[3]

$$0 < TS_{IN} < \frac{n(n-2)}{n-1} \quad . \tag{5.4}$$

Finally, (5.3) is normalized, and its complement to 1 is calculated:

$$TS^{*}_{IN} = 1 - \frac{TS_{IN}(n-1)}{n(n-2)} \quad , \quad 0 < TS^{*}_{IN} < 1 \tag{5.5}$$

so that low/high values will give evidence of a low/high degree of internal connection of the corresponding technological system.

If TS^{*}_{IN} is high, the structure of the technological system will be characterized by a high degree of innovative pervasiveness. If, on the other hand, TS^{*}_{IN} is low, the same interrelations show a high degree of polarization, which is associated with more segmented innovative processes and which reminds us of a technological system based on a series of less-connected technological trajectories.

It could be argued whether intersectoral innovation flows matrices can be used as a proper proxy to define technological systems, or whether a sectoral decomposition is a reasonable proxy for the technological system's divisions. However, the advantages of using such an approach by far outweigh its shortcomings. In fact, if, on the one hand, the results are sensitive to the sectoral disaggregation, and embodied flows are not the only ones to be relevant, then, on the other hand, different technological systems can be compared on a systematic basis, by means of consistent and integrated international databases.

Furthermore, it must be stressed that in this kind of analysis, we adopt an input perspective based on R&D and on the embodiment hypothesis, because it allows us to capture relationships of a wider nature. However, if the intersectoral innovation flows matrix is obtained by using output indicators, such as patents,[4] the relative flows might be less representative of a techno-economic kind of relationship. Indeed, a certain user sector might not have very significant economic links (in terms of goods or capital transactions or in terms of market structure) with the producer sectors to which it pays some royalties. The development of a new research programme is more likely to stimulate interactions between actors that are also close in economic terms (that is, of input–output coefficients) (DeBresson 1996; Verspagen 1997). Moreover, while output and input indicators both suffer from serious limitations (Griliches 1990; Patel and Pavitt 1995), the latter are less affected by problems of scale than the former, which are mainly significant for those economic units that can afford a patenting effort.

Finally, the embodiment hypothesis, with its use of the input kind of indicators, also suffers from serious limitations.[5] However, it has the advantage of allowing us to retain the non-intentional transmission of tacit knowledge, for which the institutional set-up, the structure of the interrelations and other extra-economic arguments are particularly relevant.[6]

2.2 The External Connection Indicator

Although quite helpful in studying internal links, an input–output analysis of the innovative activity is not suitable when facing the question of external (international) relations (Lundvall 1996). However, as we are not simply interested in the international innovative penetration of a country – for which there is an abundance of indicators (balance of technological payments, foreign direct investments in high-tech sectors, patents obtained abroad, multinationals diffusion, technological agreements and so on) – but rather in the structure of its techno-economic international openness, the input–output analysis can be complemented by an intersectoral analysis of the commercial specialization (or de-specialization) of a country. Our external indicator is thus given by the variance of the comparative advantages (disadvantages) of a certain country:

$$TS_{EX} = \frac{\sum_j (\beta_{ic} - \bar{\beta}_c)^2}{n} \tag{5.6}$$

where:

$$\beta_{ic} = \frac{EX_{ic}}{\sum_c EX_{ic}} / \frac{\sum_i EX_{ic}}{\sum_i \sum_c EX_{ic}}$$

is the Balassa index of revealed comparative advantages for sector i of country c, in which:

EX_{ic} are total exports of product i by country c;

$\sum_c EX_{ic}$ are total world exports of product i;

$\sum_i EX_{ic}$ are total exports of country c;

$\sum_i \sum_c EX_{ic}$ are total world exports; and

$\bar{\beta}_c = \frac{\sum_i \beta_{ic}}{n}$ is the average value of the competitive advantages and/or disadvantages of the *n* sectors of country *c*.[7]

Although less immediate than the internal connection indicator, the interpretation of TS_{EX} provides some insights on the degree of openness of a technological system. Very high values of (5.6) are consistent either with very high specialization or with very high de-specialization. In fact, the same value of TS_{EX} can apply for symmetrical export performances. That is, a country with generally high values of β_i and few unsuccessful sectors gains the same value for TS_{EX} of a country with generally low values of β_i and few successful sectors. In spite of this, both countries' performances can be interpreted in a similar way. Indeed, in the former case, very high comparative advantages are presumably the result of quite well-developed national capabilities. While in the latter case, the impossibility of catching-up with average world practices and of exploiting international relationships presumably results from national capabilities that are not sufficiently developed. In both cases, therefore, although for opposite reasons, an inward-oriented configuration seems appropriate. Very low values of (5.6) are instead the outcome of revealed comparative advantages that are quite aligned to the average, again with a symmetrical interpretation. Therefore, they indicate that national capabilities are neither so high as to guarantee 'self-sufficiency', nor so low as to prevent the country from penetrating foreign markets: an outward-orientated configuration is thus the most suitable in this case.

2.3 The Taxonomy

By combining the internal and the external indicators, each technological system is thus defined by means of the binomial distribution of the values of (5.5) and (5.6) which might locate it in one of the following configurations:

1. pervasive/inward oriented (high TS^*_{IN} and TS_{EX});
2. segmented/inward oriented (low TS^*_{IN} and high TS_{EX});
3. segmented/outward oriented (low TS^*_{IN} and TS_{EX});
4. pervasive/outward oriented (high TS^*_{IN} and low TS_{EX}).

We can see the same taxonomy using a scatter diagram, such as that of Figure 5.1.

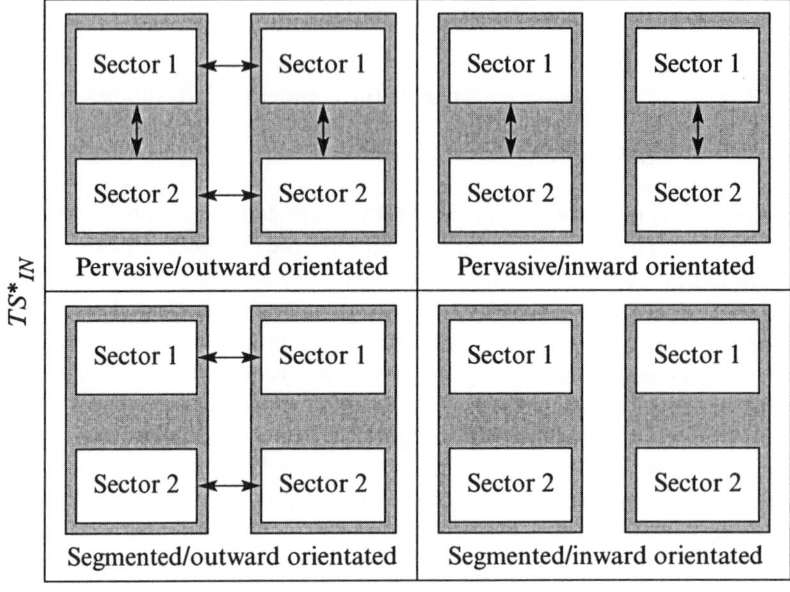

Figure 5.1 Taxonomy of technological system configurations

3 THE EMPIRICAL CONFIGURATIONS OF THE TECHNOLOGICAL SYSTEM

In this section we apply the approach described above to eight OECD countries, using a 16 manufacturing sectors disaggregation, for the following three periods: early 1980s, middle 1980s, and early 1990s.[8]

The empirical analysis is aimed at supplying a taxonomy of countries in the different periods, but it must be pointed out immediately that, by its nature, this kind of analysis describes the complexity of a system only partially. Indeed, the utilization of intersectoral innovation flows matrices to describe technological systems suffers from the shortcomings of a reductionist approach which limits the technological system's complexity to its production structure (although complemented by the R&D components) as reflected in an input–output context. If, on the one hand, we can compute a synthetic indicator and hence position each system in a single diagram, then, on the other hand, there is the necessity to complement this kind of structural analysis with another one of the institutional characteristics of the systems analysed. In order to take into account the full complexity of a technological system we thus decided to cross data sources and to parallel the synthetic treatment made available by the utilization of intersectoral innovation flows matrices, with more traditional types of data. As will be clearer in the following, certain peculiarities of the technological system configurations can be understood better in the light of structural characteristics (e.g. country size and economic dimension). For instance, small countries, characterized by a smaller population dispersed on a relatively large territory, by a higher dependency on agriculture and resource exploitation, and by a smaller percentage of manufacturing activity (Nelson 1993), all share the same basic features, in terms of their techno-economic activity, which are consistent with our more synthetic treatment. That is, a basic-oriented R&D, applied mainly to the primary sector and, only to a limited and international extent, also at industrial level; a greater pressure imposed by 'critical-mass' problems towards innovative specialization; and relatively thinner innovative sectors, dependent on few firms and few technologies (see, for example, Freeman and Lundvall 1988).

In this way, we hope to benefit from the integration of already available knowledge on systems performance with our more compact and concise methodology. We shall thus present in the following subsections a separate discussion of the two indicators, each one complemented by a discussion of more traditional indicators. The two indicators will then be drawn together in order to highlight the most important evidence and the full meaning of our analysis.

3.1 The Internal Connection Indicator

At the outset, clear clusters emerge from the analysis of the internal connection indicator (Figure 5.2): three countries (Japan, Germany and France) are characterized by values more or less higher than 0.3, with Japan consistently above 0.6, and growing over time. The rest of the sample has values around 0.1, but these are increasing; the exception is the Netherlands, which shows a decrease over time of the indicator. We shall now proceed to analyse these results in the light of the already available knowledge, in order to show how crossing data sources can improve the explanatory power of both.

The outsider characteristics of the Japanese innovative performance is, for example, a widely recognized fact, at least since the early 1980s onward, confirmed both by input and output innovation statistics (Odagiri and Goto 1993). However, our results also seem to give a quantitative confirmation of some recent qualitative studies on structural and

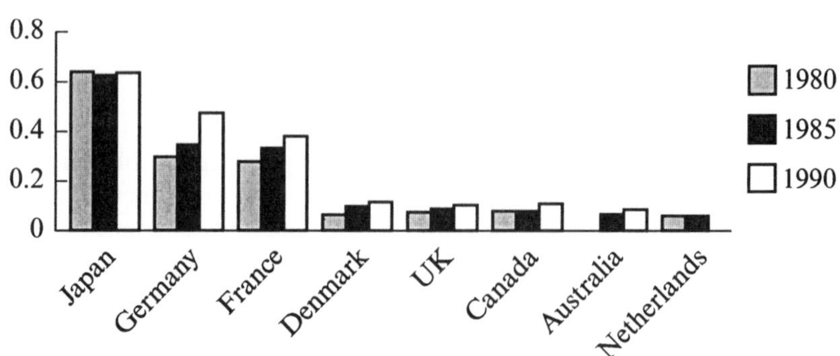

Figure 5.2 Internal connection indicator

techno-economic characteristics: the pervasive role of the Ministry of International Trade and Industry (MITI), Japan, and the diffused participation of the economic agents, either direct or indirect, to the main economic and technology policy issues (Fransman 1995);[9] the recent debate on the different models of capitalism (Dore 1992); the predominance of diffusion-oriented innovative projects with respect to the mission-oriented ones (Freeman 1988); and the characteristics of a dynamic versus myopic system (Patel and Pavitt 1991).

For France and Germany it is also possible to relate our results to structural and institutional characteristics. Several more qualitative contributions point out, for instance, the wide participation base of the Rhenish model of capitalism (Albert 1991) and the highly pervasive characteristics of their most competitive sectors (Dosi et al. 1990). It must also be noted that the high degree of internal connection of Germany seems to be the result of its superior ability in spurring innovative intersectoral virtuous circles, rather than the result of its conspicuous R&D involvement (see, for example, Leoncini et al. 1996 and Schnabl 1995). Indeed, if we consider R&D expenditure as a percentage of GDP (Table 5.1), Germany would rank at least equal to Japan (indeed it was ahead in the early 1980s), but this is not reflected in our internal connection indicator.

It is interesting to note that a large country, such as the UK, exhibits unambiguously a trajectories-based configuration. Several qualitative arguments can also be put forward for this result: the Anglo-Saxon model of capitalism is notably not very effective in spurring industrial and social cohesion (Albert 1991); the majority of the innovative projects are mainly mission oriented (Ergas 1987); moreover, the UK is one of the most 'multinationalized' countries of the OECD area (both inward and outward, Patel 1995), and this fact tends to stimulate more intra-sectoral

Table 5.1 Techno-economic indicators, 1988

	Japan	Germany	France	UK	Denmark	Canada	Australia	Netherlands*
GDP/capita (1988 PPP)	14 228	14 161	13 603	13 428	13 555	18 446	13 412	11 860
Population	122 613	61 451	55 873	57 065	5 130	25 950	16 538	14 700
Manufacturing output/GDP	29	44	27	27	25	23	18	–
Manufacturing exp/GDP	9	24	13	17	19	16	5	–
Literacy rate	>95	>95	>95	>95	>95	>95	>95	>95
Secondary level enrolment rate	96	94	92	83	107	104	98	103
University enrolment rate	28	30	31	22	30	58	29	34
Scientists and technicians (per 1000 p., 1986–90)	110	86	83	90	85	174	48	92
R&D/GDP	2.9	2.9	2.3	2.3	1.3	1.5	1.4	2.3
Business R&D/Total R&D	66.0	72.2	58.9	67.0	55.6	55.0	37.4	60.0

Note: *GDP/capita (1987), population (mid-1987), enrolment rate (1990).

Sources: Nelson (1993), World Bank (1994), OECD (1992b).

relationships with partner countries rather than inter-sectoral relationships within the country (Walker 1993). Once again, as in the case of Germany, these structural and institutional characteristics are not reflected by a performance analysis, according to which the UK is among the most innovative countries (Table 5.1).

The remaining countries of the sample show similar positions, which can be traced to the homogeneous structural characteristics of smaller countries.[10]

Interesting results finally emerge when the same variations are considered in relative terms, that is, with respect to the value assumed by the internal indicator at the beginning of each period. By simply scattering these indicators it is possible to show some peculiar features of their patterns of growth. Indeed, they grow according to a non-linear pattern (Figure 5.3): only really backward countries or those on the 'frontier' of systems complexity exhibit a negative correlation between the starting level and the rate of growth, while the same relationship does not hold for systems that are situated in-between.

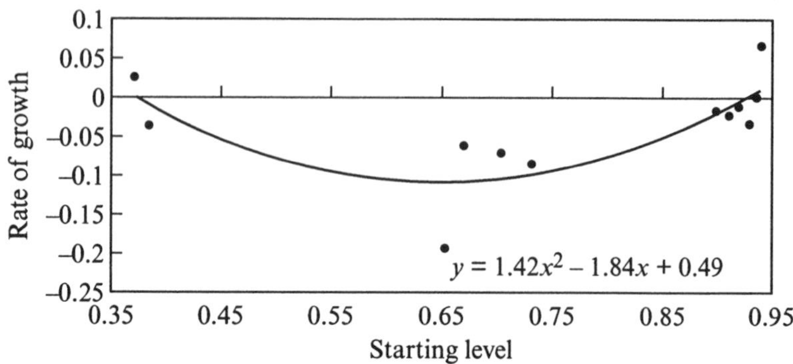

Figure 5.3 A non-linear model of convergence

3.2 The External Connection Indicator

Also, by looking at the external connection indicator (Figure 5.4), a clustering of Japan and Canada emerges, characterized by highly polarized revealed comparative advantages. Denmark shifts, in time, from this cluster to another one that comprises the rest of the countries.

A further integration with the statistics collected in Table 5.1 indicates that the present analysis is consistent with the common argument according to which dynamic small and medium-sized countries, for different reasons, exhibit a larger international involvement (Freeman and Lundvall 1988). In some cases, small countries might not find their internal market large enough to repay a considerable investment in innovation, so that they need to target highly selected sectors and exploit their results in the international markets. This is the argument that best fits the case of the Netherlands, whose R&D intensity is quite similar to the larger investors (Table 5.1), and whose international patenting activity is remarkably biased towards nationally controlled foreign subsidiaries (Table 5.2).

In contrast, in some other cases, a major degree of openness might be due to a massive injection of foreign capital which comes to control the new technologies, while more traditional sectors undertake innovations at a local level quite independently (again spurring a trajectories-based configuration). This seems to be the case for Denmark, whose inward foreign investments and new technology transfers are among the highest in the world (Mowery and Oxley 1995), while the local agro-industrial complex gradually assumes the features of an independent development bloc (Edquist and Lundvall 1993). A similar dependency account can be given

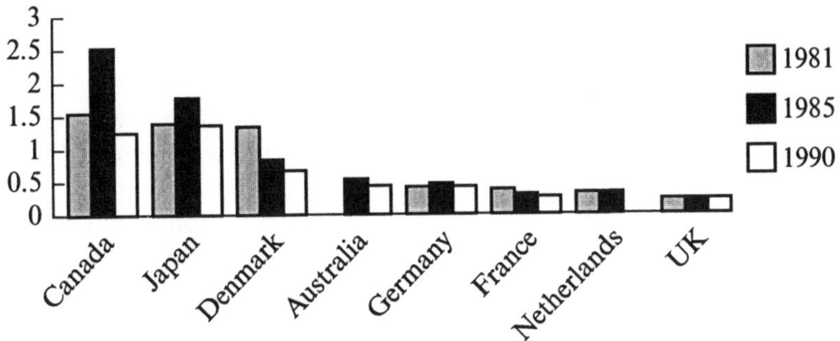

Figure 5.4 External connection indicator

for the Australian case: its high-income colonial state origin has made the development of local technological capabilities a quite recent and still maturing need, while it has also implied a certain bias towards the key sectors of the colonizers, mainly the UK (Gregory 1993).

Table 5.2 International patenting activity

	National source of US patents 1981–86* (% of total patents granted by the US office)			Source of patent applications in each country.** (% of total local patent applications 1990)***	
	National controlled	Foreign controlled	Other	Domestic	Foreign
Japan	62.5	1.2	36.3	88.5	11.5
Germany	44.8	10.5	44.2	32.6	67.4
France	36.8	10.0	53.2	16.1	83.8
UK	32.0	19.1	49.0	21.5	78.5
Netherlands	51.9	8.7	39.4	5.3	94.7
Denmark	n.a.	n.a.	n.a.	6.8	93.2
Australia	n.a.	n.a.	n.a.	24.6	73.4
Canada	11.0	16.9	72.1	6.8	93.2

Note: *West Germany, ***Differences among patents are not considered.

Sources: *Patel and Pavitt (1991); **OECD (1992b).

Quite surprisingly, the same argument does not hold for the remaining small country of the sample (indeed the smallest of the 'large' OECD economies), that is Canada, whose dominance of international linkages over domestic ones has been largely recognized by different sources (see, for example, Niosi and Bellon 1996). A recent study (Paquet 1996) on local Canadian economies seems to shed some light on this apparent contradiction, by suggesting that the balance between domestic-oriented, natural resource-based technological systems (energy, metallurgy, forestry, agriculture) and internationally oriented human resource-based ones (aeronautics, telecommunications, information technology), is not actually in favour of the latter, as has been documented by other more aggregated statistics (McFetridge 1993). However, it seems more plausible that a truer explanation comes from the very 'local' (that is, national) nature of its internationalization process, determined by the almost exclusive (that is, not dispersed) links (and therefore comparative advantages) with respect to the United States: most US subsidiaries operating in Canadian high-tech sectors deal only with parent companies and vice versa (OECD 1992, p. 220). The most active development blocs (especially the motor vehicle sector[11]) are completely integrated with, rather than simply dependent on, those in the USA, and human capital is educated and trained mainly in the USA rather than 'imported' through the US multinationals (McFetridge 1993).

As far as the larger countries of the sample are concerned, we observe that there is a clear distinction between Japan and the three large European countries, although Germany shows higher and growing values (almost twice those of the UK), and France has slightly higher values, although converging, than the UK. Again, it is possible to trace these differences to structural characteristics. In the case of UK, for example, the absence of sound and virtuous relationships between users and suppliers and between public basic and private applied R&D (Walker 1993) spurs firms, mainly large multinational corporations, to develop their research activities and safeguard the results abroad (Table 5.2).[12]

3.3 The Configurations of the Technological System

Finally, the taxonomy of the different configurations can be composed for each of the investigated periods by combining the results of the internal and external connection indicators, so that some of the comments presented above separately can be put together here (Figures 5.5 to 5.7).

The first thing to notice is that very different patterns emerge even though the group of countries analysed is more or less homogeneous and with comparable levels of development. This methodology seems to be in

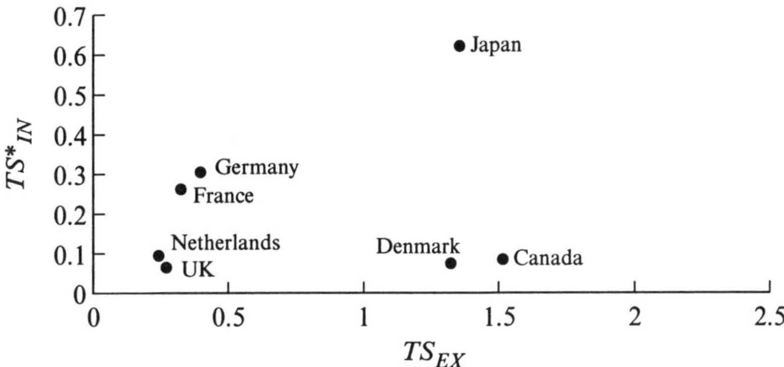

Figure 5.5 Technological system configurations, 1980

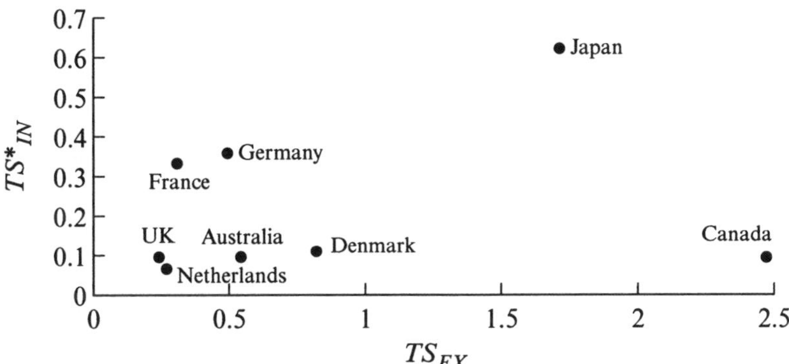

Figure 5.6 Technological system configurations, 1985

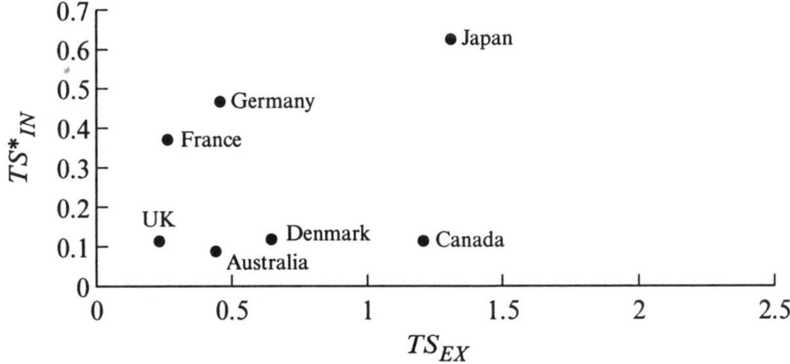

Figure 5.7 Technological system configurations, 1990

fact quite powerful in detecting structural differences and similarities among countries undergoing very similar techno-economic processes which are barely distinguishable if more traditional approaches are adopted. Second, it offers a clear clustering of technological systems within an exhaustive taxonomy encompassing the two important dimensions of a technological system, that is, the degree of coherence among its internal building blocks, and the structure of its relationships with the environment.

As far as the main results are concerned, clear clusters of technological systems are easily detected. In 1980, Japan is steadily in the top-right area, and is thus the one clearly identifiable as the most inward-oriented and pervasive technological system. Canada and Denmark, in the bottom-right area, can be defined as national trajectories-based technological systems (that is, inward-oriented and segmented). The rest of the countries are characterized by an outward-oriented configuration, with the UK, Australia and the Netherlands defined as segmented, and France and Germany as pervasive.

The whole picture is confirmed if the subsequent periods of time are taken into account. The only two relevant dynamic patterns are those of Denmark, and of France and Germany. Denmark shifts from one cluster (inward oriented and segmented) to another (outward oriented and segmented), joining the UK, the Netherlands and also Australia (for which data on 1980 were not available). France and Germany move upwards, and thus confirm their vocation of outward orientated technological systems.

An important feature of the identified clusters is that within the same configuration different 'types' of countries have very different dynamics, which seem to be diverging in time. Hence, similar clustering does not imply similar dynamic patterns.

France and Germany, for example, are both characterized by a high level of internal coherence, coupled to a strong internationalization of their external relationships. In this case, it seems likely that their internal structure is particularly suitable to exploit the benefits arising from intense international linkages. For instance, as far as the German case is concerned, its systemic characteristics actually guarantee a certain degree of self-sufficiency (Table 5.2 confirms that its foreign participation is not among the highest, although Germany is the most important foreign location for the biggest international firms, see Patel (1995)). Well-established internal virtuous circles seem to allow German firms also to tap into inferior-order centres of expertise (Cantwell 1995). A similar argument holds for France, whose intermediate position between Germany and the UK is confirmed by the external patents analysis (Table 5.2).

A different story holds for the UK and the Netherlands, which identify two cases of an apparent 'dualistic configuration', that is, with hardly any internal interaction and a high outward orientation, while Australia and Denmark are the closest countries to the national trajectories-based configuration, of which Denmark was actually a part in the early 1980s.

Finally, that internal coherence can allow one technological system to be self-sufficient appears to be demonstrated by Japan, the most 'systemic' and simultaneously the most inward oriented (that is, national).[13] However, if inward and outward innovative efforts are considered separately (Table 5.2), the importance of a coherent internal environment for the international exploitation of nationally developed technological capabilities also emerges in the case of Japan. The configurations of the Netherlands and the UK contrasts this interpretation and reveals that the condition is necessary, but not sufficient.

4 CONCLUSIONS

In this chapter we have put forward an approach dealing with two crucial levels for the analysis of techno-economically defined open systems, particularly suited to their empirical study, that is, the degree of internal and external connection of their partitions (sectors). On this basis we derived a taxonomy and identified four possible configurations for the technological system.

We also carried out an empirical application of this scheme of analysis, from which we derived interesting results, both in terms of classification of the different technological systems, and in terms of their changes along time. The need for cross-data sources has been highlighted repeatedly in order to complement qualitative studies with our more compact and concise approach.

The results are quite suggestive. In particular, Japan is the only country whose configuration is permanently pervasive and inward oriented, while Canada is the only one that is persistently segmented and inward oriented. The other countries are all classified as outward oriented, although with very different characters. Indeed, France and Germany are characterized as pervasive, while the UK-Australia and the Netherlands are segmented. The only exception is that of Denmark, for which it is possible to register a structural change of configuration from an inward- to an outward-oriented technological system.

APPENDIX 5A

Table 5A.1 Sample characteristics

	Internal connection		
	Input–output tables[1]	R&D expenditure[2]	External connection[3]
Australia	1986, 1989	1986, 1989	1985, 1989
Canada	1981, 1986, 1990	1981, 1986, 1990	1981, 1985, 1990
Denmark	1980, 1985, 1990	1980, 1985, 1990	1981, 1985, 1990
France	1980, 1985, 1990	1980, 1985, 1990	1981, 1985, 1990
Germany	1978, 1986, 1990	1978, 1986, 1990	1981, 1985, 1990
Japan	1980, 1985, 1990	1980, 1985, 1990	1981, 1985, 1990
UK	1979, 1984, 1990	1979, 1984, 1990	1981, 1985, 1990
The Netherlands	1981, 1986	1981, 1986	1981, 1985

Sources:
1. OECD, DSTI, (STAN, I-O), 1995.
2. OECD, DSTI, (STAN, ANBERD), 1994.
3. *UN, International Trade Statistical Yearbook*, various years.

Table 5A.2 Sectoral disaggregation

1	Food, beverages and tobacco
2	Textile industry, apparel and leather
3	Wood products and furniture
4	Paper, paper products, printing and publishing
5	Chemical industry, drugs and medicines
6	Energy products
7	Rubber and plastics products
8	Non-metallic mineral products
9	Ferrous and non-ferrous metalls
10	Metal products
11	Non-electric machinery, office and computing, electric appliances, radio, TV and communication
12	Shipbuilding and repairing
13	Motor vehicles and other transports
14	Professional goods
15	Other manufacturing

NOTES

* We thank Gilberto Antonelli, Giulio Cainelli, Nicola De Liso, the participants to the 10th annual Meeting of the Italian Association for Labour Economics (AIEL), to the 1997 European Association for Evolutionary Political Economy (EAEPE) Conference, and the editors of this book for useful criticisms and suggestions. We are particularly indebted to Alain Decker, who suggested the basic diagrammatic representation to one of us (RL) during his stay at the Programme of Policy Research in Engineering Science and Technology (PREST), University of Manchester.

1. It must be stressed that the empirical exercise proposed in this chapter refers to the institutional interface only implicitly because of data availability for the large number of countries it considers. However, fuller empirical applications, although limited to two countries and to a shorter time span, are in Leoncini et al. (1996) and Leoncini (1998).
2. All the details for the construction of such a matrix can be found in Leoncini et al. (1996), and in the literature cited there.
3. In the case of maximum polarization the **D** matrix is composed of n values equal to 1 (one for each column). All the remaining elements are nil, but only $(n-1)n$ of them (all except those on the principal diagonal) are considered in their deviation from those values of d_{ij} that are different from zero and equal to $1/(n-1)$. It follows that the value of the internal connection indicator is in this case given by

$$TS_{IN} = n\left(1 - \frac{1}{n-1}\right)^2 + (n^2 - 2n)\left(-\frac{1}{n-1}\right)^2 = \frac{n(n-2)}{n-1}.$$

4. To be true, the majority of studies dealing with an input–output approach to the innovative activity are based on these kinds of matrices: intersectoral flows are actually identified on the basis of the most relevant innovations 'declared' (mainly patented) by the firms of each sector, and of their presumed potential users. Some examples are collected in DeBresson (1996, Part Three), concerning the location of innovative clusters in some national economies.
5. For a specific survey of this topic, see, for example, Marengo and Sterlacchini (1990).
6. On a more analytical ground, it should also be noticed that the study of the economic structure of the innovative activity from an output perspective requires the utilization of specific techniques to compare, on the one hand, input–output tables (current flows matrices, input requirements matrices, and so on), and, on the other hand, innovative matrices as such. On the contrary, following an input perspective, it is the structure of one matrix only (that is, of the intersectoral innovation flows matrix) that has to be investigated.
7. While for the internal connection indicator the technological dimension is explicit (R&D expenditure) and the economic dimension is implicit (intersectoral structure), the reverse holds for the external indicator, where the economic side is explicit (commercial exchange) and the technological one is implicit (embodied into the transferred goods). It must be noted that the techno-economic focus, which is here the only one compatible with the definition of system's boundaries, can be released at a later stage, when the analysis is concerned with the degree and type of internal connections among the system's components (its building blocks, or subsystems) (Leoncini and Montresor 1998). In this case, from both a methodological and an empirical point of view, explicit technological indicators are necessary (Leoncini et al. 1996).
8. The countries examined are seven for the early 1980s and 1990s, and eight for the middle 1980s. For a detailed description of the dataset and of the sources see Appendix 5A, Table 5A.1. For the sectoral disaggregation used in this chapter, see Appendix 5A, Table 5A.2. The latter follows from the attempt to find a compromise between, on the one hand, the need to have as many sectors as possible (in order to represent as many components of the technological system as possible) and, on the

other, to have as many comparable countries as possible among the different years. The selected level of disaggregation is therefore crucial for the empirical application of the taxonomy proposed in this chapter.

9. For a more historically oriented account of the role of MITI, see also Brown (1980).
10. In the most comprehensive collection of case studies available at the moment (Nelson 1993), Denmark, Canada and Australia actually occupy (along with Sweden) a separate section of the book (Section II) under the heading 'Smaller high-income countries'.
11. In fact, the motor vehicle sector is quite a significant case in which the US–Canadian integration has been supported institutionally through the so-called 'Auto Pact' (McFetridge 1993).
12. However, the high percentage of external patents obtained from foreign subsidiaries follows mainly from the attempt of the superior-order centres (for example, Germany and the USA) to tap into the areas in which British expertise is greater (for example, chemicals and pharmaceuticals), rather than to extend those lines of operation on which they are focused in their home base (Cantwell 1995).
13. The national Japanese response to economic globalization is a well-known story (Fransman 1995).

REFERENCES

Albert, M. (1991), *Capitalisme contre capitalisme*, Paris: Editions du Seuil.

Brown, C. (1980), 'Industrial policy and economic planning in Japan and France', *National Institute Economic Review*, **93**, August, 59–75.

Cantwell, J. (1995), 'The globalisation of technology: what remains of the product cycle model?', *Cambridge Journal of Economics*, **19**, 155–74.

Carlsson, B. and R. Stankievicz (1991), 'On the nature, functions and composition of technological systems', *Journal of Evolutionary Economics*, **1**, 93–118.

DeBresson, C. (ed.) (1996), *Economic Interdependence and Innovative Activity*, Cheltenham: Edward Elgar.

De Liso N. and S. Metcalfe (1996), 'On technological systems and technological paradigms: some recent developments in the understanding of technological change', in E. Helmstädter and M. Perlman (eds), *Behavioral Norms, Technological Progress and Economic Dynamics: Studies in Schumpeterian Economics*, Ann Arbor, MI: University of Michigan Press, pp. 71–95.

Dore, R. (1992), 'Japanese capitalism, Anglo-Saxon capitalism. How will the Darwinian contest turn out?', Occasional Paper no. 4, Centre for Economic Performance.

Dosi, G. (1982), 'Technological paradigms and technological trajectories: a suggested interpretation of the determinants and directions of technical change', *Research Policy*, **11**, 147–62.

Dosi, G., K Pavitt and L. Soete (1990), *The Economics of Technical Change and International Trade*, New York: Harvester Wheatsheaf.

Edquist, C. (1997), *Systems of Innovation. Technologies, Institutions and Organisations*, London: Pinter.

Edquist, C. and B. Lundvall (1993), 'Comparing the Danish and Swedish systems of innovation', in Nelson (ed.) (1993), pp. 265–98.

Ergas, H. (1987), 'The importance of technology policy', in P. Dasgupta and P. Stoneman (eds), *Economic Policy and Technological Performance*, Cambridge: Cambridge University Press, pp. 51–96.

Fransman, M. (1995), 'Is national technology policy obsolete in a globalised world? The Japanese response', *Cambridge Journal of Economics*, **19**, 95–119.

Freeman, C. (1979), 'The determinants of innovation', *Futures*, **11**, 206–15.

Freeman, C. (1988), 'Japan: a new national system of innovation?', in G. Dosi, C., Freeman, R. Nelson, G. Silverberg and L. Soete (eds), *Technical Change and Economic Theory*, London: Pinter, pp. 330–48.

Freeman, C. (1994), 'The economics of technical change', *Cambridge Journal of Economics*, **18**, 463–514.

Freeman, C. and B.A. Lundvall (eds) (1988), *Small Countries Facing the Technological Revolution*, London: Pinter.

Gregory, R. (1993), 'The Australian innovation system', in Nelson (ed.) (1993), pp. 324–52.

Griliches, Z. (1990), 'Patents statistics as economic indicators: a survey', *Journal of Economic Literature*, **28**, 1661–707.

Leoncini, R. (1998), 'The nature of long-run technological change: innovation, evolution and technological systems', *Research Policy*, **27**, 75–93.

Leoncini, R., M. Maggioni and S. Montresor (1996), 'Intersectoral innovation flows and national technological systems: network analysis for comparing Italy and Germany', *Research Policy*, **25**, 415–30.

Leoncini, R. and S. Montresor (2000), 'Network analysis of eight technological systems', *International Review of Applied Economics*, **14**(2), May.

Lundvall, B.A. (1988), 'Innovation as an interactive process: from user–producer interaction to the national system of innovation', in G. Dosi, C. Freeman, R. Nelson, G. Silverberg and L. Soete (eds), *Technical Change and Economic Theory*, London: Pinter, pp. 349–69.

Lundvall, B.A. (ed.) (1992), *National Systems of Innovation. Towards a Theory of Innovation and Interactive Learning*, London: Pinter.

Lundvall, B.A. (1996), 'National systems of innovation and input–output analysis', in DeBresson (ed.) (1996), pp. 356–63.

Malerba, F. and L. Orsenigo (1995), 'Schumpeterian patterns of innovation', *Cambridge Journal of Economics*, **19**, 47–65.

Marengo, L. and A. Sterlacchini (1990), 'Intersectoral technology flows. Methodological aspects and empirical applications', *Metroeconomica*, **41**, 19–39.

McFetridge, D. (1993), 'The Canadian system of industrial innovation', in Nelson (ed.) (1993), pp. 299–323.

McKelvey, M. (1991), 'How do national systems of innovation differ? A critical analysis of Porter, Freeman, Lundvall and Nelson', in G. Hodgson and E. Screpanti (eds), *Rethinking Economics: Markets, Technology and Economic Evolution*, Aldershot: Edward Elgar, pp. 117–37.

Mowery, D. and J. Oxley (1995), 'Inward technology transfer and competitiveness: the role of national innovation systems', *Cambridge Journal of Economics*, **19**, 67–93.

Nelson, R. (ed.) (1993), *National Innovation Systems. A Comparative Analysis*, New York: Oxford University Press.

Nelson, R. (1995), 'Recent evolutionary theorizing about economic change', *Journal of Economic Literature*, **33**, 48–90.

Niosi, J. and B. Bellon (1996), 'The globalisation of national innovation systems', in J. de la Mothe and G. Paquet (eds), *Evolutionary Economics and the New International Political Economy*, London: Pinter, pp. 138–59.

Odagiri, H. and A. Goto (1993), 'The Japanese system of innovation: past, present, and future', in Nelson (ed.) (1993), pp. 76–114.

Organization for Economic Cooperation and Development (OECD) (1992a), *Technology and Economy. The Key Relationship*, Paris: OECD.

Organization for Economic Cooperation and Development (OECD) (1992b), *Main Science and Technology Indicators*, Paris: OECD.

Paquet, G. (1996), 'Technonationalism and meso innovation systems', http://iirl.uwaterloo.ca/MOTW96/summer96/GillesPaquet.html.

Patel, P. (1995), 'Localised production of technology for global markets', *Cambridge Journal of Economics*, **19**, 141–53.

Patel, P. and K. Pavitt (1991), 'Europe's technological performance', in C. Freeman, M. Sharp and W. Walker (eds), *Technology and the Future of Europe*, London: Pinter, pp. 37–58.

Patel, P. and K. Pavitt (1995), 'Patterns of technological activity: their measurement and interpretation, in P. Stoneman (ed.), *Handbook of the Economics of Innovation and Technological Change*, Oxford: Blackwell, pp. 14–51.

Pavitt, K. (1984), 'Sectoral patterns of technical change: towards a taxonomy and a theory', *Research Policy*, **13**, 343–73.

Schnabl, H. (1995), 'The subsystem – MFA: a qualitative method for analysing national innovation systems – the case of Germany', *Economic Systems Research*, **7**, 383–96.

Verspagen, B. (1997), 'Measuring intersectoral technology spillovers: estimates from the European and US Patent Office databases', *Economic Systems Research*, **9** (1), 47–65.

Walker, W. (1993), 'National innovation systems: Britain', in Nelson (ed.) (1993), pp. 158–191.

World Bank (1994), *World Development Report*, Oxford: Oxford University Press.

6. The effects of productive and technology specialization on inter-firm technological cooperation: evidence from joint patent activity

Frederico Rocha*

1 INTRODUCTION

Richardson (1972) has emphasized the role of asset dissimilarity in the shaping of organization of industry when complementary, specific assets are involved. He has argued that dissimilar assets are unlikely to be internalized and therefore should be acquired through alliances. The decision to collaborate in technology should then be understood as a strategic choice to avoid high sunk costs and the delays in accumulating competencies in a rapidly changing environment. This chapter argues that in this sense the propensity to cooperate should be a positive function of the firm's specialization level and of the rate of technical change of the environment. This line of argument sheds some light on the debate over the substitutive or complementary character of technological cooperation to intramural R&D efforts by giving an alternative interpretation of the relationship between these two different modes of organizing technological activities.

This chapter attempts to contribute to the debate using a database built from patents jointly filed by two or more firms in the European Patent Office. Section 2 is dedicated to the display of some theoretical ideas that illuminate the chapter. Section 3 presents the data that is going to be used in the analysis. Section 4 presents the main results of the chapter and links them to the main propositions made by the literature on the subject. Section 5 concludes.

2 THEORETICAL AND EMPIRICAL BACKGROUND

2.1 Evidence for the Growth of Technological Cooperation

Chesnais (1988) has surveyed the literature on technological cooperation and has collected evidence that shows that the number of technological agreements between firms has grown during the 1980s. Hagedoorn and Schakenraad (1990) confirmed this trend for a wide range of modes and motives of agreement. Figure 6.1 uses data drawn from a database composed by joint patents filed by two or more firms in the European Patent Office. It confirms the results of the literature, but reveals a stabilization in the number of joint patents at the end of the 1980s. This evidence suggests that the 1980s witnessed some transformations in the organization of technological activities. Apart from being carried out in intramural R&D labs (Mowery 1983; Teece 1988), technological activities appear to be undertaken through external linkages.

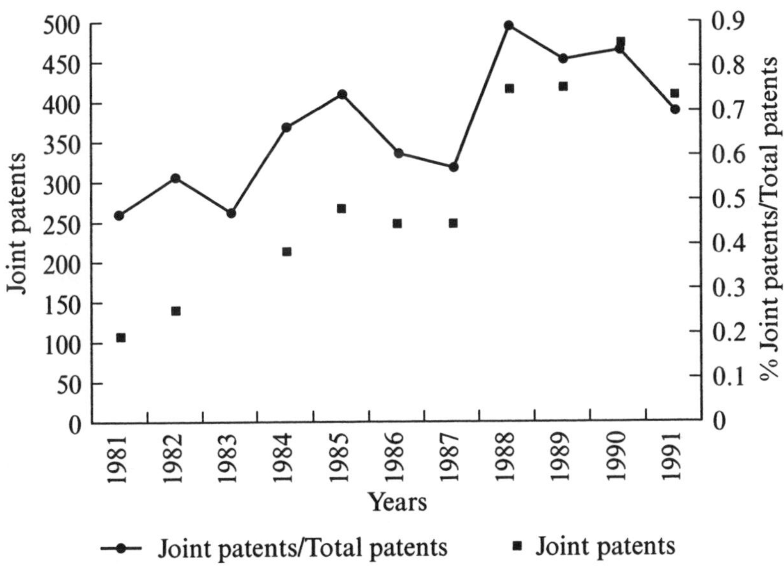

Sources: Own manipulation of European Patent Office (1993) and European Commission (1994).

Figure 6.1 Evolution of technological cooperation

Mowery and Rosenberg (1989) and Chesnais (1988) have argued that the reorganization of technological activities during the 1980s was related mainly to recent developments in technology, such as: (i) the continued growth in development costs; (ii) technological convergence or cross-fertilization of technologies; (iii) shorter product cycles; and (iv) faster rates of technical change. The growth in the R&D costs implies that the minimum efficient scale of R&D projects has increased. This can create obstacles for the development of innovative activities by small firms that would not reach the minimum necessary scale. This is particularly important if the time to profit from innovation – the product cycle – has been reduced, because the fixed costs of innovation will have to be covered in a shorter period. Technological convergence has demanded that a firm's competencies be spread over a greater number of technological fields, that is, the number of technologies that a firm has to deal with has increased.

2.2 The Importance of Asset Specificity, Complementarity and Dissimilarity

This scenario is perfectly consistent with the framework elaborated by Richardson (1972), based on the importance of asset specificity for the organization of industry. According to his arguments, coordination problems may arise when assets are transacted in small numbers, that is, when the demand for a certain asset is associated with specific characteristics so that *ex ante* specific investments from one or both parties are necessary. In the case of technology, two characteristics may cause coordination problems. First, the uncertainty present in innovative environments requires investments to gather and develop information that is not available (Milgrom and Roberts 1993, p. 93). This is particularly important because it is difficult to determine *ex ante* the value of the results of the efforts to be undertaken and even the tasks to be performed by each party involved in the transaction (Teece 1988). Second, the tacit attribute of knowledge generates situations in which important pieces of information are held by the client and not by the innovating party. This may require interaction between the two parties and some problems may arise if costly investment from either party is required or if the non-achievement of the right relationship among variables or parties is very costly.[1]

These problems could be solved through the internalization of the technological activities. According to Richardson (1972), this kind of solution will be successful only if the competencies (capabilities) required to develop the knowledge asset belong to the firm's technology base. However, if the innovation requires the development of (*new*) dissimilar knowledge, then internalization will not be the way to develop the required asset and cooperation may be an alternative solution. In this case, one can see at least three drawbacks to internalization: (i) the firm may lose coherence; (ii) when cumulative attributes of knowledge are very important, the development of new competencies may require very costly and time-consuming investments to overcome technology gaps; and (iii) due to the time requirements to develop the new competencies, the firm will be left at a disadvantage to its competitors if the pace of technical progress is very rapid.

Therefore, two factors should be considered. First, one should take into account the attributes of the firm. Its level of technological and productive diversification will influence its reaction to new technology requirements. A more diversified firm will be more likely to internalize the activity because of its technological portfolio, which should be spread over a greater number of technical fields, and because the diversification into new fields should belong to its long-term strategy, that is, it may be less risk averse to diversify into new productive and technology areas. The size of the firm may also affect its propensity to internalize complementary assets. Smaller firms may have tighter financial constraints and may be risk averse. Besides, size may also affect the probability of firms holding a more diversified technological and productive portfolio.

Second, one should consider the characteristics of technical progress with respect to the importance of accumulated knowledge and the pace of technical change. In the first case, if cumulative attributes are not important, late-comers or new firms may face low barriers to develop new competencies. In the latter case, slow changing technical fields are less likely to present problems of uncertainty, and delays in developing technology should not compromise the positioning of the firm relative to its competitors.

Figure 6.2 summarizes the propositions made above. In the case of more diversified firms, there is a higher propensity to adopt internalization strategies, while in the case of more specialized firms, the externalization strategies should be more common. On the other hand, in industries where technical change is very intense and knowledge is more likely to be cumulative, firms should be inclined to adopt cooperative strategies, while in a non-cumulative and slowly changing environment, firms may be prone to adopt either internalization or market-dependent strategies.

2.3 The Importance of Absorptive Capacity

One can derive from the above arguments that cooperative organization is a substitute for internal organization. Thus, one could expect a reduction of intramural R&D in those firms that engage more frequently in technological alliances. Yet, Cohen and Levinthal (1989) argue that R&D has

Technologically and productively:	Cumulative knowledge and fast-changing environment	Non-cumulative knowledge stable environment
Diversified firm	Internalization/ Cooperation	Internalization/ Market transaction
Specialized firm	Cooperation	Market transaction

Figure 6.2 Modes of organization of the production of specific, dissimilar assets

indeed two different functions – it is undertaken to generate innovations and/or to increase the firm's absorptive capacity – and therefore it should be considered complementary to external sources of knowledge, that is, an increase in the intensity of use of external sources of knowledge should be accompanied by an increase in intramural R&D expenditures.

2.4 Trust and Nationality

Still, although cooperation may solve some of the problems associated with the *ex ante* coordination of the production and acquisition of complementary assets, it cannot end suspicions about possible *ex post* opportunistic behaviours in collaborative alliances, that is, Richardson's (1972) interpretation about the cooperative problems solves only partially the problems posed by the transaction costs paradigm. A transaction costs theorist would hold that one of two beliefs should underly the above developed approach: (i) contracts in collaborative arrangements are self-fulfilling; or (ii) there is always a contractual design to overcome hazard problems in partnerships.

For a self-enforcing agreement to take place, it would require 'a sequence of transactions over time such that ending date is unknown and uncertain' (Williamson 1985, p. 169).[2] The infinite horizon approach has been used by many studies dealing with collusive behaviour in the product market (Jacquemin and Slade 1989). According to them, an infinitely repeated non-cooperative game may have a cooperative solution. Williamson (1985, p. 169) avoids this kind of argument because 'the transactions I consider can be (indeed, normally are) finite'. Kreps (1990) suggests an alternative vision. He argues that the fundamental role of the firm is to be the long-lasting party in a sequence of short-run (finite) transactions. It should then have a reputation that ensures the satisfactory completion of contracts. He therefore states that the role of corporate culture is to act as a commitment for the maintenance of contractual clauses, that is, for not acting opportunistically. As a possible extension of his theory, Kreps (1990) proposes to analyse a situation where the two parties are long-lived. Their corporate culture or the reputation in sustaining certain kinds of behaviour is their *ex ante* commitment for the agreement to take place. In this case, another type of influence may be important: corporate culture or held reputation.

None the less, the measurement of the effect of corporate culture would be very difficult. Here, it is proposed that nationality may gather some of the characteristics of corporate cultures. Lazonick (1991) states that there are some important differences between the managerial capitalism, practised in the USA and spread across many Western economies, and the collective capitalism, present in Japan. The distinguishing factor of managerial capitalism would be the separation between the management and the ownership and heavy investments in managerial structure, while the distinguishing factors of the collective capitalism would be: (i) the organizational integration of a number of different firms; (ii) long-run links between employees and the enterprise; and (iii) state cooperation. Caves (1989) surveying the literature on firm behaviour also registers some important international differences. He calls attention to differences in firm organization between French corporate groups and North American diversified firms. French corporate groups have weaker ownership links and their control is maintained through interlocking directorate rather than strict administrative linkages. He also argues for the differences in Japanese firm organization, such as the *keiretsu*. Imai and Itami (1984) argue for the relevance of the differences of market and organization penetration between Japanese and US industrial structures. In this sense, it would be licit to use nationality of a generalization for certain corporation practices.

Williamson (1985) has argued that to avoid cheating in agreements some kind of commitment, such as the provision of hostages, from transacting parties may be necessary. The literature on technological cooperation is rich in works trying to relate modes of cooperation to technological requirements for the transaction. Contractor and Lorange

(1988) relate the types of incentives that different modes of collaborative arrangement have to the level of dependence required in the transaction. Arora and Gambardella (1990) show that each type of cooperative agreement may be associated with different requirements of knowledge transference or sharing. According to this line of work, there is a vast type of contractual arrangement that can be regarded as a settlement for the carrying out of the collaborative arrangement. Our database – as will be clear in the next section – cannot account for different contractual situations. Therefore, the rather cynical view that for each specific situation there is a different contractual arrangement will be adopted in this chapter or, at least, that the probability of adapting contracts to technological requirements does not vary across firms.

It should be acknowledged that nationality involves characteristics other than trust. Levels of firm productive and technological specialization vary across nationalities. Firm size, anti-trust legislation and general legal environment may also differ across countries. Therefore, nationality may be capturing many more effects than corporate culture.

3 METHODOLOGY

3.1 The Variables

Technological cooperation

This chapter attempts to increase the knowledge about technological alliances by using an alternative indicator composed by patents jointly filed by the firm with one or more firms (in the European Patent Office) divided by the total number of patents filed by the firm as a measure of the propensity to cooperate. The use of patents to measure technological activity has some important drawbacks: (i) smaller firms have a greater propensity to patent; (ii) sector and technologies differ in their propensity to patent; (iii) the presence of a firm in a market covered by a specific patent office affects its propensity to patent; and (iv) the value of patents have great variability.[3]

One can pose further problems in the use of patents as an indicator of cooperation: (i) it does not capture non-innovative technological cooperation. Efforts such as technology transfer and cross-licensing will remain out of the analysis; (ii) it does not perform uniformly across different modes of cooperation. For instance, joint-venture contributions should be largely underestimated by the indicator, due to the fact that they may be filed in the name of the joint venture, which is a sole firm; (iii) it is mainly a measure of output, so it can underestimate cooperative inputs that are not turned into patents; and (iv) it may capture some effects that do not result from cooperative relationships, such as the use of joint patents as hostage for the establishment of long-term collaboration in productive activities.

Unlike most of the literature on technological cooperation that uses the number of agreements as a measure of intensity of cooperation and tests it against the R&D intensity (Colombo and Garrone 1996), this chapter uses a relative variable – joint patents divided by total patents – as a measure of the propensity to cooperate.[4] Colombo and Garrone suggest the use of the rate of the number of agreements divided by R&D expenditure. However, the use of the rate of joint patents to the amount of R&D expenditures would cause problems, because of uneven patenting and R&D propensities across sectors.[5] In contrast, the use of the rate of joint patents to total patents is useful to control for some of the above-mentioned biases. It may correct for differences in the patenting propensity of firms of different nationality, size and sector.

Dissimilarity and absorptive capacity

As traditionally used in the literature, the firm's absorptive capacity will be measured by the R&D to sales rate.

Dissimilarity will here be approached in three different ways. First, one should expect the probability of an asset being dissimilar to the firm's productive base to be a positive function of the level of the firm's technological specialization. The greater the firm's level of specialization, the higher the level of cooperation should be. Technological specialization is here measured by the rate of patenting in the firm's main sector of activity divided by the total number of patents. However, as there may be differences across sectors in the pace of technical progress and in the propensity to patent of technologies that have different importance levels across sectors, sectoral dummies will be used for all science-based sectors.

The probability of the occurrence of asset dissimilarity should also be an inverse function of the firm's level of productive specialization. This could be an inverse function of the firm's level of productive integration. The literature has used the rate of total added value to total gross output. Unfortunately, data at the firm level was not available. An alternative measure used in this chapter as a proxy for the level of integration is the sectoral rate of total added value to total gross output at the national level. In this case, some biases may arise: (i) multinational firms that participate in many economies will be represented by their home country's data; and (ii) some firms may have activities outside their main sector of production.

A third variable used to measure the effect of asset dissimilarity on its propensity to cooperate is the firm's size represented by its total sales.

Bigger firms are more likely to be more technologically diversified and therefore more able to access complementary technologies (Patel and Pavitt 1994). Some side-effects may also be captured by this variable. Bigger firms are more likely to invest in sunk costs associated with technology and should have easier access to financial funds.

3.2 Data Sources

Data was obtained for 81 firms – 31 Japanese, 23 European and 27 North American – with activities in high-technology sectors (see Table 6.1). The sectoral distribution of firms is: 21 in chemicals, 12 in pharmaceuticals, 1 in rubber and plastics, 1 in food, 2 in glass and building materials, 5 in machinery, 18 in electronics, 5 in computers, 3 in instruments and 13 in motor vehicles. Data on joint patents filed by two or more firms in the European Office were obtained by the Bulletin CD-ROM of the European Patent Office for the years 1988 to 1992. Patents were selected by the firm name. Patents that were filed in the name of the firm's subsidiaries were also included. Joint patents between parent and subsidiaries or subsidiaries of the same group – according to the 'who owns whom' criteria – were excluded. As a consequence, there should be no ownership linkages between firms jointly patenting in the database and the firms are not expected to have patents filed by subsidiaries that are not part of the database.

Data on the total number of patents and patents in the firm's main sector of activity between 1988 and 1992 and R&D data for 1992 were obtained from the European Commission. Data on sectoral rate of value added to gross production for each country was obtained from OECD, (1994). Firms were classified according to the nationality of its parent firm.

Table 6.1 Main characteristics of the firms in the database

Variable	Mean	Standard deviation	Minimum	Maximum
R&D Intensity	6.89	3.3950	1.12	15.56
Sales*	16 365.58	19049	1412	101 916
Joint Patents	16.96	21.56	1	99
Total Patents	1 070.85	1189.61	19	5 529

Note: *Billions of ECUs 1992.

Sources: European Commission (1994). The number of joint patents is obtained by own manipulation of the European Patent Office (1993).

4 EMPIRICAL ANALYSIS

4.1 The Model

According to the variables that have been referred to above, one can define a model in which

$$PROPCOOP = f(RDINT, SIZE, PRODSPEC, TECHSPEC, NATION),$$
$$LNPROPCOOP = LN\,\alpha + (LNRDINT + \gamma\,LNSIZE + \varphi\,LNPRODSPEC + \lambda\,LNTECHSPEC + \mu\,EUDUM + \upsilon\,JPDUM$$

where *PROPCOOP* is a firm's propensity to cooperate, represented by the number of joint patents divided by the total number of patents filed between 1988 and 1992, *RDINT* is the firm's R&D intensity in 1992, *SIZE* is represented by sales in 1992, *PRODSPEC* is the firm's productive specialization here associated with a proxy measure denoted by the firm's national sector rate of added value to gross production (*LN AVGO*), *TECHSPEC* is the firm's level of technological specialization expressed by the number of patents in its main sector of technological activity divided by its total number of patents between 1988 and 1992. Nationality is represented by two dummies, *EUDUM* for European firms and for *JPDUM* Japanese firms. Sectoral dummies are introduced in those equations where *TECHSPEC* is present.

4.2 The Results

Table 6.2 shows the results of the regressions. All equations have acceptable *R*-square and *F* levels. Equation (1) has the presence of all the variables introduced in the model. However, it does not include firms outside science-based and motor vehicles sectors. Equation (2) introduces firms that belong to specialized suppliers (machinery, rubber and instruments) and other scale-intensive (food and glass) sectors; data on technological specialization was not available. Equation (3) excludes the nationality dummies and shows some important differences in the behaviour of the other variables.

Absorptive capacity

The R&D intensity variable has a negative sign in all equations, though it is not significant in any of them. There seems to be no evidence to support the hypothesis that a firm has to increase its level of R&D in order to acquire externally produced knowledge. On the contrary, the general results seem to show that technological cooperation is not influenced by the firm's R&D intensity. It should be stressed that cooperation is not totally external to the firm's activities and thus it should be understood as a particular case. We should be careful to make any further assertion due to the characteristics of the sample. The chapter gathers data only on large, technology-intensive firms (see Table 6.1). There could be a threshold R&D level for firms to absorb external knowledge.[6] In this case, the technological cooperation would increase with R&D intensity when the level of R&D was very low until it reached intermediary intensity.

Dissimilarity, specialization and the choice of partners

In the two equations where the technological specialization measure is included, it holds a positive sign, though it is only significant when the equation is not controlled for nationality. This shows that the level of specialization may be slightly affecting the firm's propensity to cooperate technologically. Technological cooperation may be an alternative strategy to overcome constraints of narrow competence formation. This result may conflict with some of the literature which asserts the need for firms to be technologically diversified in order to access complementary assets (Granstrand et al. 1997). Some remarks should, however, be made in this respect. First, the analysis undertaken here does not assess a firm's technological base fully. Some of the more specialized firms may still hold some technological competencies in those technical fields where they are cooperating. Furthermore, tests should still be held with other measures of technological specialization in order to have a robust result. Some further work has still to be done in this respect.[7] Second, not all technological cooperation may involve the absorption of knowledge. Some of the firms in the sample may be suppliers of inputs to other firms and in this respect their greater technological specialization may be an important asset, though they still need some information on the specification of the product to be produced.

The sectoral rate of added value to total production (*LNAVGO*) has a negative and significant sign in all equations. As this variable is a proxy for the level of integration of the firm, it is negatively correlated to the level of productive specialization. Therefore, the result shows that more productive specialized sectors tend to cooperate more than those sectors where firms are in average more integrated. There is some evidence to support the idea that firms that adopt lean production strategies are more inclined to cooperate. The higher coefficient in equations (2) and (3) in Table 6.2 demonstrate also that there are important differences in the level of specialization across countries.

Firm size shows a negative sign and is significant in all equations, suggesting that smaller firms are more likely to cooperate than bigger firms.

Table 6.2 *OLS regressions – dependent variable* LN *(number of joint patents/total patents)*

Variable	Equation (1)	Equation (2)	Equation (3)
Constant	10.300	5.41	22.633***
	(1.597)	(1.64)	(6.629)
LNRDINT	–0.229	–0.01	–0.156
	(–0.689)	(–0.04)	(–0.459)
LNSIZE	–0.273*	–0.31**	–0.314**
	(–1.896)	(–2.82)	(–2.159)
LNTECHSPEC	0.250		0.395*
	(1.063)		(1.720)
LNAVGO	–2.902*	–1.49**	–6.232***
	(–1.728)	(–2.45)	(–7.916)
CHEMICAL	–0.141		0.335
	–0.289		(0.750)
PHARMA	1.098		2.533**
	1.214		(3.975)
ELECTRO	0.595		1.799**
	0.812		(3.616)
COMPUTER	0.763		1.752**
	(0.921)		(2.703)
EUDUM	0.289	0.21	
	(0.927)	(0.82)	
JPDUM	1.282**	1.52***	
	(2.377)	(4.94)	
R-squared	0.61	0.58	0.57
Adjusted R-squared	0.54	0.55	0.51
F	9.05***	20.48***	9.99**
N	69	81	69

Notes
t-statistics in parentheses.
* Significant at the 10% level.
** Significant at the 5% level.
*** Significant at the 1% level.

This result was predicted by the examination of the literature. Smaller firms tend to be productively and technologically less diversified. However, the maintenance of its significance even when technological and productive specialization variables are included suggest some side-effects of size not directly related to the specialization level. In this case, financial constraints may be playing a role in the above equations. As bigger firms have easier access to external financial sources and may have greater amount of internal funds, they could be less likely to engage in cooperation. Some risk aversion considerations may also apply.

The results on the levels of productive and technological specialization and on firm size seem to suggest the importance of the acquisition of complementary and dissimilar assets for the undertaking of cooperation. More-specialized, smaller firms that cannot internally access some knowledge-intensive asset because either they lack competencies or they are not sufficiently productively diversified show a greater propensity to cooperate than more diversified, bigger firms.

Figure 6.3 shows data on joint patenting activity of 536 firms belonging to 14 different two-digit sectors. Joint patent among firms with ownership relations were eliminated. About two-thirds of the inter-firm technological alliances involve partners that belong to different productive sectors at a very high level of aggregation. For the same set of firms, Patel and Pavitt (1994a) show that at this level of aggregation firms of different sectors are most likely to hold dissimilar competencies. This suggests that most of the technological linkages in the patent sample occur

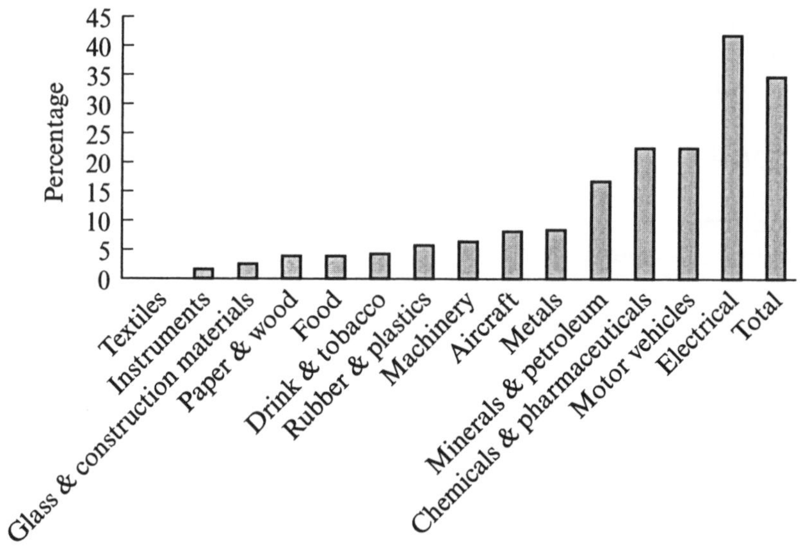

Source: Own manipulation of European Patent Office (1992) and European Commission (1994).

Figure 6.3 Percentage of intra-sectoral patents

between firms that hold dissimilar competencies. Furthermore, even when the patents are filed by firms that belong to the same two-digit sector, there is still room for differences in competencies. The analysis of the firms involved in most of the intra-sectoral alliances reveals that in most cases they are associated with vertical relationships. In the motor vehicle sector, for instance, the alliances established inside the two-digit sector almost without exception have the involvement of an auto part producer and an auto assembler. In the electrical–electronics sector, producers of electrical cables have connections with firms in the telecommunications sector, and so on. This finding gives support to Richardson's (1972) interpretation of cooperation linkages, that is, the need to access dissimilar competencies in more specialized firms may be an important element in the determination of the level of technological cooperation.

The four sectoral dummies, representing the four science-based sectors included in the database, have the right sign in the two equations where nationality dummies are included and three of them are significant in the equation where they are not included. It should be stressed that pharmaceuticals, electronics and computers are the most R&D intensive sectors in the sample. One can then surmise a relationship between the pace of technical change and the rate of cooperation. However, first, the loss of significance when nationality dummies are included is probably related to the loss of some degrees of freedom in the limited sample of the database. Second, having no variable controls for effects of cumulativeness in the propensity to cooperate and the empirical verification of its importance should be an aim of further work. Third, according to Figure 6.3, patterns of choice of partners vary across sectors, that is, in some sectors the frequency of choice of an intra-sectoral partner is greater than others. One explanation for this fact is the complexity of the intra-sectoral division of labour. Another explanation can be associated with the importance of technology flows. This may relate technological cooperation with sectoral patterns of technical change (Pavitt 1984). We can then speculate that the type of knowledge that should be exchanged across firms' borders through cooperation will depend, on one hand, on the asset specificity and dissimilarity and, on the other side, on the technology flows across sectors.

Nationality

The European dummy is not significant, although it is positive, but the Japanese dummy shows a positive and significant sign in both equations where it is included. Japanese firms seem to be more likely to cooperate than Western ones. Similar results have been achieved in a number of case studies (Hamel 1991 and Chesnais 1988), although most of the analysis undertaken using big databases shows no particular tendency for bigger firms to have higher levels of cooperation, due to biases in their information sources (Chesnais 1988; Hagedoorn 1995). Thus, the result does not present great novelty.

The literature on technological cooperation is very rich in justifications for the greater level of cooperation in Japan. Chesnais (1988) gives three main reasons: (i) the existence of business groups in Japan; (ii) the state interference in the formation of inter-firm technological alliances; and (iii) the very early recognition in Japan of the advantages of cross-sectoral horizontal cooperation.

Cooperation inside business groups does not seem to be a useful explanation for the results obtained here. Table 6.3 presents information on

Table 6.3 Number of patents and leading partners per business group

Business groups[1]	Number of joint patents	Intra-group joint patents (%)
Mitsubishi	380	17
Mitsui	176	9
Sumitomo	434	12
Fuyo	124	4
DKB	304	11
Sanwa	132	4
Tokai	40	0
IBJ	67	1
Nippon Steel	72	0
Hitachi	196	3
Nissan	122	11
Toyota	235	43
Matsushita	68	0
Toshiba-ihi	98	3
Others	748*	–
Not identified	300**	–
Total	1934	15

Notes

1. Industrial Groupings in Japan (1987), lists 17 Japanese business groups. Three of them have been excluded from the table because of lack of relevance in total joint patenting.

*Others are firms that had their capital structure identified but do not belong to any of the 14 groups listed in the table.

**Not identified are firms that did not have their capital structure identified.

Source: Own manipulation of the European Patent Office, 1993.

joint patenting activities of 139 Japanese firms. It shows that only about 15 per cent of the total domestic partnerships involved firms inside the same business group. The only exception is Toyota, which is well known for the verticalization of some of its intra-group linkages.

The state intervention should also provide limited explanation. Technological cooperation stimulated by the Japanese government takes the form of consortia (Mowery and Rosenberg 1989 and Chesnais 1988), which are usually associated with a large number of firms. None the less, less than 5 per cent of the total patents in the database were filed by more than two firms. The third reason given by Chesnais (1988) still requires refinement.

Aoki (1988) adds some insights to the explanation of Japanese behaviour. He claims that the ownership structure of the Japanese firm – bank and employee control – has many consequences for the functioning of firms. One of these differences is that the protection of incumbent employees' interests in Japanese firms imposes restrictions for the hiring of new labour. As a consequence, Japanese firms spin-off a large number of labour-intensive activities. One outcome of this spinning off is the greater specialization inside the same production chain. In Table 6.2, where the nationality dummies are not included, there is an increase in the elasticity of the rate of added value to gross production. This is confirmed in Table 6.4, which shows that the *AVGO* rate is lower in Japan than in any other country/region. This table also demonstrates that Japanese firms are generally smaller than European and North American firms. This suggests that Japanese firms are less integrated and therefore need to rely more on external built capabilities and on an external supply of specific assets. The need to share information may induce cooperation.

Second, Aoki (1988) emphasizes the role played by labour market rigidities in Japanese firms in the establishment of cooperation. According to him, Japanese firms are not allowed to hire mid-career scientists because this would be contrary to organizational rules for contract labour.[8] Therefore, the acquisition of competencies in fast-moving technological fields would depend on the hiring of inexperienced, young scientists on external sources or on firm acquisition. This situation is similar to Richardson's (1972) interpretation, and cooperation should be a solution for this kind of dilemma. This should be represented by a greater technological specialization of Japanese firms. Although the *TECHSPEC* variable in Table 6.2 is significant only in equation (3) where there is no nationality dummy, Table 6.4 shows that although Japanese firms have a greater level of specialization than those in North America, European firms appear to be more specialized. However, this result could be biased by the sectoral composition of the national samples of firms and further analysis should be done with larger samples.

Table 6.4 National differences in key variables

Nationality	Variable	Mean	Standard deviation	Minimum	Maximum	Cases
Japan	R&D Intensity	6.67	2.79	1.12	13.48	31
	Sales	14831.12	15 604.67	1 488.24	54 924.48	31
	TECHSPEC	0.47	0.27	0.04	1.00	27
	AYGO	31.53	6.64	21.32	42.08	31
	Patents	918.87	1 064.82	28.00	3 224.00	31
Europe	R&D Intensity	6.96	4.11	1.57	15.57	23
	Sales	16946.82	13 763.97	2 194.92	50035.4	23
	TECHSPEC	0.59	0.22	0.13	0.93	21
	AVGO	38.87	6.88	27.18	52.17	23
	Patents	1 374.48	1 542.24	158.00	5 529.00	23
United States	R&D Intensity	7.57	3.31	2.83	13.66	27
	Sales*	19 132.24	25 788.56	1 411.80	108 916.08	27
	TECHSPEC	0.43	0.22	0.10	0.75	21
	AVGO	43.21	6.16	30.04	55.22	27
	Patents	986.70	957.64	19.00	3 964.00	27

Note: *Billions of ECUs, 1992.

Sources: R&D Intensity, Sales, Technological Specialization and Total Patents are from European Commission (1994). Added value to gross output ratio (*AVGO*) is from OECD (1994).

5 CONCLUSIONS

The chapter raises three issues. On the theoretical level, it gives support to Richardson's (1972) interpretation about the organization of industry. It also emphasizes the role played by strategic positioning of assets and technology characteristics of sectors in the determination of the degree and type of partnering. On the empirical level, the chapter stresses the importance of specialization in the determination of a firm's propensity to cooperate. There are robust results over the sign and significance of the variables associated with specialization of a firm's technological and productive base. In this case, the intensity of cooperation seems to be higher in smaller, technologically less diversified and productively specialized firms. This finding suggests that technological cooperation may be complementary to lean production strategies. The chapter also relates the intensity of cooperation to nationality, suggesting that trust linkages may play an important role.

Finally, the chapter emphasizes the need for further empirical and theoretical analysis. In this respect, one line of analysis could be to compare the competencies firms develop inside their boundaries and those they acquire from external sources in order to confirm the hypotheses suggested here. A second line of analysis is to relate sectoral technology characteristics to technological cooperation.

NOTES

* This chapter was mostly written during the author's stay at the Institute for High Technologies, United Nations University (INTECH/UNU). The author is grateful for the financial help provided by the United Nations Development Programme (UNDP) and the Conselho Nacional de Desenvolvimento Cientifico e Tecnologico, Ministeria da Ciência e Tecnologia, Brazil (CNPq/MCT). The chapter has benefited from comments made by Nagesh Kumar, Pari Patel and the participants of the INTECH/UNU internal seminar. The usual disclaimers apply.

1. In the Words of Milgrom and Roberts (1993, p. 91), these problems are created due to *design attributes* of the transaction.
2. Quoting Lester Telser.
3. Some useful surveys on the use of patents as economic and technology indicators may be found in Griliches (1990) and Pavitt (1988).
4. Some problems may arise when an absolute variable is tested against a relative variable. For instance, firms with higher levels of R&D expenditure may cooperate more though in lower proportion than less R&D-intensive firms. Bigger firms may have greater linkages than smaller ones and be less intensive in technological cooperation.
5. When *ln Patents* are regressed against *ln R&D*, $R_2 = 0.34$, the coefficient is positive and significant at 0.0001 per cent.
6. Though the coefficient in the test of a quadratic form showed no significance.
7. A more thorough analysis comparing the technical fields where firms patent jointly with other firms with those where they hold competencies should be done.
8. Either associated with corporate culture and the discipline in intra-firm labour market or with inter-firm agreements against raids on each other's labour force.

BIBLIOGRAPHY

Aoki, M. (1988), *Information, Incentives and Bargaining in the Japanese Economy*, Cambridge: Cambridge University Press.

Arora, A. and A. Gambardella (1990), 'Complementarity and external linkages: the strategies of the large firms in biotechnology', *Journal of Industrial Economics*, **38**(4), June, 361–79.

Bound, J., C. Cummins, Z. Griliches, B. Hall and A. Jaffe (1984), 'Who does R&D and who patents?', in Z. Griliches (ed.), *R&D, Patents, and Productivity*, Chicago: University of Chicago Press, pp. 21–54.

Caves, R. (1989), 'International differences in industrial organization', in R. Schmalensee and R. Willig (eds), *Handbook of Industrial Organization*, Amsterdam: North-Holland.

Chesnais, F. (1988), 'Technical co-operation agreements between firms', *STI Review*, **4**, 51–119.

Cohen, W. and D. Levinthal (1989), 'Innovation and learning: the two faces of R&D', *Economic Journal*, **99**, 569–96.

Colombo, M. and P. Garrone (1996), 'Technological cooperative agreements and firm's R&D intensity. A note on causality relations', *Research Policy*, **25**(6), 923–32.

Contractor, F. and P. Lorange (1988), 'Why should firms cooperate?', in F. Contractor and P. Lorange, *Cooperative Strategies in International Business*. New York: Lexington Books.

Dosi, G. (1988), 'The nature of innovative process', in G. Dosi, C. Freeman, R. Nelson, G. Silverberg and L. Soete (eds), *Technical Change and Economic Theory*, London: Pinter, pp. 221–38.

European Commission (1994), *The European Report on Science and Technology Indicators*, Luxembourg: Office for Official Publications of the European Communities (OOPEC).

European Patent Office (1993), 'Bulletin Patent Database in CD-Rom', European Patent Office, Vienna.

Foray, D. (1991), 'The secrets of industry are in the air: industrial cooperation and the organizational dynamics of the innovative firm', *Research Policy*, **20**, 393–405.

Granstrand, O., E. Bohlin, C. Oskarsson and N. Sjöberg (1992), 'External technology acquisition in large multi-technology corporations', R&D *Management*, **22**(2), 111–33.

Granstrand, O., P. Patel and K. Pavitt (1997), 'Multi-technology corporations: why they have "distributed" rather than "distinctive core" competencies', Brighton, March, mimeo.

Griliches, Z. (1990), 'Patent statistics as economic indicators: a survey', *Journal of Economic Literature*, **27**, 1661–707.

Hagedoorn, J. (1995), 'The economics of cooperation among high-tech firms – trends and patterns in strategic technology patterns since the early seventies', Paper for the Conference on the Economics of High Technology Competition and Cooperation in Global Markets, Hamburg.

Hagedoorn, J. and J. Schakenraad (1990), 'Interfirm partnerships and cooperative strategies in core technologies', in C. Freeman and L. Soete (eds), *New Explorations of Technical Change*, London: Pinter, pp. 3–37.

Hamel, G. (1991), 'Competition for competence and inter-partner learning within international strategic alliances', *Strategic Management Journal*, **12**, 83–103.

Imai, K.K. and H. Itami (1984), 'Interpenetration of organization and market', *International Journal of Industrial Organization*, **2**, 285–310.

Jacquemin, A. and M. Slade (1989), 'Cartels, collusion and horizontal mergers', in R. Schmalensee and R. Willig (eds), *Handbook of Industrial Organization*, Amsterdam: North-Holland, pp. 415–73.

Jaffe, A. (1989), 'Characterizing the "technological position" of firms, with application to quantifying technological opportunity and research spillovers', *Research Policy*, **18**, 87–97.

Kreps, D. (1990), 'Corporate culture and economic theory', in J. Alt and K. Shapsle (eds), *Perspectives on Positive and Political Economy*, Cambridge: Cambridge University Press, pp. 90–143.

Langlois, R. (1992), 'Transaction-cost economics in real time', *Industrial and Corporate Change*, **1**(1), 99–127.

Lazonick, W. (1991), *Business Organization and the Myth of the Market Economy*, Cambridge: Cambridge University Press.

Mariti, P. and R. Smiley (1983), 'Co-operative agreements and the organization of industry', *Journal of Industrial Economics*, **31**(4), June, 437–51.

Milgrom, P. and J. Roberts (1993), *Economics, Organization and Management*, Englewood Cliffs, NJ: Prentice-Hall.

Mowery, D. (1983), 'The relationship between intrafirm and contractual forms of industrial research in American manufacturing, 1900–1940', *Explorations in Economic History*, **20**, 351–74.

Mowery, D. and N. Rosenberg (1989), *Technology and the Pursuit of Economic Growth*, Cambridge: Cambridge University Press.

Organization for Economic Cooperation and Development (OECD) (1994), *The OECD Input–Output Database*, Paris: OECD.

Patel, P. and K. Pavitt (1994a), 'Technological competencies in the world's largest firms: characteristics, constraints and scope for managerial choice', Sussex Policy Research Unit, STEEP Discussion Paper no. 13.

Patel, P. and K. Pavitt (1994b), 'The continuing, widespread (and neglected) importance of improvements in mechanical technologies', *Research Policy*, **23**, 532–45.

Pavitt, K. (1984), 'Sectoral patterns of technical change: towards a taxonomy and a theory', *Research Policy*, **13**, 343–73.

Pavitt, K. (1988), 'Uses and abuses of patent statistics', in A. van Raan (ed.), *Handbook of Quantitative Studies Science and Technology*, Amsterdam: North-Holland.

Richardson, G.B. (1972), 'The organization of industry', *Economic Journal*, **82**, 883–96.

Rocha, F. (1995), 'Competências Tecnológicas e Cooperção Inter-firma: Resultados da Análise de Patentes Depositadas em conjunto' (Technological competencies and inter-firm cooperation: results from joint patents filed at The European Patent Office), PhD, Instituto de Economia Industrial, Universidad Federal do Rio de Janeiro (in Portuguese).

Rocha, F. (1997), 'Sectoral patterns of technological cooperation', Universidade federal fluminense, Rio de Janeiro, mimeo.

Teece, D. (1988), 'Technological change and the nature of the firm', in G. Dosi C. Freeman, R. Nelson, G. Silverberg and L. Soete (eds), *Technical Change and Economic Theory*, London: Pinter, pp. 256–81.

Veugelers, R. (1997), 'Internal R&D expenditures and external technology sourcing', *Research Policy*, **26**, 303–15.

Williamson, O. (1985), *The Economic Institutions of Capitalism*, New York: Free Press.

7. Organizational dynamics and the evolutionary dilemma between diversity and standardization in mission-oriented research programmes: an illustration

Emmanuelle Conesa*

1 INTRODUCTION

The growing globalization of technology and the deeper integration of the world economy have generated an ongoing debate in economics focused on the role of the policy makers in the dynamics of technological change and questioned the efficient design of national technological policies (Branscomb and Florida 1997; Fransman 1995). As Branscomb (1993) stated: 'Economic competitiveness will no longer be left to a *laissez-faire* economic policy; government will share costs of base technology development with commercial firms' (Branscomb 1993, p. 7).

This prompts a rethinking of the rationale of mission-oriented research programmes. These are generally initiated by the policy maker, in partnership with industry, in order to invest in R&D and advance technical innovation for high-technology sectors. They are traditionally linked to the post-Second World War technology policies in industrialized countries such as the United States or France.

In the economic literature on technical change, mission-oriented research programmes undertaken to develop cutting-edge technologies have usually failed (Ergas 1987). It has frequently been suggested that they are too costly and tend to survive in spite of failure. The US SuperSonic Transportation (SST) programme and the much more controversial Concorde project[1] are some famous examples of failures. Most of the economists criticize heavily (i) their centralized organization (Cohen et al.1991); (ii) their high cost, which encourages a narrowing in the range of the options explored (Collingridge 1991); and (iii) their technical complexity which restricts participation in programme execution to a few, technologically sophisticated firms, that is, the 'national champions' (see Ergas 1992, p. 3).

The economist faces an interesting paradox between, on the one hand, the policy makers' unanimity to adopt mission-oriented programmes as an instrument of technological policy during the last half-century, and, on the other hand, the economists' unanimity to ignore the high probability of expensive failure that can persist for a long time (Cohen et al. 1991). Are these national technology programmes doomed to failure or can they be justified? This chapter is an attempt to examine this issue critically.

The discussion is based on an account of the US National AeroSpace Plane (NASP) programme (1986–94), a joint effort by the National Aeronautics and Space Administration (NASA), the Department of Defense (DoD), and five major industrial contractors to explore the development of technologies for a new generation of aerospace vehicles. More precisely, the NASP research programme was devoted to the exploration and demonstration of hypersonic airbreathing propulsion technologies for aerospace transportation. Reviewing the NASP programme gives insights to the challenges entailed by mission-oriented technology policy in a context of technological discontinuities.

Looking at mission-oriented programmes through evolutionary glasses reveals that their organizational management is challenged by a tension between variety and standardization through the learning process (Cowan and Foray 1995). The tradeoff between variety and standardization corresponds to the evolutionary dilemma between exploration and exploitation (March 1991). It also refers to the distinction between static and dynamic efficiency (Klein 1977). Exploration is linked to experimentation, flexibility, discovery, innovation and dynamic efficiency, while exploitation corresponds to choice, selection, implementation and static efficiency. A pure strategy of exploration causes the system to suffer the costs of experimentation without benefiting from the diversity generated, simply because of a lack of coordination while, adopting a pure strategy of exploitation, the risk is to become locked into a suboptimal technological path. Thus, to keep the dynamic efficiency in the long run, the organizational stake is to maintain an appropriate balance between exploration and exploitation.

Challenged by the risk of technology lock-in, the policy maker, aiming to develop a radically new technological system, should rather favour some parallel research efforts through decentralized experiments, that is, adopt a network approach. It would guarantee the simultaneous exploration of technological variety in the framework of a mission-oriented programme.

Usually mission-oriented programmes deal with complex technology challenges. They require basic research activities to be carried out, so as to displace the knowledge frontier. In this case, research programmes are characterized by radical uncertainty. To decrease the degree of radical uncertainty, common pools of knowledge about the radically new technology area have to be created. The US NASP programme, with the aim of designing a hypersonic space plane, was concerned with such a precondition. Preliminary production of codified knowledge was first required so as to enable the exploration of the technological variety. Consistently with the context of radical uncertainty, the need to produce pieces of standardized knowledge called for a centralized organizational management of R&D in a club involving together all the research partners committed in the programme. Basically, research carried out in a single entity makes it possible to scan only sequentially the potentially wide spectrum of possible technological paths. Still, considering the irreversibility increased tenfold by the radical uncertainty, a centralized organization does not protect the programme from the risk of a lock-in into an inefficient technology.

The tension between variety and standardization relates to the well-known debate on the comparative advantage of what Chesbrough and Teece (1996) called 'virtual organization'[2] versus integrated organization. The former yields in principle more variety, the latter more integration and coordination. Yet, when comparing those organizational forms, one must keep in mind that even if the 'club', a design which has often been used, has been proved to exhibit a number of defects, the untested network design, which in principle shows greater promise, can only be advocated on theoretical grounds.[3]

On a management perspective, the challenge faced by policy makers consists in selecting the organizational design that best matches the type of innovation they are pursuing in the framework of a mission-oriented research programme. In other words, to solve the organizational dilemma between learning from diversity and learning from standardization, one has to take account of both the idiosyncratic nature of the technology at stake and the specific characteristics of the research outputs.

2 THE NATIONAL AERO-SPACE PLANE PROGRAMME

At the beginning of the 1980s, the US NASA studied various ways to lower the economic barriers to space and concluded that one of the most promising avenues was to develop a fully reusable launch vehicle. That craft would take off as an aeroplane, attain orbit with a single propulsion system and fly back to the Earth, landing horizontally at a conventional airport.

The NASP Programme was a US mission-oriented programme undertaken by NASA, the DoD, and five major industrial contractors in the mid-1980s to accept the challenge. It consisted of a multi-year technology demonstration effort based on a three-phase research, development, test and evaluation programme (RDTE) to explore a range of technologies – including airbreathing propulsion, advanced aerodynamics, materials and structure, fuel systems, avionics and the computational fluid dynamics – that could lead to the development, fabrication and flight testing of an experimental flight vehicle called X-30. The demonstrator would have paved the way for the building of a fully reusable single-stage-to-orbit (SSTO) vehicle, needed to achieve a dramatic cut in space launch costs (Augenstein and Harris 1993). Basically, the idea underlying the NASP research programme was that technological advances in both basic and applied knowledge might allow space launchers of the next century to operate much as aircraft do today. This change might cut the cost of reaching orbit tenfold.

Hence, low-cost – or at least affordable – access to space which would satisfy both NASA and US strategic plans, was the primary goal of the NASP programme. Thus the NASP supported national security and would have facilitated commercial applications in aerospace transportation.

It focused on the reduction of space access costs to hundreds of dollars per pound and the provision of core research and technology needed for the next 25 years, that is, the demonstration of advanced technologies in aeronautics (transatmospheric) and space transportation systems (Barthélémy 1989) so as to enable the US aerospace launch vehicle industry to compete in a global market. In this perspective, main critical technologies concerned the airbreathing propulsion system. Therefore, the NASP aimed to explore and the demonstrate hypersonic and transatmospheric airbreathing propulsion technologies, and more specifically to design a supersonic combustion ramjet, that is, the scramjet.[4]

A propulsion system consisting in the integration of airbreathing propulsive engines, including a scramjet for hypersonic flight, would allow the design of a 'space aircraft' that could reach satellite speeds and attain orbit, while taking off like a regular aeroplane, and returning to the Earth once its mission is accomplished. In this case, space station resupply would be far cheaper than with rockets that cannot be recovered or with a space shuttle that cannot be reimplemented without any refurbishment (see Section 3.2 below). By contrast, such an aerospace vehicle could be flown repeatedly. Another foreseen application for hypersonic propulsion technologies would be the design of a civil transatmospheric transport vehicle travelling at a minimum of Mach 6, that is, six times the speed of sound, at altitudes of 20 miles or higher. The starting point was that a hypersonic vehicle might prove an attractive option for long-distance transportation in the next century. Travellers would travel between distant destinations within a very few hours (for instance two hours from Paris to Tokyo).

A key feature of the NASP programme was the building of a flight research vehicle, the X-30, for investigating flight conditions that cannot be simulated in ground facilities. It would have served as a pathfinding vehicle for hypersonic transportation systems.

At the beginning, support for the NASP programme was strongly optimistic: 'We have the capability to integrate these technologies in the experimental X-30, which should begin validation in actual flight by the early 1990s' (Executive Office of the President, OSTP 1987, p. 10). But the main difficulty was precisely the critical integration of the airframe with the propulsion systems. In this respect, the development of the X-30 lifting body vehicle entered the category of systemic rather than stand-alone technologies and placed additional focus on system technology demonstration and not just on innovation in separate components.

2.1 Dealing with the Organizational R&D Management for a Systemic Innovation

Basically, for hypersonic vehicles such as the X-30, there are a number of factors which affect configuration design, including vehicle flight mechanics and control. The most basic factor is the requirement for satisfactory operation of several propulsion systems over the Mach range from 0 to 25. The vehicle must fly a high dynamic pressure trajectory in the airbreathing corridor to ensure that adequate volume of air is supplied to the propulsion system This is quite critical for proper operation of the ramjet and scramjet above Mach 4. The forebody, which needs to be long and wide to provide adequate air capture at high Mach numbers, is an integral part of the inlet. Therefore the building of the forebody must be carefully integrated into the propulsion system design, which may require changes in the forebody to ensure satisfactory performance.[5] All these requirements explain why the NASP vehicle design was a major technical challenge.

But some people were against this ambitious enterprise, stating that the development of technologies involving large-scale shifts away from the status quo cannot permit low-cost control of technology and ensure successful performance in policy without increasing the risk of technology lock-in. 'The national aerospace plane will do nothing to promote the technical diversity of the space programme. It is likely to repeat the errors of the Shuttle, having such large sunk costs that it will have to be used intensively' (Collingridge 1990, p. 197).

Initially, the NASP programme consisted of three phases, but it never reached Phase 3 because Congress ended funding at the end of 1994. Is this sufficient for considering the NASP as one more costly failure?

Inflexible technology and the risk of costly failure

According to Collingridge, the NASP had all the indicative properties of what he described as complex inflexible technologies, namely large-unit size; long lead-time; high capital intensity; and dependence on specialized infrastructure (Collingridge 1992). The degree of technology inflexibility makes it possible to account for its non-incremental nature. Therefore, the development of inflexible technology is peculiarly prone to costly failure.

Following the doctrine of 'Incrementalism' (Collingridge 1990), technology should be developed in a piecemeal, experimental way involving a series of trials. This demand is most easily met when decisions are made in a decentralized, pluralistic way. By contrast, the development of an inflexible technology involves highly centralized decision making dominated by large firms, able to transfer risk in some way to government, thereby excluding many legitimate stakeholders. To avoid such a drift, the policy maker should look for more flexible technical alternatives, some of which may be developed in a more decentralized way.

Here, the doctrine of incrementalism should be distinguished from incremental innovation. Incremental innovation is usually adopted when a technological paradigm has already been established and a technology has become standardized and possibly locked in. Therefore, incremental innovation cannot be used to introduce a radical innovation. Moreover, a single technical breakthrough alone is not sufficient for progress in high technology. Rather, it is only through the organic fusing of several technical breakthroughs in a number of different fields that a radically new technology can be created (Kodama 1991, p. 3). Incrementalism, on the other hand, represents an alternative organizational arrangement to introduce a radical innovation. This means that dealing with radical innovation rather than incremental innovation requires a step-by-step approach, with different paths explored in parallel for a while, so as not to be locked into an inefficient technology option.

In the case of the NASP programme, the technology at stake possessed all the four characteristics mentioned by Collingridge. But, one should add the systemic character of hypersonic transportation technology for aerospace vehicles. Here, we refer to the distinction made by Teece (1986) between autonomous and systemic innovations. Following Teece we can talk about an innovation as systemic if change in one part of the system necessitates corresponding change in other parts; by contrast, an innovation is autonomous if change in one part can proceed without materially affecting the rest of the system. This means that autonomous innovations can be pursued independently of other innovations. In contrast, for innovations that are fundamentally systemic, their benefits can be realized only in conjunction with some related, complementary innovations.

Adequate organizational design

Hypersonic propulsion technology can be considered as a systemic innovation because it requires interrelated changes in engine and airframe designs, as well as in supply technology, simulation and test technologies, and so on. This is why to profit from hypersonic propulsion, aerospace industrial companies involved in the NASP needed to develop both new airbreathing engine technology and new airframe design.

The distinction between autonomous or stand-alone innovation and systemic innovation is fundamental to the choice of organizational design. For stand-alone innovation, a decentralized network organization – what Chesbrough and Teece (1996) called 'virtual organization' – can manage the development task quite well. For systemic innovation, members of a virtual organization are dependent on their research partners, over whom they have no control. Chesbrough and Teece concluded that in either case, the wrong organizational choice can be costly. Regarding hypersonic technology, two distinct characteristics of innovation affect the appropriateness of an organizational design, namely, the systemic character of innovation and its radicalness. These two factors often ride together, while in principle they are separable. As Robertson and Langlois (1995) showed:

> decentralized networks of innovation do well under conditions of autonomous innovation but . . . systemic innovation calls for integration of both ownership and coordination in order to surmount adverse power relationships and avoid information impactedness. Moreover, one would typically think of systemic innovation as more 'radical' than autonomous, since changing many parts of a system is clearly a relatively drastic procedure, whereas adjusting only a part seems to be necessarily an incremental business. (pp. 554–5)

Complexity itself is not a problem if it concerns stand-alone innovation. But problems arise when different components hang together and need to be developed in step and in a coordinated fashion, as for the design of the X-30 forebody, which requires a high level of engine–airframe integration.

Despite some unique technical and functional merits, such as recovering satellites from orbit while at the same time being reusable, NASP transportation technology development might have been proved to be a hugely expensive failure because (i) it represented a large shift from anything that had existed earlier,[6] and (ii) the risk of failure resulted in its development and operation under a decision-making process of considerable centralization. Cohen et al. (1991) regarded such an organizational R&D management style as positively inimical to fostering the flexibility that, according to them, would have been necessary to search for a technology 'optimal' design.

2.2 Organization of the NASP Programme: A Matter at Issue

The NASP programme was drawn out of the US industrial–government structure to form a new single entity that incorporated both public and private sector expertise, previously dispersed throughout the national research system. Innovative partnership strengthened the alliance between government and industry, thus enabling costs and risks to be shared between them. Cohen et al. (1991) compared the NASP organizational design with the great Japanese government–industry collaborations of the 1970s and the 1980s (see Fransman 1990).

Five major industrial contractors collaborated with Academia and leading committed government agencies in the NASP programme. The two engines contractors, Pratt & Whitney and Rocketdyne, were involved in the design of the engine systems including ground tests of engine modules. The airframe contractors, General Dynamics, McDonnell Douglas, and Rockwell, took charge of the design of the X-30 configurations, the conduct of wind tunnel tests and the building of sample vehicle sections. In the technology maturation area, government laboratories, NASA research centres (Langley, Lewis and Ames), Air Force Wright Research Development Center, Johns Hopkins University, and others worked closely with the contractors to improve critical technologies to reduce the risk in developing the X-30. The National Program Office, managed by NASA and the DoD, assumed the function of programme coordinator.[7]

According to Cohen et al. (1991), the NASP organizational management design – described as an 'innovative team' approach – exhibited two drawbacks. First, the grouping of all available expertise into a single entity narrowed the range of alternative development paths to be explored. Consideration of a wider range of technological options that could have catalysed broader industrial involvement in the research programme was, therefore, inhibited. Second, they early stated that:

> The innovative team approach makes the program more difficult to kill if NASP becomes nothing more than an expensive toy. Involving all of the important players in the aerospace industry eliminates short-term sources of political attack because it picks no winners and has no competitive external R&D effort. Involving multiple government agencies creates a stable support coalition within government. (Ibid, p. 53)

Cohen et al. suggested that integration in the management of NASP relied more on political pressure rather than on any attempt to achieve technical and economic optimization. They claimed that this organizational design might have increased the likelihood of failure, that is, it raised both the risk of missing the best technology design and the cost of the event of failure concerning the selected technological trajectory.

Integrated organizational structure through 'clubbing' might a priori have strongly impeded the exploration of the requisite diversity in the NASP programme. Actually, from an economic point of view, such a curtailment of diversity might have impeded resource-allocation efficiency in the relevant research area and increased the cost of a possible failure, as, in principle, organizational integration increases the risk of missing the best design. None the less, integration can be justified, as long as one takes into account the effective tasks performed in the course of a research programme leading to a paradigmatic shift. Therefore, clubbing together the research partners involved in the NASP programme was consistent with the need to conduct basic research activities on which the introduction of the radical innovation concerned relied.

3 DESIGNING A HYPERSONIC SPACECRAFT: TOWARDS A PARADIGMATIC SHIFT

To understand the organizational dilemma of the NASP programme management, it is useful to refer to the evolutionary economic approach of technological change dynamics.

In parallel with Kuhnian notions of scientific evolution, evolutionary economics of technical change has provided a treatment of the technological evolution of an industry, underlying the succession of disruptive scientific breakthroughs entailing technology discontinuities and periods of incremental development (Dosi 1984). Basically, two stages in the evolutionary development of a given branch of scientific and technological knowledge can be distinguished: the preparadigmatic stage when there is no single generally accepted conceptual treatment of some phenomena in a specific technology area, and the paradigmatic stage which begins when a body of theory appears to have passed the canons of scientific and engineering acceptability. This cumulative and path-dependent process, in which 'small historical events' play a crucial role in the orientation of technological trajectories (David 1985), is relentlessly exposed to irreversibility as a result of the market-driven diffusion process of technologies in the presence of increasing returns to adoption (Arthur 1988,1989; David 1987). The emergence of a dominant paradigm

(Abernathy and Utterback 1978) signals scientific maturity and the acceptance of agreed-upon 'standards' by which what has been referred to as 'normal' scientific and technological research can proceed. Cooperation in R&D activities emerges more spontaneously when uncertainty over technology design is overcome. Shared experience reduces persistent uncertainty and fosters a process of cross-fertilization which accelerates the innovation design. The 'standards' remain in force unless or until the paradigm is overturned. Revolutionary technology is what overturns normal technology, that is, the dominant design.

Here, it is important to stress that the need to build an aerospace plane induced the breaking of the performance criteria which have hitherto governed the technological evolution of space launch engines (McLean 1985). More specifically, the technical challenge consisted in the combination of potentially conflicting sets of performance requirements that were previously applied exclusively either to aeronautics or to space systems.

3.1 Technological Paradigms and Conventions of Technical Change

Economists like Dosi emphasized the existence of paradigms to explain the observed regularity in the development of a technological path (Dosi 1984, 1988). The technical change dynamics is reinforced by the emergence of self-fulfilling prophecies shared by the technologists, that is, what may be called a 'convention of technical change' (CTC). As common knowledge, such a convention consists of a structure of mutually consistent expectations about the future course of technical change (see David 1994).

A CTC can be described as a set of design parameters, which embodies the principles that will generate both the physical configuration of the product and the process and materials from which it is to be constructed. It refers to the notion of technological regime as described by Georghiou et al. (1986, p. 34): 'The basic design parameters are the heart of the technological regime, and they constitute a framework which is shared by the firms in the industry'. It also corresponds to the definition of a self-reinforcing institution stressed by Vanberg (1994, p. 7): 'as configurations of interconnected and mutually-stabilising behavioural routines. They are constituted by routines practices of number of persons that are functionally interlaced and reinforce each other in a mutually-stabilising manner'.

As an informal commonly agreed conventional institution, this set of technical and functional parameters of design and behavioural routines guides the expectations of designers about the future development of technologies, like the famous Moore's Law[8] which has regulated the technological path in the industry of integrated circuits since the 1960s.[9] For instance, the pattern of technical change in the semiconductor industry mixes four main directions of progress: increasing miniaturization, increasing speed, increasing reliability and decreasing costs.[10] For integrated circuits, increasing miniaturization is a function of increasing *density*, that is, an increasing number of components on a single chip. This evolutionary dynamics is consistent with what Dosi (1984) called a 'technological paradigm', defined as the specific set of knowledge, associated with the exploitation of selected physical/chemical principles and the development of a given set of artefacts.

The CTC plays the role of shared cognitive maps, based on structures of mutually consistent expectations about the course of technological dynamics. The paradigm gives the direction of technical change while the convention specifies its orientation by giving precise hints about the *rhythm* and, possibly, about the *timing* of technical change. The convention operates as a perceived exogenous constraint that, afterwards, can be assimilated into a sort of *self-fulfilling prophecy*: its institutional nature reduces the uncertainty about the future path of technical change. As a focal point, its salience allows the emergence of coordination among the engineers participating in the innovation process in a specific industrial sector. Therefore, it appears to be a powerful driving force for technological standardization.

But when an unexpected technological breakthrough appears, such a disruptive evolution results in a paradigmatic transition. The reference to the old cognitive framework is no longer possible. A new set of cognitive and social practices has to emerge, as the previous practices associated with the old paradigm have by and large become obsolete.

From the foregoing, when a mission-oriented research programme is dedicated to the development of a radical innovation that *de facto* induces a paradigmatic shift, the economic agents have to deal with what may be called 'radical uncertainty'. Concerning the NASP programme, the radical uncertainty resulted from structural change in the CTC's requirements regarding both the technical and service characteristics of the technology.

3.2 New Priorities for Space Launch Vehicles: A Drift in the Existing CTC

Since the beginning of the 1960s, the development of space propulsion technologies has been shaped through the emergence of a 'steady' orientation, that is, a trajectory, in technical change based on a set of specific criteria regarding costs and industrial implementation. These criteria emphasized the need to deal primarily with the acceleration speed and orbital access requirements.

The existing orientation, rendering the existence of 'standard operating procedures' to generate technological change, has long been compatible with the technological option of rocket engines. Moreover, the incredible magnitude of federal R&D expenditures in aerospace industries combined with the concentration on military-oriented R&D performed by government agencies, like NASA in partnership with the DoD, made it possible to neglect the industrial potential for aerospace transport systems (Pace 1990; Macauley 1986). Therefore, designers could ignore issues of reutilization, operability and payload mass for space launch vehicles.

Looking for a more affordable space transportation system: mismatching rocket

To understand how the building of X-30 was supposed to pave a less expensive road into space, one needs first to appreciate why current launch costs are so high. Some of the difficulties arise from the unalterable laws of physics.

To place a satellite in orbit, a rocket must rise above almost all of the atmosphere and has to give its payload sufficiently horizontal velocity so that, when it falls back towards the Earth, the Earth's curved surface 'falls' away at the same rate. For instance, the energy expended, in little more than eight minutes, to place a space shuttle into orbit could power a typical automobile for millions of miles. Most of the expenses result from some of the basic principles of rocketry. The payload that can be carried by a launch vehicle depends in large part on the performance of its engines and the ratio of propellant to structural weight. A rocket designer thus has two key tasks: to maximize propulsion efficiency and minimize the amount of mass to be accelerated. 'The problem here is to attain orbital velocity with one set of rocket engines which requires high-thrust engines and large tanks of propellant to feed them. This feat requires something like 90 per cent of the weight of the vehicle to be allotted to propellant. Only by using two or more separate stages, each with its own engines and propellant, have designers been able to build practical launch vehicles. Such 'staging' works because it allows segments of the vehicle to be jettisoned en route.

Besides, the introduction of multiple stages – which continue to be employed by all launch vehicles, including the space shuttle – allowed practical rockets to be constructed from the same materials then used to build aircraft. Staging, in essence, made space transportation feasible. Moreover, about half the cost of an expendable launch vehicle (ELV) can be attributed to the many careful inspections and tests required to ensure that its one and only flight goes exactly as planned. In other words, each flight mission for such an ELV must be the culmination of a labour-intensive process during which engineers check and recheck every single component needed to fly.

When compared with the fleets of airliners serving commercial aviation, the space shuttle and the various ELVs now flying appear fragile, inflexible and extraordinarily costly. Aircraft land, exchange cargo, refuel and return to flight in hours, whereas the space shuttle needs months to do so, and newly ordered expendable rockets require a year or so to fabricate. Therefore, the advanced technology demonstration at stake with the NASP clearly marked the end of the existing technical change convention.

Mixing aeronautic and orbital launching CTCs to build a reusable single stage space aircraft

The challenge was to combine in a single transportation system the distinctive and potentially contradictory sets of performance requirements that were previously applied exclusively either to aeronautics or to space systems. The new propulsion concept was supposed to mix the economic advantages of aeroplane engines – cost, ratio to mass, operability, maintenance–with the performance criteria of rockets, in terms of flight speed and orbital access. At that time, the development of reusable launch vehicles appeared to hold great promise as the key to unlocking the vast potential of space business exploitation. The NASP programme answered precisely that purpose in planning the development of a fully reusable single-stage-to-orbit (SSTO) launch vehicle that could be flown much like an airliner, using air-breathing primary propulsion as well as horizontal takeoff and landing. Aerospace engineers could plan such an ambitious project because improvements in propulsion efficiency and lightweight composite materials appeared to be just then bringing within reach a fully reusable single-stage spacecraft for hypersonic cruise in the Earth's atmosphere. The programme expected such a vehicle to fly at Mach 25, that is, 25 times the speed of sound.

Unfortunately, while this would be a great improvement over current systems, the cost per pound delivered to orbit for currently proposed systems remains greater than that needed to exploit space for many business uses. One of the limiting factors in potential reductions for chemical rockets is the Ips (specific impulse) limit. As a consequence of change in the performance criteria, the technology portion of the programme had to focus on propulsion technologies. Therefore the ultimate success of the project depended mainly on solving the propulsion issues arising from the implementation of airbreathing engines.[11]

The need for hypersonic airbreathing propulsion technology

To develop technology for a new class of fully reusable aerospace vehicles that could take off from conventional runways and fly directly into Earth orbit or cruise at hypersonic speeds in the Earth's atmosphere, NASA and

DoD decided to implement airbreathing propulsion systems. These systems were visualized as operating in an air-like mode with significantly reduced operating costs relative to conventional rocket space vehicles.

Actually, airbreathing propulsion technology offers substantial advantages for hypersonic flight, notably the ability to take off and land horizontally as well as the elimination of auxiliary solid fuel rockets and other types of launch support. Implementation of airbreathing engines holds potential for very significant increases in Ips which could result in a significantly lower cost per pound to orbit: an improvement in mass ratio of the order of 3 to 5 in comparison with the mass placed into orbit by non-airbreathing propulsion systems – rocket engines – is possible because airbreathing engines utilize air as the combustive agent, removing the need for mass-loaded oxygen as for combustion in rockets.

Moreover, possible vehicle reutilization would eliminate costly replacement or, in the best case, the recovery of the space vehicle which means – as for the US space shuttle – reconfiguring and refurbishing the vehicle after each flight – the so-called refurbishment phase. In this respect, maintenance practices would be closer to those of aeroplanes than to those of rockets. Less retraining of the ground staff is needed, which is consistent with a cut in maintenance costs. By contrast, a rocket-powered vehicle has some operational penalties, such as large infrastructure requirements, and needs to transport its own oxidant for combustion, exacting large payload penalties.

As a result, airbreathing propulsion is an essential ingredient for sustained endo-atmospheric hypersonic cruise implementations such as 'global reach' vehicles, and can improve both economic and functional performances of space launch vehicles significantly.

At the sought-after flight speeds (beyond Mach 5) supersonic combustion becomes necessary.[12] In other words, the air-breathing hypersonic technology is certainly the scramjet engine – supersonic combustion ramjet. The tricky problem of developing the existing propulsion technology (called 'ramjet') into a fully functioning supersonic combustion ramjet (called a 'scramjet') was then thought to require a critical combination of the airbreathing principle with the sought-after hypersonic speeds. An integrated airbreathing ramjet/scramjet engines system could thereby improve mission effectiveness by reducing on-board propellant load in favour of payload and by increasing operational flexibility.

3.3 Presumptive Anomaly as the Dynamo of Paradigmatic Shift

According to Constant (1973), a paradigmatic shift may be speeded up with the intuition of a probably occurring 'presumptive anomaly'. In the midst of an existing paradigm, a presumptive anomaly arises from the growing scientific evidence or a widely shared conviction that the conventional technology cannot perform some new missions and/or reach a higher level of performance. It means that attempts to extend the existing paradigm to a new set of service characteristics[13] are expected to fail in providing 'satisfactory' answers, thus leading with time to the 'presumption' that the existing paradigm is fundamentally flawed and then stimulating search for new ways of looking at things, that is, radical innovation.

Considering it was the building of an SSTO at stake in the NASP programme, the anomaly was expressed in terms of a growing conviction that, in any case, existing conventional airbreathing propulsion systems would not function at hypersonic speeds. This presumption prompted the expectation of further possible technical/functional anomalies in design, and then the search for a new paradigm

Historically, the fastest airbreathing engine-powered aeroplane, the SR-71, can cruise just above Mach 3, about 60 per cent of the Mach 5 transition to the hypersonic regime. Ramjet powered vehicles have flirted with the hypersonic threshold. History's only hypersonic aeroplane, the Mach 6.7 X-15 of the 1960s, used rockets – as have all space flight launch vehicles to date, the expendable ones and the reusable Shuttle alike.

For some people, there was no point in trying to design a hypersonic jet on the basis of a technology (airbreathing propulsion) when existing science suggests that the principle cannot be applied to hypersonic speeds. The presumptive anomaly, though it arose primarily within science, brought in its train a technical/functional anomaly that, adversely, affected progress in design and development of a reusable space transportation system. Although there was no empirical observation of any functional failure in ramjet implementation to reach hypersonic speeds, the presumption of a 'theoretical' anomaly spilled over to the design dimension and constrained the full momentum of development in the hope that analytical work would, one day, clarify the situation.

3.4 Exploring the Technological Diversity: The Need for Organizational Decentralization

On the basis of the above-mentioned considerations, the NASP programme was clearly facing a situation of structural change as the building of a hypersonic space plane rendered the transition to a new technical change convention incompatible with the existing technological knowledge. In turn, this transition through new performance criteria has induced a paradigmatic change regarding the technological basis. Yet, a paradigmatic transition calls for the exploration of the technological diversity so as to avoid the risk of lock-in into an inefficient technological design. Structural change results in structural uncertainty.

The structural uncertainty[14] arose not only from the unspecified performances and functions that hypersonic and transatmospheric technologies would one day be able to carry out. Structural uncertainty in the NASP programme arose also from the inability of the research partners to anticipate both the ways in which the research should be carried out as well as the concrete research finality, that is, the expected research outcomes.

Consistently it rendered the necessity to scan the technological variety, that is, a large number of possible 'design candidates', before any commitment would be made to any particular system design. To face up to structural uncertainty, organizational NASP management should have led rationally to incremental development and flexible management schedules, consistent with the exploration of different technical paths (Cohen et al. 1991, p. 53).

So as to benefit from the virtues of diversity, an efficient organizational design has to promote option generation and facilitate experimentation along different trajectories. Option generation refers to the process by which alternative design approaches are developed and tested before any selection procedures can take place (Ergas 1994). Efficiency in option generation consistently depends on the range of alternatives being explored. It is also affected by the effectiveness in the knowledge diffusion process, that is, both the velocity and integrity with which created knowledge resulting from exploratory efforts is transmitted throughout the technological community involved in the process of innovation. Both contribute to the learning by exploring the technological variety.

As for the NASP programme, it could have been advisable to create a system that was likely to handle multiple, decentralized projects. Such a procedure of investigation is one way to explore a broad range of all the conceivable technological and functional spectra. The final focus of the NASP programme towards the single, originally predetermined area of research in hypersonic airbreathing propulsion technologies, would have been validated after the completion of several pilot projects and a broad base of experiments (Cohen and Noll 1991, p. 42; David and Rothwell 1996). Yet such a decentralized organization may be difficult to manage, for two main reasons: first, because it needs to arrange some procedures of financial compensation regarding the projects finally not selected – which can be considered as technological orphans (David 1987) – and second, decentralization does not ensure that information generated through different experiments will *actually be shared*. In other words, such a decentralized organization should include mechanisms and procedures providing the incentives to transfer widely, towards the research partners, all the information produced in the course of each individual project, while, at the same time, it has to be centralized enough to ensure an efficient research management. Organizational design influences options assessment, decision-making procedures about the timing, as well as the selection process of standard configurations to be implemented in the overall programme. In this regard, decentralization is likely to be concerned with moral hazard and free-riding when some agents adopt a non-cooperative strategy so as to avoid the cost of participating in any of the experiments. Yet, non-cooperative strategies are not compatible with the requirements of producing knowledge 'at the frontier'.

4 TECHNOLOGICAL DISCONTINUITIES AS A SOURCE OF STRUCTURAL UNCERTAINTY

The technical challenge in the NASP depended not only on the allocation of financial or human resources, important as they were. The main difficulty arose from the lack of a sufficiently robust analytical framework to guide both research activity (mixing basic and applied research activities – BR&AR) and technological design. As a matter of fact, designing a hypersonic aircraft required the exploration of a new paradigm to solve the airbreathing propulsion issues.

Basically, such a complex innovation implied discontinuities, both scientific and technological. Resort to traditional supports in the elaboration of new technological designs – both existing scientific models and design know-how gathered through preceding technological generations, that is, the supersonic 'ramjet' – was no more effective, as they provided only very limited hints. These limits applied to hypersonic flight – first, in the difficulty of developing predictive models and second, in the inability of previous experience with ramjets to compensate for the absence of these models. Progress, apparently blocked on both fronts, spurred the search for a new paradigm.

The following paragraphs explain in greater detail why the design of a scramjet required a paradigmatic shift. The reported discontinuities had implications for both the very nature of the research outcomes as well as for the appropriate circumstances for knowledge production.

4.1 Critical Properties of Basic Research Outcomes

The required paradigm change stood out clearly. Basically, most of the scientific outcomes observed at the threshold of Mach 5 are no longer valid beyond Mach 5. For instance, some physic-chemical laws reverse as velocities pass from the supersonic to the hypersonic flight areas (Barthélémy 1989). Besides, beyond Mach 5, air no longer behaves as a perfect gas; and beyond Mach 8, some properties dependent upon temperature and even dissociation phenomena become dominant.[15] 'As a result of kinetic chemical phenomena of increasing significance, simple extrapolation parameters no longer exist which can be applied to the domain of supersonic combustion' (Barthélémy 1989, p 2).

Here it is useful to refer to the notions of homotopic and non-homotopic mappings – or connections – analogic links, and technological lumpiness, developed by David et al. (1992). These notions allow one to determine the distinctive properties of the research outcomes in a specific area, and to assess the potential 'transferability' of knowledge, when generated by one basic research programme, to another research area, not necessarily basic. This framework helps in rendering the complexity of the hypersonic flight area (Conesa 1997). It emphasizes the interaction between basic and applied research activities 'as the ultimate source of the economic benefits of basic research' (David et al. 1992, p. 80).[16]

According to David et al., the number and wealth of links between the corpus of knowledge generated through some basic scientific projects and other scientific and applied research endeavours are important determinants of the potential economic returns resulting from discoveries occurring in some specific disciplines.

Two types of links are distinguished: 'homotopic mappings' and 'analogic links'. The former refers to scientific information potentially applicable to problems quite far removed from those of concern in the original inquiry. Such information is reputed to be homotopically mapped to different scientific or applied research issues: 'The extension of the domain of "application" proceeds by validation of the hypothesised regularities, subject to the parametric change' (ibid., p. 70). The conclusion is that once a theory exhibits such homotopic mappings, progress in other fields of basic and applied research can focus on practical implementation issues rather than on the discovery of new phenomena. This notion helps to anticipate the pace and impact of progress within a scientific field in which examination of a single portion of an entire system composed of interrelated phenomena provides useful generalizations and applications in other areas. The analogic links between knowledge from basic and applied research 'are based on the surmise that nature is conservative in the use of concepts and structures, and posit that physical regularities in one field underlie other natural phenomena' (ibid., p. 85).[17]

Thus, existing 'homotopic mappings' or 'analogic links' show or reveal the potential implementation area of the basic research outcomes in question. Such an economic assessment highlights the existence of discontinuities, beyond some thresholds, in the validity of knowledge produced in the study of some physical phenomena. Inside a particular discipline, the lack of scalability[18] resulting from missing homotopic connections means that the observed outcomes cannot be extrapolated to another range of size. Each of these notions has implications in the empirical examination of mission-oriented programmes, including basic research activities.

An additional characteristic of basic research projects is the indivisibility in R&D, what David et al. call the property of 'lumpiness'. It means that additions to the knowledge base 'may be more or less "lumpy", i.e. subject to indivisibilities in the resource inputs required to expand the scale of experimentation or research'[19] (ibid., p. 71). The degree of lumpiness is at work in the formation or lack of homotopic mappings and analogic links. The lumpiness may be either informational (minimum of subproblems to solve through R&D in the course of innovation) or material (minimum of required experimental installations). Strong R&D indivisibility means that the production of new knowledge depends on the prior resolution of a great number of subproblems in the research area. This property of lumpiness is no doubt particularly pronounced where the homotopic connections are weak (see David et al. 1992).

4.2 Difficulties with Experimentation and Lack of Scientific Models: The 'Research Lumpiness'

As for the NASP, concerning supersonic combustion (Mach 5-6), the main difficulty arose from the impossibility of producing the ground-based scientific data needed for the validation of the 'scramjet' concept and to predict its performances for specific vehicle designs. Indeed, ground-based test capacities and experimental installations, that is, technological infrastructures, do not yet exist for vehicles flying beyond Mach 8. Installations capable of reproducing the combination of speeds, pressures and temperatures necessary to simulate hypersonic flight are not available.

In addition, possible ground-based experiments with existing test facilities are of extremely short duration. For example, hypersonic wind tunnel tests generally last less than a few seconds because of the great quantities of energy required. It swiftly became apparent that new suitably sized installations were needed for the experimental verification of propulsion and aerodynamics concepts beyond Mach 8 (US GAO 1988; Sullivan 1991; Piland 1991). Weaknesses in the experimental apparatus could partially be overcome by using computational simulation methods.

In principle, computer simulations enable researchers to limit wind tunnel tests to the precise areas where simulations alone are either too difficult to conduct or do not provide sufficiently precise results. With computer simulations, the cost of experimental effort decreases. But if computer simulations allow some 'virtual experiments' to be carried out, they do not eliminate altogether the need for 'real' experiments.

Regarding the NASP programme, the scientists faced two impediments: the absence of predictive laws for the modelling of turbulence in the study of laminar flows and the difficulties encountered in the solving of supersonic combustion equations (Harsha and Waldman 1989; Bogue and Erbland 1993). The latter required substantial computing power because of the quite long calculation times it implied. All simulations included a significant number of approximations, but, even when justified, they did not eliminate the need for experimental tests. Available scientific knowledge was not sufficient to provide predictive models on which a design configuration might have been based. 'The observed difficulties in ensuring synergy between computer simulations and real tests testified to that.

If further research was blocked by lack of theoretical guidance, could not this weakness have been, at least partially, overcome by using other sources of information, such as concepts and design ideas inherited from previous technological generations?

4.3 The Gap Between the New and the Previous Technical Regimes

Identified discontinuities between hypersonic and supersonic flight conditions account for the gap between the new and the previous technical regimes. Originally, it was due to some critical properties of the research outputs in this knowledge area. First, in the transition to the hypersonic domain, the homotopic connections[20] between the concepts developed at different velocity levels are weak. This means that extending existing scientific concepts cannot bridge the discontinuity between supersonic and hypersonic propulsion areas, merely by making additional, modest improvements in existing facilities and human resources. Further, the analogic links between older rockets and the newer airbreathing propulsion technologies are relatively insignificant (Conesa 1997). There are only a limited number of opportunities to transfer design know-how and practical development experience from one domain to the other: Harsha and Waldman (1989, p. 2) emphasized that 'The installation requirements for aerodynamic experimentation and propulsion systems appear to be quite different depending upon whether they concern the development of a shuttle or a scramjet demonstrator'.

Finally, weaknesses in both homotopic connections and analogic links implied that new facilities should be built. It increased the lumpiness in

R&D, and also it potentially decreased its productivity – basically, the rate of return from research activities increases with the rate of fall in R&D costs. Essential large-scale investments in new facilities and basic knowledge production were likely to alter the expected economic returns from the NASP programme.

5 THE NEED FOR TECHNOLOGICAL INFRASTRUCTURES AND ORGANIZATIONAL INTEGRATION

With the paradigmatic change, both existing science and accumulated experimental know-how became inadequate to provide guidance about how to proceed (to push the 'knowledge frontier') in the hypersonic flight area. Consequently, before any demonstration of hypersonic technologies could be implemented, there was a need to build up a new technical knowledge base – that is, to gather data and develop search methods.

Experts had to agree on and confirm the research agenda prior to the selection of any particular design configuration. Top priority was given to the production of both the research infrastructures and the technology instrumentation required for tests and experiments. In this way, it effectively put a block on both further experimentation and organizational innovation.

5.1 The Very Nature of the NASP's Research Outputs: The Production of Infratechnologies as a First Priority

During the preparadigmatic stage, a new set of absolutely essential instrumental knowledge has first to be produced before R&D can start. The situation stressed in the previous sections regarding the NASP programme highlights the strong indivisibility in the hypersonic research area. De Meis, a project leader responsible for the NASP programme, stated: 'Lots of things need to be measured that we do not know how to measure' (De Meis 1990, p. 34). This means that there was a need for new infratechnologies, that is, experimental methods and instrumentation to design a hypersonic aerospace plane. Following Tassey (1991), infratechnologies include: the scientific data necessary for operations of measurement, test control and trial; methods and research instruments, techniques and knowledge. Infratechnologies are the basis of technological development in that they enable precise measurements and furnish scientific and technical data, evaluated and organized, necessary to the understanding, characterization, and interpretation of pertinent research results. Infratechnologies are linked to the basic units of measure. In addition, infratechnologies incorporate the concepts and techniques of measurement and testing which allow for increased quality. Infratechnologies are the instrumental basis of R&D activities. According to Rosenberg (1992, p. 385), 'Scientific instruments may be usefully regarded as the capital goods of the scientific research industry. That is to say, the conduct of research requires some antecedent investment in specific equipment for purposes of enhancing the ability to observe and measure specific categories of natural phenomena'. The infratechnologies consist in standards of measurement, experimental methods and shared modes of comparing and checking research results that are produced collectively and underlie collective experimentation. When this set of knowledge and instruments is lacking, it becomes extremely difficult to identify what particular kinds of design issues need to be addressed. In turn, it creates what may be referred to as a situation of structural uncertainty.

It is clear that the NASP programme with its strong technological composition has been, effectively, in just such a preliminary phase (Bogue and Erbland 1993). How then does one produce infratechnologies and instrumentation that will constitute the new knowledge base, given the weakness of scientific support, and the technology discontinuities marking the transition from supersonic to hypersonic propulsion? The weakness of homotopic connections and analogic links both precluded this and, at the same time, revealed the need for lumpy technological *and* research projects.

5.2 Creating Common Pools of Knowledge: Standardization as a Priority

Because, the argument runs, infratechnologies constitute the basic procedures and routines that enable measurements to be collected and compared across projects carried out at different sites, a high degree of standardization is essential. Basically, it is only on the basis of such data that the conventionally described 'research and development' phases can be undertaken with a minimum of acceptable efficiency in resource allocation.

In the framework of a precompetitive research programme, the difficulty is to produce efficiently infrastructural knowledge of two types: first the infratechnologies and instrumentation required because of the paradigmatic shift; and second, the information generated through the parallel experiments of alternative technology options. The infrastructural character of knowledge relies on its collective production and use. To be infrastructural, knowledge has to be widely diffused in 'codified' form (which is tricky for the first type of knowledge), whereas it must be shared by the research partners despite its partially tacit characteristics (which is critical for the second type of knowledge). The problem here is to combine private ownership of technology with the public character of infrastructural knowledge. In principle, infrastructural knowledge is persistent,[21] and exhibits at the same time non-rival and non-exclusive characters. Yet, in spite of its strong 'public good' aspect, the second type of infrastructural knowledge can be kept private by some free-riders, that is, be concerned with partial exclusivity. Actually, if it is known that such infrastructural knowledge (Steinmueller 1995) has to be widely shared, there will be an incentive for agents to become 'free-riders', and thus avoid the cost of participating in any of the parallel experiments of alternative designs. In other words, there is a risk of information being retained by the competing projects and teams. Yet wide diffusion and adoption of this structural knowledge is essential for a particular programme to go forward. This is consistent with the collective mode of knowledge production in the specific context of application of a mission-oriented research programme.

Besides, the experience cannot easily be shared as it can exhibit some tacit character. Actually, the degree to which knowledge is tacit or codified affects ease of imitation. Codified knowledge is easier to transmit and receive, and is more exposed to industrial espionage and the like. Tacit knowledge by definition is difficult to articulate and so transfer is hard unless those who possess the know-how in question can demonstrate it to others (Teece 1981).

Such obstacles can be offset either directly by means of government investments in programme infrastructure and/or indirectly via the formation of a club – similar to the consortia gathering together industrial firms with public agencies. The latter is consistent with the integration and the coordination of dispersed public and private sources of expertise.

5.3 Existing Network Externalities and Organizational Integration

What is the most appropriate type of organization to generate such infrastructures and infratechnologies, given that the chosen organizational form needs to reflect the 'collective' nature of infratechnologies and research infrastructures?

Investments in infrastructural knowledge production exhibit a strong technological character. Since these preliminary investments enable both subsequent research and design considerations, they hold the collective and lumpy dimensions of the innovative project. Therefore, the appropriate organizational form should be subject to the requirements of collective production of the infratechnologies and research infrastructures needed to create what amounts to the 'conditions of feasibility' of taking the project forward.

Yet, infratechnologies are more than the sum of the experimental routines developed by the participants and their production cannot be left to the participants alone. They have no significance outside their collective use in the research process. They are collective goods in that they require investments that none of the participants individually will feel inclined to pay for. Actually, their public good aspects inhibit some, mainly private sector, participants from investing resources in innovation from which they cannot capture direct benefits. As technological standards, they promote collective research.

The need for new infratechnologies and research instruments suggests the desirability of forming a single entity to produce the required structural knowledge and facilitate its diffusion throughout the programme. The adoption of an integrated organizational design helps to deal with the structural uncertainty and to create a shared cognitive framework, among the partners, regarding the new research area.

5.4 Infratechnologies and Organizational Integration: Lessons from the NASP

Considering the content of the research carried out, the NASP programme was clearly oriented towards the production of adequate infratechnologies and scientific instruments as a preliminary requirement for the exploration of technology variety. This phase of research has been crucial and had to precede whatever basic and applied research activities might be undertaken.

The first objective was to push back the frontier of experimentation 'on the ground' rather than 'in the air',[22] so as to produce the experimental infrastructures required to pursue the ground tests. Second, it was necessary to develop computer simulations – numerical simulation and computer modelling – for fluid dynamics to enable the prediction of the performance and flight characteristics at speeds beyond ground-experimentation capacities. However, vehicle performance calculated in this way can vary and is greatly dependent on the theoretical assumptions embodied in the computer codes. Thus, a first task was to verify vehicle design methods, using the correlation between simulation and experimentation[23] (Bogue and Erbland 1993). As a result, the eventual NASP engine (the experimental vehicle X 30) would not – yet – have been a prototype or even an 'R&D instrument'. Rather, it was accurately considered as a demonstration vehicle or a 'basic' research instrument enabling the production of infratechnologies and instrumentation necessary for further research and development. It was impossible to follow incremental research and step-by-step approaches.

Basically, the production of infratechnologies and, hence, of the related structural flight data, had to be based on an experimentation–simulation relationship relying on the design of a demonstration vehicle. Consistently, a set of factors would support the formation of a single entity, in the spirit of organizational integration. One is derived from the need to produce a collective technology infrastructure gathering the research instrumentation and the infratechnologies necessary to support the R&D activity in the hypersonic scientific area. A mixture of public and private investments was used jointly to develop the technological infrastructure (Kandebo 1990). Setting up such an entity for the NASP aimed at facilitating the sharing of technical results and enabling the formulation of a single technical design, drawing as much as possible on the research experience of a variety of individual firms (Cohen et al. 1991, p. 51).

It was also intended to establish NASP's identity clearly and quickly, making it extremely difficult for the new entity to dissolve into its former, dispersed state. Formally, this was accomplished by producing specific codes and developing specific communication channels to guide flows of information in the nascent organization of NASP. For example, an electronic communications team was created from the beginning with the objective of developing networks within the contracting system composed of subgroups of independent firms.[24] This action brings to mind Arrow's idea (Arrow 1974) that codes and information channels are forms of irreversible organizational capital. Indeed, this strategy imposed an irreversible character upon organizational investments.

One of the main organizational features in the NASP was the unprecedented level of commitment of public agencies in the research enterprise. For those government agencies, the integration process – known as 'mainlining' – involved going beyond traditional generic tasks to include research and experimental instrumentation: 'mainlining brings the government-run facilities into positions often played by contract research labs or subcontractors' (Kandebo 1990, p. 37).

Finally, the NASP programme allowed important progress towards its key milestone (extensive testing of engine modules and airframe structural components). However, despite acceleration of efforts in government and industry to reduce the risk in the critical technologies required to build the X-30, the NASP programme was cancelled by Congress in November 1994, before the decision to build and test the X-30 flight research vehicle scheduled for Phase 3 was taken.

Despite the unsuccessful development of a full-scale demonstrator,[25] these technologies and the programme's work with supersonic-combustion ramjet propulsion, including flight mechanics controls, flight management and flight test considerations for the X-30, have already proved useful to subsequent US aerospace efforts in the hypersonic area (Mattingly 1997).

6 CONCLUSION

The aim of the NASP programme was to explore and demonstrate hypersonic and transatmospheric SSTO airbreathing technologies that would have supported future national security and commercial applications and could have provided dramatic cuts in space launch costs. It consisted of a diverse range of specialists to work in teams on problems in a complex applications-oriented environment. The organizational challenge was to reduce the technological risk – to become locked in the wrong technological path – without impeding the experimentation on the technological designs envisaged. Consistently, the rationale behind NASP's particular choice of an organizational design reflected an attempt to resolve the arising organizational dilemma, trying to balance the two imperatives of diversity and standardization in learning and knowledge production.

This chapter shows that before any organizational and managerial issues could be addressed, the very nature of the research and technical challenges facing the NASP programme has to be analysed in more detail. As far as such an assessment is done regarding research priorities of the programme, the discussion suggests what kind of organizational design to adopt.

To understand why it appeared necessary to scientists and technologists to first develop infratechnologies, one has to specify in what respect the existing scientific knowledge base and evidence break down when dealing with hypersonic transportation systems. For the achievement of the technological programme's goal, problems raised by the lack of homotopic and analogic connections had to be solved. It increased the high degree of lumpiness in both basic and applied research. Structural change, that is, radical innovation envisaged in the NASP, caused structural uncertainty. Scientific and technological discontinuities caused the need for the production of a new infrastructural knowledge base. New infratechnologies including new scientific instruments were required so as to identify and surmount the existing gap between supersonic and hypersonic propulsion technologies as well as to decrease the uncertainty about the new technology.

Because of the cutting-edge technology goal, there was a need to first produce the standardized knowledge as a precondition for the exploration of technology diversity. Therefore, an integrated organizational design was adopted so as to favour a coordinated management in the first stages of the research agenda. At the time the decision was taken regarding the organizational design, the supporters of NASP focused on the production of the collective research infrastructure. Standardization in knowledge was a preliminary to the following broad exploration of the technological and functional dimensions of possible design configurations for the demonstrator. Investments in the production of the research infrastructure were critical, despite the great uncertainty attached to the potential returns from the following individual projects conducted on the basis of the infrastructural knowledge first generated. In that case, the option of a 'club' as a research consortium grouping all the partners committed in the programme in a particular location – in a central laboratory – was proving consistent with the need for the rapid creation of irreversible organizational capital. It was also consistent with the strong commitment of federal public agencies in the production of infratechnologies and technological infrastructures. As a consequence, there was a preference for the 'innovative team approach' within a unified organizational form, prior to the management of multiple decentralized and 'distributed' experimental and exploratory projects with a research network that could have been required later to scan the technology diversity.

NOTES

* This work was undertaken at the International Institute for Applied Systems Analysis, as part of the Technological and Economic Dynamics project. An earlier version of this chapter was presented at the EAEPE Conference on 'Institutions, Economic Integration and Restructuring', at the University of Athens, Greece, November 1997. My thanks to G. Dosi and D. Foray for inspiring suggestions and to B. Nooteboom and P.P. Saviotti for constructive criticisms. I also gratefully acknowledge the financial support of the European Commission (TMR Programme) to this research.

1. Nevertheless, the case of Concorde should be distinguished from the American SST failure, since the project has been recognized to be a commercial failure, not a technical one.
2 This notion is close to the concept of 'industrial district', which goes back to Alfred Marshall.
3. Chesbrough and Teece (1996) notice that the popularity of networked companies and decentralization arises, in part, from observations over a time horizon that remains far too short to be conclusive.
4. At that time, several hypersonic vehicle programmes were being pursued by other nations with goals similar to the NASP programme. The United Kingdom studied the horizontal takeoff and landing (HOTOL), a single-stage space vehicle that used a combination of airbreathing and rocket propulsion. West Germany was investigating the Sänger vehicle which featured airbreathing propulsion for the first stage and rocket propulsion for its second stage. The first stage was also proposed as a developmental vehicle for a future high-speed, Mach 5, civil transport. The Japanese had shown both single- and two-stage conceptual hypersonic vehicles similar to the NASP and the Sänger. These programmes were discussed at several international meetings and conferences organized mainly in the United States in the early 1990s.
5. For instance, the position of the shock wave and the occurrence of the boundary layer at the entrance of the combustor are critical for required thrust at high speeds. The same forebody-combustor must also operate efficiently at low and intermediate speeds. In a similar way, the aft body of the configuration is the nozzle, and proper design is also critical for required performance. As the vehicles proceed through the atmosphere at higher Mach numbers, the aerodynamic forces become more complex with an expanding boundary layer transitioning from laminar to turbulent flows with attendant effects on propulsion efficiency and heating.

6. 'There is an inherent risk in building a vehicle that departs radically from all its predecessors and whose design cannot be fully validated *before flight testing*' (OTA 1989, p. 74, emphasis added).
7. The NASP programme was managed by a team of DoD and NASA personnel located in the Joint Program Office (JPO) at Wright Paterson Air Force Base and the NASP Inter-Agency Office (NIO) in the Pentagon, Washington, DC. The overall national team numbered more than 5000 people and involved some 200 companies in 40 states.
8. Moore's law is not based on any scientific demonstration. It has been inferred by Gordon Moore from his observation of the technological change trend in the semiconductor industry between 1959, when the integrated circuit was patented by two engineers from Fairchild Semiconductor Corporation, and 1965. He noticed that the number of individual components printed on integrated circuits had been doubling each year since 1959. In an article, published in April 1965 in *Electronics Magazine*, he extrapolated this exponential growth for another decade and came up with an astounding projection: that the circuits of 1975 would contain some 65 000 devices. Now enshrined as Moore's Law, his prediction has continued to hold true for more than three decades, although the doubling period has grown to about eighteen months. The most advanced chips today contain millions of transistors – each with typical dimensions of less than half a micron. According to Moore, this trend has been reinforced by the specific properties of the technology: 'A unique aspect of the semiconductor industry is that prices for products tend to decrease over time. . . . Not only does the price fall for a given integrated circuit, but as the complexity of the chip increases, the price per electronics function decreases from product generation to generation as more and more functions are integrated into a single structure' (Moore 1996, p. 56).
9. The semiconductor industry formally began in 1947 with the invention of the transistor at the Bell Telephone Laboratory. A transistor is based on the building block of digital logic and memory circuits (Moore 1996, p. 55). An integrated circuit (IC) is a device performing more than one function on a single chip, that is, it embodies more than one component, either active or passive; for example, several transistors connected through patterns 'written' on the chip (see Dosi 1984, p. 23).
10. '[B]ecause of the unique nature of the technology, by making things smaller the speed of the circuits increases, power consumption drops, system reliability increases significantly, and, most importantly, the cost of the electronic system drops' (Moore 1996, p. 56).
11. The aerobe or airbreathing principle is distinguished by the utilization of oxygen from air – taken up from the atmosphere – as the combustible agent, whereas rocket propulsion requires the loading of both fuel and combustible agents – anaerobe or non-airbreathing principle.
12. Hypersonic speed is obtained from supersonic combustion, just as supersonic speed is obtained from subsonic combustion in a ramjet.
13. Here, we refer to Saviotti and Metcalfe (1984), who developed a typology of products based on the mappings of two sets of characteristics, one relying on their internal structure, that is, technical characteristics of the products, and the other on the functions or 'services' performed for their users, that is, service characteristics.
14. Drawing from Shackle (1972), Langlois (1984) established a typology of a different type of uncertainty. He distinguished the parametric uncertainty from the structural one. Parametric uncertainty refers to a situation of parametric change, that is, change of certain known variables within a known framework. At the most radical extreme, structural change designates change in the very structure leading to an indeterminate problem whose set of potential solutions (that is, states of the world) is unknown, that is, structural uncertainty.
15. Viscous effects become significant at hypersonic speeds and at very high Mach numbers and skin friction drag can account for 50 per cent of the total drag. At Mach numbers above 10, real gas effects begin to appear and can affect the performance of aerodynamic control effectors.
16. David et al. (1992) attempt to assess the economic payoffs of basic research outcomes. They analyse the role played by spillovers in basic research, by exploring the implica-

tions of existing R&D externalities in the symptomatic area of physics of high energy particles. Their framework focuses on the informational outputs of basic research. The aim is to identify potential connections among these outputs both with other scientific areas and with applied research and innovation issues.

17. The concept of symmetry, applied in mathematics and physics, as well as chemistry and crystallography, is a good example of an analogical link allowing for the extension of theoretical results from one domain to another (see David et al. 1992).
18. Scaling is a way of dealing with different levels of aggregation. The main implication of scalability is that it is possible to move between different levels of aggregation.
19. David et al. (1992) focus on stock of information. Here, we emphasize the knowledge characteristic of the research outputs, that is, tacitness, cumulativeness and specificity.
20. The methods and results may or may not be extrapolated to every size of range. The notion of 'homotopic correspondence' comes from topology: two phenomena have homotopic connections if one of them can be deformed continually within the other. Thus, in mechanics, a theory predicting the reaction of a physical object to attraction by an external force will be true for any object of greater mass. The relationship between force and mass is unaffected by changes in the mass parameter (see David et al. 1992).
21. Knowledge that plays an infrastructural role in industry needs to persist long enough that it can be recognized and exploited by the organizations not directly involved in its creation (see Steinmueller 1995).
22. After determining that existing Air Force, NASA, industry and university engine test facilities were not capable of testing scramjets above speeds of Mach 8 for sustained periods, the NASP programme awarded two contracts in October 1986 totalling US$9.6 million for two engine test facilities. These facilities were expected to provide the capability to test full-scale scramjets up to speeds of Mach 8 (US GAO 1988).
23. For example, government efforts led by NASA–Ames have provided an understanding of how to safely contain hydrogen, especially during the NASP's high-temperature flight (Korthals-Altes 1987).
24. 'The team has already developed an unclassified network to develop scheduling and other plans, and is now working on a classified system to handle electronic transfer of drawings and other data' (Kandebo 1990, p. 45).
25. A full-scale aircraft was never built because Congress ended funding in 1994: only a one-third-scale concept demonstrator was built, 'flown' only in a high-temperature tunnel.

BIBLIOGRAPHY

Abernathy, W.J. and J.M. Utterback (1978), 'Patterns of industrial innovation', *Technology Review*, **80**(7), January/July, 40–47.

Andrews, E.H. and E.A. Mackley (1994), 'NASA's hypersonic research engine project – a review', NASA TM 107759, October.

Arrow, K. (1974), *The Limits of Organization*, New York and London: Norton.

Arthur, W.B. (1988), 'Competing technologies: an overview', in G. Dosi, C. Freeman, R. Nelson, G. Silverberg and L. Soete (eds), *Technical Change and Economic Theory*, London and New York: Pinter, pp. 590–607.

Arthur, W.B. (1989), 'Competing technologies, increasing returns, and lock-in by historical events', *Economic Journal*, **99**(394), March, 116–31.

Augenstein, B.W. and E.D. Harris (1993), 'The national aerospace plane (NASP): development issues for the follow-on vehicle', Report R-3878/1-AF prepared for the US Air Force, Santa Monica, CA: RAND.

Barthélémy, R. (1989), 'The national aero-space plane program', American Institute of Aeronautics and Astronautics first NASP Conference, Dayton, 20–21 July.

Bimber, B. and W. Popper (1994), 'What is a critical technology?', DRU-605-CTI, RAND, February.

Bogar, T.J., J.F. Alberico, D.B. Johnson, A.M. Espinosa and M.K. Lockwood (1996), 'Dual-fuel lifting body configuration development', AIAA 96-4592, Norfolk, VA, November.

Bogue, R.K. and P. Erbland (1993), 'Perspective on the National Aero-Space Plane Program Instrumentation Development', Report Number NASA-TM-4505, NASA Dryden Flight Research Facility, Edwards, CA, May.

Branscomb, L. (1993), 'The national technology policy debate', in L. Branscomb (ed.), *Empowering Technology: Implementing a U.S. Strategy*, Cambridge, MA: MIT Press, pp. 1–35.

Branscomb, L. and R. Florida (1997), 'Challenges to technology policy in a changing world economy', in L. Branscomb and J.H. Keller (eds), *Investing in Innovation: Creating a Research and Innovation Policy that Works*, Cambridge, MA: MIT Press.

Chesbrough, H.W. and D. Teece (1996), 'When is virtual virtuous? Organizing for innovation', *Harvard Business Review*, **74**(1), January–February, 65–73.

Cohen, L.K, S. Edelman and R.G. Noll (1991), 'The national aerospace plane: an American technological long shot, Japanese style', *American Economic Review, Papers and Proceedings*, **81**(2), May, 50–53.

Cohen, L.R. and R.G. Noll (1991), 'Efficient management of R&D programs', in L.R. Cohen and R.G. Noll (eds), *The Technology Pork Barrel*, Washington, DC: The Brookings Institution, pp. 37–52.

Collingridge, D. (1990), 'Technology, organization and incrementalism: the space shuttle', *Technology Analysis and Strategic Management*, **2**(2), 181–200.

Collingridge, D. (1991), 'Undemocratic technology: big organizations, big government, big projects, big mistakes', in *Proceedings of the MASTECH Conference*, vol. 2, Thème IV: 'Maitrise sociale de la technologie', Lyon, 9–12 September, pp. 89–96.

Collingridge, D. (1992), *The Management of Scale: Big Organizations, Big Decisions, Big Mistakes*, London: Routledge.

Conesa, E. (1997), 'Contribution a l'analyse des politiques technologiques dans une perspective évolutionniste: Le cas des grands programmes de recherche "lointains"' (Contribution to the study of technology policies in an evolutionary perspective: the case of government 'remote' research programmes), Thèse de Doctorat en Sciences Economiques, Université Lumière Lyon 2.

Conesa, E. and D. Foray (1992), 'Economie et organisation des programmes de recherche "lointains": Au-delà de la frontière des connaissances' (Organization and management of remote research programmes: beyond the knowledge frontier), *Revue d'Economie Industrielle*, **61**, 3ème trimestre, 99–110.

Constant II, E.W. (1973), 'A model for technological change applied to the turbojet revolution', *Technology and Culture*, **10**, June, 132–45.

Cowan, R. and D. Foray (1995), 'The changing economics of technological learning', International Institute for Applied Systems Analysis (IIASA) Working Paper 95–39, Laxenburg: IIASA, May.

Curran, E. (1990), 'The potential and practicability of high speed combined cycles engines', in Advisory Group for Aerospace Research & Development (AGARD), *Hypersonic Combined Cycle Propulsion*, AGARD Conference Reprint no. 479, May, Neuilly-sur-Seine: North Atlantic Treaty Organization, pp. K1–K9.

David, P.A. (1985), 'Clio and the economics of QWERTY', *American Economic Review, Papers and Proceedings*, **75**(2), May, 332–7.

David, P.A. (1987), 'Some new standards for the economics of standardization in the information age', in P. Dasgupta and P. Stoneman (eds), *Economic Policy and Technological Performance*, Cambridge: Cambridge University Press, pp. 206–39.

David, P.A. (1994), 'Why are institutions the "carriers of history"? Path-dependence and the evolution of conventions, organizations and institutions', *Structural Change and Economic Dynamics*, **5**(2), 205–20.

David, P.A., D.C. Mowery and W.E. Steinmueller (1992), 'Analyzing the economic payoffs from basic research', *Economics of Innovation and New Technology*, **2**, 73–90.

David, P.A. and G.S. Rothwell (1996), 'Standardization, diversity and learning: strategies for the coevolution of technology and industrial capacity', *International Journal of Industrial Organization*, **14**(2), 181–201.

De Meis, R. (1987), 'An Orient-Express to capture the market', *Aerospace America*, September, 44–7.

De Meis, R. (1990), 'Many means to NASP', *Aerospace America*, February, 32–8.

Dosi, G. (1984), *Technical Change and Industrial Transformation: The Theory and an Application to the Semiconductor Industry*, London: Macmillan.

Dosi, G. (1988), 'Sources, procedures and microeconomic effects of innovation', *Journal of Economic Literature*, **26**, September, 1120–71.

Ergas, H. (1987), 'The importance of technology policy', in P. Dasgupta and P. Stoneman (eds), *Economic Policy and Technological Performance*, Cambridge: Cambridge University Press, pp. 50–96.

Ergas, H. (1992), 'A future for mission-oriented industrial policies? A critical review of developments in Europe', Working Paper, OECD.

Ergas, H. (1994), 'The "new approach" to science and technology policy and some of its implications', mimeo, RAND/CTI, 20 April.

Fransman, M. (1990), *The Market and Beyond: Cooperation and Competition in Information Technology Development in the Japanese System*, Cambridge: Cambridge University Press.

Fransman, M. (1995), 'Is national technology policy obsolete in a globalised world? The Japanese response', *Cambridge Journal of Economics*, **19**(1), February, 95–119.

Georghiou, L., J.S. Metcalfe, M. Gibbons, T. Ray and J. Evans (1986), *Post-Innovation Performance: Technological Development and Competition*, London: Macmillan, pp. 29–43.

Guy, R.W., R.C. Rogers, R.L. Puster, K.E. Rock and G.S. Diskin (1996), The NASA Langley Scramjet Test Complex 32nd AIAA/ASME/SAE:ASEE Joint Propulsion Conference, Lake Buena Vista, FL, 1–3 July.

Harsha, P. and B. Waldman (1989), 'The NASP challenge: testing for validation', American Institute of Aeronautics and Astronautics first NASP Conference, Dayton, 20–21 July.

Hunt, J.L. and C.R. McClinton (1997), 'Scramjet engine/airframe integration methodology', AGARD Conference, Palaiseau France, 14–16 April.

Kandebo, S. (1990), 'Lifting body design is key to single-stage-to-orbit', *Aviation Week and Space Technology*, 29 October, pp. 36–47.

Klein, B.H. (1977), *Dynamic Economics*, Cambridge, MA: Harvard University Press.

Knezo G.J. (1993), 'Critical technologies: legislative and executive branch activities', Congressional Research Service Report for Congress, 93-734SPR, August.

Kock, B. (1993), 'HYFLITHE III: Concept study results', NASP Paper No. 1275, presented at Monterey, CA, 13 April.

Kodama, F. (1991), *Analyzing Japanese High Technologies: The Techno-Paradigm Shift*, London: Pinter.

Korthals-Altes, S. (1987), 'Will the aerospace plane work?', *Technology Review*, January, 43–51.

Langlois, R.N. (1984), 'Internal organization in a dynamic context: some theoretical considerations', in M. Jussawalla and H. Ebenfield (eds), *Communication and Information Economics: New Perspectives*, Amsterdam: Elsevier Science, pp. 23–49.

Macauley, M. (1986), 'Economics and technology in U.S. space policy', National Academy of Engineering, Washington, DC.

March, J. (1991), 'Exploration versus exploitation in organizational learning', *Organization Science*, **2**(1), 71–87.

Mattingly, T.K. (1997), 'A simple ride into space', *Scientific American*, October, 88–93.

McIver, D.E. and F.R. Morrell (1989), 'National aeroSpace plane: flight mechanics', 75th Symposium of the Flight Mechanics Panel on Space Vehicle Flight Mechanics, Luxembourg, France, Paper No. 20, November.

McLean, F.E. (1985), *Supersonic Cruise Technology*, Washington, DC: National Aeronautics and Space Administration, Scientific and Technical Information Branch.

Moore, G.E. (1996), 'Intel – memories and the microprocessor', *Daedalus*, Special Issue on Managing Innovation, **125**(2), Spring, 55–80.

Nooteboom, B. (1998), 'Innovation, learning and industrial organization', Paper for the Schumpeter Conference, Vienna, 13–16 June.

Office of Science and Technology Policy (OSTP) (1987), 'National aeronautics R&D goals: technology for America's future', Executive Office of the President, Washington, DC, 20500.

Office of Technology Assessment (OTA) (1989), 'Round trip to orbit: human space flight alternatives – special report', Washington, DC: US Government Printing Office.

Pace, S. (1990), 'U.S. Access to Space: Launch Vehicles Choices for 1990–2010', A project Air Force report prepared for the United States Air Force, Rand Publication Series R-3820-AF, Santa Monica, CA: RAND Corporation, March.

Piland, W.M. (1991), 'Technology challenges for the national aero-space plane', 38th congress of the International Astronautical Federation, Brighton, 10–17 October.

Rausch, V.L., C.R. McClinton and J.W. Hicks (1997), 'NASA Scramjet flights to breath new life into hypersonics', *Aerospace America*, July.

Robertson, P.L. and R.N. Langlois (1995), 'Innovation, networks, and vertical integration', *Research Policy*, **24**(4), 543–62.

Rosenberg, N. (1982), 'Learning by using', in N. Rosenberg, *Inside the Black Box: Technology and Economics*, Cambridge: Cambridge University Press, pp. 120–40.

Rosenberg, N. (1992), 'Scientific instrumentation and university research', *Research Policy*, **21**(4) August, 381–90.

Saviotti, P.P. (1998), 'On the dynamics of appropriability of tacit and of codified knowledge', *Research Policy*, **26**(7–8), 843–56.

Saviotti, P.P. and J.S. Metcalfe (1984), 'A theoretical approach to the construction of technological output indicators', *Research Policy*, **13**, 141–51.

Shackle, G.L.S. (1972), *Epistemics and Economics*, Cambridge: Cambridge University Press.

Stalker, R.J., A.P. Simmons and D.J. Mee (1994), 'Measurement of Scramjet thrust in shock tunnels', 18th AIAA Aerospace Ground Testing Conference, AIAA 94-2516, Colorado Springs, CO, June.

Steinmueller, W.E. (1995), 'Technology infrastructure in information technology industries', MERIT Research Memorandum 95-010.

Sullivan, W. (1991), 'Conducting the NASP ground test program', AIAA Third NASP Conference, American Institute of Aeronautics and Astronautics, Washington, DC.

Tassey, G. (1991), 'The functions of technology infrastructure in a competitive economy', *Research Policy*, **20**(4), August, 345–61.

Tassey, G. (1996), 'Infratechnologies and economic growth', in M. Teubal, D. Foray, M. Justman and E. Zuscovitch (eds), *Technological Infrastructures Policy: An International Perspective*, Dordrecht: Kluwer Academic, pp. 59–86.

Teece, D. (1981), 'The market for know-how and the efficient international transfer of technology', *The Annual of the American Academy of Political and Social Science*, **458**, November, 81–96.

Teece, D. (1986), 'Profiting from technological innovation: implications for integration, collaboration, licensing and public policy', *Research Policy*, **15**, 285–305.

US House of Representatives (1989), 'The national aerospace plane', Committee on Science, Space and Technology, 101st Congress, Washington, DC: USGPO, 2 August.

United States General Accounting Office (US GAO) (1988), 'National aero-space plane: a technology development and demonstration program to build the X-30', Report NSIAD-88-122, April,

Vanberg, V.J. (1994), *Rules and Choice in Economics*, London and New York: Routledge.

Whitmore, S.A. and T.R. Moes (1994), 'Measurement uncertainty and feasibility study of a flush airdata system for a hypersonic flight experiment', NASA Technical Memorandum 4627, June.

Zander, U. and B. Kogut (1995), 'Knowledge and the speed of the transfer and imitation of organizational capabilities: an empirical test', *Organization Science*, **6**(1), January–February, 76–92.

8. Surviving technological discontinuities through evolutionary systems of innovation: Ericsson and mobile telecommunication

Maureen McKelvey and François Texier*

1 INTRODUCTION

Evolutionary economics stresses the importance of accessing, developing and using knowledge and technology (Dosi et al. 1988). Innovations and technical change are motors of economic change, in the sense that they create industrial and economic dynamics. Moreover, high-tech sectors involving product innovations have been identified as having high growth in productivity and employment (Edquist and McKelvey 1998; Edquist and Texier 1996; Edquist et al. 1998).

At the same time, divergent opinions exist about what types of firms can take advantage of different types of innovation opportunities. The two heroes of the debate are typically large, existing firms with resources and routines versus small, new flexible firms not bound by tradition. This can be termed the Schumpeterian tradeoff (Nelson and Winter 1982, ch. 14). However, there is some sort of idea that large firms are good at changes which are technically incremental and/or involve economies of scale (like distribution of consumer products) whereas small firms can move into radically new technologies quickly. This may be related to a time dimension, where small firms can initiate a new industry (sector) based on a new technology, but after a while, the large ones dominate.

This chapter will challenge this view by developing a dynamic systems of innovation perspective to analyse how a large firm can survive a technological discontinuity.[1] Such a discontinuity leads to a situation where small start-up firms are expected to have an advantage over incumbent firms. The focus here is on radical innovation in high-tech, or research and development (R&D)-intensive, sectors. An innovation is defined here as new knowledge or technology of economic value, used in products or production processes. One example of a dynamic system of innovation is given, that of the Swedish telecommunication firm Ericsson and its move from being a firm selling switches (exchanges) for public or fixed/wireline telecommunication to national postal, telegraph and telephone (PTT) administrations to being a firm specializing in all aspects of mobile telecommunication (network and phones) to various types of customers. Our perspective is based explicitly on evolutionary economics, in that it emphasizes the importance of firm diversity, knowledge and routines, and selection environments (Nelson and Winter 1982; see also discussion in McKelvey 1998a).

Arguing that large firms can survive technological discontinuities runs in many ways against ideas about how sectors change. The theoretical discussion of industry life cycle stimulated by Abernathy and Utterback (1978) and more recently in Utterback and Suarez (1993) strongly suggests that, initially, a number of different, and often small, firms develop competing technological designs, particularly product design. As the industry matures, the small firms disappear and a fewer number of larger firms remain. The range of products offered becomes limited around a dominant design. R&D also becomes more centred on price reductions and on improvements to production processes. Thus, one basic insight of the life-cycle models is that the focus of innovation as well as type of firms innovating changes over time in an industry.[2] This implies that large firms would increasingly dominate as time progresses and that they have advantages in large-scale production and production innovations rather than in phases of turbulent product innovations.

One challenge to the industry life cycle is based on the idea of markets in motion and lots of smaller firms entering and exiting a sector. This includes research by Klepper and Graddy (1990), Malerba and Orsenigo (1996), Audretsch (1997), and Malerba et al. (1997).[3] As Audretsch (1997, p. 76) puts the basic question, 'Why is the general shape of the firm-size distribution not only strikingly similar across virtually every industry – that is, skewed with only a few large enterprises and numerous small ones but has persisted with tenacity not only across developed countries but even over a long period of time?'. Thus, rather than arguing that there are different phases over time, the common argument in this research approach is that there are two groups of firms, namely a core group of large firms which is fairly stable over time and a fringe group of small firms which enter and exit.

Based on this work, two types of research questions arise: under what conditions can the small ones become part of the core group and under what conditions can large ones at the core remain there? The first has generated a lot of interest while the second is relatively neglected. This chapter therefore focuses on the latter question, in relation to radical innovation opportunities and the development and use of technology and knowledge. Contrary to expectations, can large firms 'bridge' technological discontinuities in the sense of accessing and developing the necessary new knowledge and thereby survive? If so, how can we explain their success?

The particularly relevant aspects of industrial dynamics are thus the relationships between shifts in knowledge/technologies and the type(s) of firms able to innovate and survive over time. To address such questions which unite micro and macro levels of analysis, it is necessary to go down to the micro level of evolutionary economics – namely the diversity of firms (Nelson 1991; McKelvey 1998a).

Our basic argument here is that the firm has a set of relationships relevant for innovation opportunities, or a system of innovation (SI). The different sets that a firm has, or can create, over time – in combination with their internal diversity in terms of strategy, structure and core competencies (Teece et al. 1994; Nelson 1991) – determines the ability of the firm to respond to radical shifts in technology. Here, these arguments are first made theoretically and then used to discuss the shift to mobile telecommunication, particularly the innovative success of Ericsson.

Thus, having ditched the neoclassical assumptions that choices can be weighed rationally and calculated based on adequate information (Nelson and Winter 1982; Hodgson 1988), the basic challenge to understand at the firm level is how firm decisions about search and innovation processes are made. For any radical and uncertain technology, arguments can be made both for waiting and for investing early on. Waiting means that an existing firm can see whether the new technology has the promised growth potential before spending its own resources. Moving in early means the possibility of becoming a technological leader and capturing early, and hopefully durable, market share. Both strategies may seem equally rational to different persons in one firm or to different firms at one point in time. Thus, different persons within the same firm – or different firms – can answer the question of whether to exploit, explore, or ignore new technologies differently. How they answer will depend on their evaluation of the firm's current and potential capabilities as well as of innovative opportunities of the technologies (Dosi et al. 1997). The early history of genetic engineering in pharmaceuticals clearly demonstrates this (McKelvey 1996a, ch. 8). Having said that different decisions are possible, this does not, however, imply that chance is the only thing that matters in an evolutionary perspective. Instead, perceiving and acting upon innovation opportunities requires a combination of factors which are internal and external to the firm.

2 HIGH-TECH FIRMS USING SYSTEMS OF INNOVATION

Focusing only on the question of when a new industry (sector) arises and gives possibilities for new firms to enter, then many researchers argue that radical technological change is involved.[4] Tushman and Anderson (1986) argue that in cases of radical technological discontinuities, existing firms have trouble assimilating new core technical knowledge and so they are replaced by new firms. In short, new firms are thought to develop new industries when there are radical discontinuities. This is so even if the new product is a substitute for the previous one. In other cases where technical change is radical but continuous with (or close to) old competencies, then existing firms are able to adapt, develop their core knowledge, and remain the innovators. In some senses, Tushman and Anderson's arguments could be interpreted as defining when a new industry life cycle will start.

However, there are serious questions about the general validity of the argument, hence focusing our attention on circumstances under which this general pattern does not hold. Ehrnberg and Sjöberg (1995) include different examples of what can happen to new and incumbent firms during technological discontinuities. They argue that the disruptiveness depends mainly on structural factors – especially the character of the technological discontinuity, changes in barriers to entry or mobility, diversity in strategic action, and speed of diffusion – as well as on firms' visions of the future. Ehrnberg and Jacobsson (1997) then go on to develop a four-level model to analyse how well an incumbent firm performs during a technological discontinuity. The four are the industry, firm, local technological system and technology levels, and they argue that variables at each level help determine the outcome.

Rather than dividing the analysis into different levels, the approach here is to place the firm in a dynamic relationship to an SI. McKelvey (1996b) uses the system of innovation approach to question the idea that discontinuities lead to replacement of incumbent firms with new innovators. In the case of genetic engineering used in the pharmaceutical industry, the knowledge and technology was clearly radical and discontinuous compared to the competencies of existing pharmaceutical firms, and yet the pharmaceutical firms have not been replaced by the start-up biotechnology firms. Instead, a system of innovation can be observed which links the existing pharmaceutical firms to new biotechnology firms and to other sources of new competencies and ideas. The firms live in a symbiosis of information relations, and while the large pharmaceutical firms continue to exist (independently or through mergers), many small ones enter and exit.

McKelvey (ibid.) thus not only argues that this example can shake up notions of a well-behaved industry life cycle but that it also implies that the identification and development of innovative opportunities occurs as a process between actors. So, instead of having a dichotomy where we analyse whether innovation is concentrated in one type of firm or another (small–large, old–new), the analytical perspective shifts to focus on how and why different types of firms, other organizations and institutions may in symbiosis help a firm identify and act upon innovation opportunities. This requires re-thinking what innovation processes are and how they proceed.

The analytical framework and research questions must therefore move beyond the firm or sector *per se* to see the firm in relation to firms, other organizations and individuals as well as the relevant institutions which make innovations possible, that is, the SI. Systems of innovation is an approach which Edquist (1997a, pp. 15–29) argues has a number of common features such as inclusion of institutions, holistic and historical perspective, emphasis on dynamics and the importance of innovation to the economy.

In addition to the above common characteristics, most approaches to SI define the boundaries of the system based on either regional or technology/industrial sector delineations, and the vast majority have developed out of evolutionary economics and innovation studies. However, there are a number of competing definitions, based on different research questions and different epistemological starting-points.[5]

The boundaries used to define an SI are usually based on a geographical region, technological area and/or industrial sector. The geographical boundaries used range from global to national to local. The best-known approaches come from (heterodox) economists and look at national systems of innovation (NSIs) in order to examine the specialization of public and private knowledge production – often in relation to questions about its importance for long-term national economic growth (Nelson (ed.) 1993; Lundvall 1992). With the exception of Lundvall (ibid.), much of the early work on NSIs, however, tended to be mostly descriptive as has much of the regional work (Texier et al. 1998). The local or regional system approaches often come from economic geography and try to analyse whether local patterns of interaction matter and if so, how (Cooke et al. 1997). Saxenian (1994), for example, addresses the question of whether small firms in Silicon Valley and large firms outside of Boston are those developing microelectronics, and if so how does it matter. Some of the basic ideas within the local or regional studies relate back to the discussion about post-Fordism. In other words, there is a sense that geographically close relationships and flexible production can substitute for an international and Fordist (Taylorist) approach to production (Storper 1995).

The technological system approach examines the economic competence, knowledge infrastructure (network) and institutions relevant for an area of technological knowledge like microelectronics, regardless of which sector uses the technology. The approach has so far been used mainly in connection with specific industries within specific countries (Carlsson 1995). The sectoral system approach has a similar emphasis on the importance of knowledge conditions but further develops the arguments in terms of technological regimes, and innovation opportunities. Proponents argue that the SIs exist in specific industrial sectors, and these should be visible in the same sectors in different nations (Malerba and Orsenigo 1997).

Our definition of systems of innovation is explicitly based on evolutionary economics and developed in full in McKelvey (1997a, p. 201).

> [Systems of innovation] are here defined as a network involving individual and collective processes of searching, learning and selection among different innovation opportunities, including technical and economic dimensions.
>
> The perspective . . . is that systems of innovation (SIs) are constituted by innovative activities. Innovations are seen as having two dimensions, namely technical novelty and market selection.[6] Innovative activities are therefore here defined as knowledge-seeking activities to develop novelty of economic value. The SI approach as developed here argues that interactions among agents, institutions and environmental conditions define technical and economic opportunities, which taken together define innovation opportunities. Firms have to identify and act upon these technical and economic opportunities in order to innovate (Dosi et al. 1990).[7] It is further assumed that innovative activities in SIs involve different types of agents making decisions and acting based on their perceptions of opportunities.

Thus, the evolutionary SI definition used here is general in the sense that it is not restricted to national or regional or sectoral variables. This is an important point because instead of trying to define SI as something existing 'out there', it is here defined dynamically from an evolutionary economics framework.

Systems of innovation are thus defined in terms of how and why relations with others help, or hinder, firms to identify and act upon innovation opportunities. Relations may, for example, lead to the transfer of technical information relevant for a new technological area or specify a market demand of which an incumbent or new firm otherwise might not be aware. Firms are active in their own right in such a definition, but the SI is important to the extent that relationships and institutions influence the firm's perceptions of opportunities. That influence can either be positive in the sense of helping them identify a new one or negative by, for example, locking them into a relationship which gives information which does not represent wider trends in markets or technology.

Thus, the main actors driving innovations are seen to be firms, while firms' relations with others, as well as institutional conditions, constrain the range of the possible but also enable firms to envisage and act upon opportunities to innovate (Johnson 1992). Evolutionary SI is explicitly dynamic because the basic definition is based on innovation processes, and because it is assumed that the SI is of changing importance to the firm over time, so the relevant SI can change. SI so defined focuses our attention on how knowledge production, exchange and selection occur in an interactive process.

The basic idea that innovation is an interactive process of identification, learning and acting involving relationships between a producing firm and others is thus a more generalizable statement about how modem innovation processes work. The general ideas are supported by research on collaboration, joint R&D and other network relations (Håkansson 1989; Hagedoorn and Schakenraad 1990; Caroll and Harman 1995; Powell et al. 1996). Much research on innovation networks has examined high-tech firms and/or industries like biotechnology and information technology. At the firm level, R&D is a formal commitment to search and learning about new possibilities and about improving existing products.

As such, McKelvey (1998a) forwards the hypothesis that existing firms in R&D-intensive industries, as well as in services with a high level of education among personnel, should have the greatest potential to exploit innovative opportunities – rather than being replaced by new firms – due precisely to this formal commitment to search and learning. One implication is that the general industry life-cycle model, and technological discontinuities discussion may no longer hold, especially in high-tech sectors, due to the nature of modern innovation processes. Existing firms in R&D-intensive sectors are able to survive, and even thrive, because they develop a system of innovation, or network of relationships with others and institutions, which provides them with information and incentives to move into new technological areas.[8]

However, having said that firms in high tech may have the greatest potential to succeed in a move into new technological areas, this does not imply that individual firms will (automatically) succeed. What it means instead is that we need more research which contrasts successes with failures, not only between firms but also inside one and the same firm over time. The next section does this in an illustrative example of the firm Ericsson and of mobile telecommunication more generally.

3 ERICSSON AND MOBILE TELECOMMUNICATION

For existing firms producing telecommunication equipment, mobile telecommunication has represented quite a radical change in the types of technical and market knowledge necessary to survive.[9] There was a move in the late 1970s to digital switches for fixed telecommunication which began one shift in core competencies from electromechanical to microelectronics and computer programming. Then mobile telecommunication added the additional requirement of competence in radio communications and signal processing. Here, mobile telecommunication refers to the hardware aspects of producing the infrastructure (consisting of switches and radio base stations, RBS) and cellular phones. The current study is thus limited to goods products and does not address the services around its use.[10]

Mobile telephony is that part of the telecommunications industry which has globally expanded at an extremely high rate, particularly during the 1990s. For example, the OECD market for mobile communication services doubled between 1993 and 1995 (to US$65 billion) as have the number of subscribers (users) in most OECD countries (OECD 1997, pp. 37–41, 49–51). This is at a time when fixed telecommunications has not only had lower sales and expansion but also lower profits. Large-scale commercial mobile telecommunication networks were first installed in the late 1970s, early 1980s, so these are quite recent products. Moreover, international data shows that the same large firms such as NEC, AT&T (later Lucent), Mobira (later Nokia), Motorola and so on have had market shares from close to the beginning in both networks and phones.[11] Although small firms are often suppliers of components, they are not the major competitors. The market for network equipment has been more concentrated than the market for phones. It was initially fragmented but has become increasingly dominated by a few large firms.

This section will argue that knowledge relations in a dynamic system of innovation, combined with internal firm factors, can explain why a large firm remains in the core group – despite radical shifts in technology and market. To do so, the Swedish firm Ericsson, which has been highly successful in mobile networks and phones, will be discussed, with some comparison with its competitor Motorola.

Ericsson is a truly multinational company, with 104 000 employees at the end of 1998, located in more than 100 countries (LME 1998). Even so, they are important to their home country Sweden because Ericsson's exports of mobile telecommunications accounted for about 13 per cent of total Swedish exports in 1996 (or about the same as paper and pulp). Only 3 per cent of their sales were in Sweden in 1996, compared to 60 per cent of their R&D (LME 1996).[12] Moreover, Ericsson directly created 14 000 new jobs in Sweden between 1993 and 1996, on top of which there are second-order employment generation effects through supplier firms and so on (*Ny Teknik* 1997). In cellular phones, Ericsson has gone from having about 4 per cent of the global market share to 20–25 per cent internationally and 55 per cent of the US digital market, whereas in mobile networks they have held a market share of about 40–45 per cent from the beginning. In phones globally, Motorola has fluctuated between about 10 and 35 per cent whereas their market share for network equipment has gone down from 30 per cent in 1990 to 15 per cent in 1996.[13]

The twenty-year history of large-scale civilian mobile telecommunication has already seen two points of technological discontinuity. From the firm's perspective, the first was that it moved to mobile telecommunication at all and the second the move from analogue to digital standards, with three major competing standards internationally. Each shift will be discussed in turn, focusing on Ericsson and a dynamic system of innovation, and based on an evolutionary economics perspective.

3.1 To Mobile Telecommunication

First, there is the move to mobile telecommunication in the late 1970s and early 1980s. At that time, Ericsson was predominately a traditional telecommunication firm, selling switches to PTTs for public networks. They were, however, just developing a new generation digital switch (later called AXE) in collaboration with the Swedish PTT Televerket. Televerket would produce it for the Swedish market, and Ericsson for the international one. They often sold it to third world countries (*Ny Teknik* 1997) as telecommunication sales were under monopoly and regulation in the Triad countries of the USA, Japan and Europe. Ericsson as a total corporation had little technical competence in radio communication, which is the foundation of mobile telecommunication, and little market experience with either mass consumers or private operators.

However, Ericsson owned a company called Svenska Radio AB (SRA), together with Marconi. SRA mainly developed land mobile radio equipment for the Swedish army, although part was involved in the development of mobile radio for private use in cooperation with Televerket (Mölleryd 1996, p. 34). By the mid-1960s, SRA had an explicit policy to shift its business from military to civilian areas (Lundqvist 1997). SRA also had its own in-house entrepreneur, Åke Lundqvist, who worked hard to convince others of his vision, to use resources to develop the necessary technologies, and to find orders for mobile networks. In other words, he believed that civilian mobile telecommunication offered many innovation opportunities. Moreover, Televerket had an in-house entrepreneur in Östen Mäkitalo, head of Televerket's R&D laboratories, who was very involved in the development of the initial Nordic, then European, cellular systems. SRA acted very much like a small entrepreneurial firm, always ready to solve problems when they arose rather than making sure they had a solution before making an offer, and military equipment was sometimes used fairly directly for civilian orders in the earliest period.

Although SRA could develop one component of the network, namely the radio base stations, Lundqvist thought Ericsson's strategy should be to provide a whole mobile telecommunication infrastructure. In order to sell a whole system, however, SRA needed RBS and switches. Although it did do some prototypes, SRA very quickly relied on other parts of the larger Ericsson corporation to be able to deliver such a system. Over time, as cellular communication became more successful, these various units of the corporation were integrated and reorganized so that the boundaries became transparent. In other words, although SRA could act fairly independently in some respects and this was important initially, it drew upon the competencies and products of the rest of the larger Ericsson corporation. This integration was necessary if, as the SRA entrepreneur believed, Ericsson was to deliver a whole system from RBS to switches to phones.

With relatively minor modifications at the software level, the AXE switch could be used for cellular systems. From those working in the radio systems part of Ericsson, this was seen as a very small niche application. However, Lundquist's arguments and visions were supported by a very close-to-home market developing in the late 1970s, namely the Nordic one. This was driven by the development of technical standards for an early analogue system called NMT-450 by the Nordic PTTs (especially Televerket). The bidding to provide the three components of the system was international and not based on explicit policy of helping Nordic firms (Mäkitalo 1997). Nevertheless, Nordic ones were successful, particularly the smaller independent competitors of SRA/Ericsson that existed at this time. The NMT standard was later used in a number of countries, with Ericsson winning orders which enabled it to reap some scale economies.

Although mobile telecommunication was a niche application for AXE, general characteristics of the AXE switch are quite crucial for explaining Ericsson's early success in mobile telecommunication. Without AXE, there would be no cellular success. But ironically enough, both Ericsson and its competitors thought that offering AXE was initially a disadvantage because it had more processing power and hence was more expensive than competing switches developed explicitly for mobile network use.[14] Network operators, however, wanted the most up-to-date technologies (like AXE), and more importantly, growth in mobile telecommunication had been greatly underestimated. This meant that within a few years, large switches which had previously seemed so overdimensioned and expensive suddenly seemed so obviously necessary. Demand and capacity were greatly underestimated – by all players.

Throughout the 1980s, SRA and Ericsson gained competencies both by expanding in-house and by buying up small competing firms in Sweden. They did so to obtain technical competencies, to bring in fresh entrepreneurial air, and also to 'tidy up the market' (Lundqvist 1997). Three things are quite striking about Ericsson's behaviour within Sweden: first, that Ericsson bought up small competitors; second, that there are only a few of these companies; and third, that they were not spin-offs of Ericsson. Ericsson's expansion was initially Swedish.[15]

Over time, moreover, the civilian applications were developed in different directions from military ones, and Ericsson as a whole had to actively develop new competencies and reorganize. Cellular telecommunication started out as a very small business area in a jointly owned company, but a product area which grew rapidly and became the core of the Ericsson corporation. The deregulation of telecommunication in the 1980s also sparked growth in international markets, partly because now operators could be not only PITs but also private firms and partly because previously closed national markets were now open to foreign competitors.

This does not mean, however, that mobile telecommunication was considered important to Ericsson during the 1980s. Quite the contrary! Most analysts and most telecommunication firms, including Ericsson, had decided that their future lay in the 'paperless office', an integration of telecommunication and information technology. From 1982, Ericsson's Chief Executive Officer (CEO) Björn Svedberg put enormous resources into its Ericsson Information Systems (EIS), with up to 20 000 employees and 30 per cent of the total corporation in terms of sales. This was at a time when radio communication (mobile telecommunication) was about 8 per cent of the total corporation.

Although EIS was large, it was never profitable. Ericsson failed for a number of reasons both external and internal to the firm, then sold EIS to Nokia in 1988. Major problems were that the markets for firm IT systems never expanded as expected; the market for personal computers more generally developed in a different direction than Ericsson had identified; and there were technical problems with its linchpin technology, the switch MD 110.

Thus, there were two visions of innovation opportunities within Ericsson during this crucial time. On the one hand, top management believed in the paperless office and allocated resources accordingly. Ericsson's move to mobile telecommunication was not an innovation opportunity which top management had identified and acted on thereafter. On the other hand, cellular telecommunication was something being developed as a small product and niche application in one jointly owned firm, which, however, had the independence to sink profits into further technical developments. Nevertheless, with great expansion every year, this was soon integrated into the core of Ericsson through reorganizations, and small firms were bought.

In explaining why Ericsson succeeded in moving to mobile telecommunication at all, the most important factors seem to be

- in-house entrepreneurs and entrepreneurial spirit;
- Nordic PTTs/Televerket's role in developing an initial technical standard and hence opening a market close by;
- Ericsson's ability to supply whole networks based on a powerful switch; and
- a market which expanded much more explosively than anyone expected. Forecast demand was sometimes only 10 per cent of actual![16]

3.2 From Analogue to Digital and Competing Standards

The second important technological discontinuity from the firms' perspective has been from analogue to digital standards in the late 1980s and early 1990s, with three major competing standards internationally. Europe, then the USA and Japan, each developed a set of standards for digital mobile telecommunication. So far, the European standard GSM has been by far the most successful internationally, used in at least 105 countries by 1996 (*America's Network* 1996).

The challenges here for firms can generally be stated as needing to monitor rapidly changing bodies of knowledge and techniques and to bet on which standards to develop equipment for, under conditions of uncertainty. Moreover, having identified that in the phone market, the ratio of price to performance (services as well as size and weight) was important for different customer groups, the firm had to try to reduce costs and increase performance. Improving the overall ratio of price to performance has demanded improvements in areas where the firm has had to rely on suppliers and their specialized competencies, such as microelectronics.

The problem for Ericsson was, first, whether to pursue all three major standards and, second, how to find resources to do so. The task fell to Lars Ramqvist, who became president and CEO of Ericsson in 1990. Ramqvist's had previously been head of Ericsson Radio System and had seen the explosive growth of mobile communication, at about 40 per cent per year (Skidé 1994, p. 13). His strategy was to pursue mobile telecommunication for the corporation as a whole and to go for all the international markets and hence develop equipment for the three major standards.

> 'I had to explain to share-holders that I was proposing to increase R&D costs very considerably, possibly by as much as 50% a year over two years,' Ramqvist says. 'We would be spending 15 times as much on R&D as on dividends at a time when recession had hit the industry and we were faced with falling profits'...
>
> Ramqvist's first priority was to focus his efforts. 'We had to realize that we could not do everything,' he says. 'Five or 10 years ago, Ericsson even produced its own nuts and bolts. Now we were going to spend more money on development than anyone else in the industry, but we had to concentrate that effort on our primary goals in telecoms systems, and in particular on mobile communications.
>
> 'We also needed partners in areas outside our core competence. So we established alliances with Texas Instruments in microelectronics, General Electric in a marketing company in the US, and Matra in France.' (*International Management* 1994, p. 27).

Thus, Ramqvist's new strategy for Ericsson to be a major player in mobile telecommunication in the 1990s was threefold. He decided to: (i) dramatically increase in-house R&D, (ii) refocus activities to central ones, and (iii) develop alliances with firms to gain different types of competencies or access to marketing.

It is useful to compare Ericsson with one of its competitors that has had more ups and downs, namely Motorola.[17] Motorola was seen as a very strong company moving into the 1990s, voted company of the year and seen to expand like Intel or Microsoft (Tetzeli 1997). In relation to cellular telecommunication, Motorola had competencies in components related to microelectronics and in radio base stations. However, unlike Ericsson, Motorola did not have any particular competence in switches, in fact, its technical performance began lagging behind the leaders. By 1992, Motorola had stopped producing switches altogether and was instead cooperating with Northern Telecom, Siemens and NEC. Thus, unlike Ericsson which has followed Lundquist's vision of developing and supplying a complete infrastructure and phones in-house, Motorola did not develop all three elements. At that time, Motorola's CEO was George Fisher, and he was sceptical of Ramqvist's strategy for the 1990s: Ericsson was spending too much money without demanding enough profitability; it could not continue to do everything by itself; and so on (Ahlbom 1992). Motorola has since had trouble in a number of its markets, from chips (designed for the Macintosh PowerPC) to cellular phones. Although the global mobile telecommunication market is expanding rapidly and although cellular phones are Motorola's largest revenue source, the company had a 35 per cent decline in earnings between 1995 and 1996 (Tetzeli 1997, p. 124).

Ericsson's strategy of betting on the three international digital standards has meant it has become increasingly important to monitor technical developments outside the firm in order to decide what to do in-house. Its expansion has occurred internationally, with about 40 R&D centres globally by 1996. Moreover, expansion has been expensive and sparked in-house competition over the three technical standard areas (and future ones), but the firm has so far been successful in terms of innovations and sales.[18] Ericsson's shift to producing equipment for the digital standards has entailed a number of other changes relative to its SI as well:

- More direct contacts with university researchers, in Sweden and internationally.
- Hiring more engineers with Master's degrees in computer science and electronics. Basically, Ericsson hires the same number, 900, as graduates with such degrees in Sweden each year (*Ny Teknik* 1997)
- Continuing to evaluate which activities are core and should be done in-house or else contracted out.
- Continuing to develop different types of collaboration and agreements with firms with complementary competencies.
- Taking an increasingly active role in both developing and trying to influence decisions about standards all over the world (McKelvey et al. 1997).[19]

Thus, for Ericsson, the shift from analogue to digital standards and of internationally competing digital standards are together a period both of intensive investment of in-house R&D (throughout the world) and of more extensive contacts with universities and with firms having complementary competencies. Both technological developments and market expansion have been explosive, requiring the firm to be flexible to identify and act upon new opportunities. Moreover, it has had to be flexible to access – at each new point of potential difficulties – the technical and market information crucial for identifying and acting upon new innovation opportunities. For Ericsson, monitoring and accessing external information relevant for innovation opportunities has required enormous investment in in-house capabilities.[20]

Mobile telecommunication has completely changed the nature of what Ericsson does and how it carries out its organizational tasks. Mobile telecommunication has increased from 8 to 60 per cent of the firm between 1985 and 1996 while public telecommunication (switches) has decreased from 34 to about 20 per cent (McKelvey et al. 1997, Table 3. 1). Ericsson has also changed the way it organizes business activities. It has gone from a firm which produces most things in-house to a firm which concentrates on all three components of cellular telecommunication (for example, radio base stations, switches and phones) in-house; which spends extensive resources on R&D (15–18 per cent of sales); and which engages in strategic alliances when necessary.

4 CONCLUSIONS

The theoretical arguments and illustrative examples of two technological discontinuities of mobile telecommunication which challenged Ericsson and telecommunication firms more generally have focused our attention on the importance of analysing a firm in relation to a dynamic SI. The analysis is explicitly evolutionary, not only because it focuses our attention on innovation processes but also because it leads us to ask, 'What is it in an SI that can aid a firm in generating novelty, in inheriting necessary routines and knowledge, and in selecting among alternatives?'. In other words, the analytical focus shifts to understanding the interaction of internal firm procedures and external interactive relationships for obtaining and choosing among market and technical information. Thus, the micro level of the firm is explicitly placed in relation to the macro level of overall innovation opportunities, including market and technical developments.

There are three main topics in this conclusion, each of which can be discussed as specific for the case and more generally. They are elaborations of the ideas of (i) dynamic systems of innovation, (ii) the importance of interaction for size of firm, and (iii) mechanisms for creation, inheritance and selection of knowledge about innovation opportunities.

Dynamic systems of innovation The first topic is that a firm's relationship to a dynamic SI influences its ability to survive technological discontinuities. SIs do not exist but are instead created, as is implied by the definition of SI – that is, those relationships and institutions relevant for identifying and acting upon innovation opportunities.

In this case, Ericsson moves from being heavily influenced by an SI located mainly nationally to being at the centre of an international and flexible SI. Over time, the types of relationships which are important change (for example, from early state customers with technical competencies to universities) as does their location (for example, increasingly international as well as local/national). Ericsson does not succeed because there is the 'correct' SI on the outside which gives relevant impulses about innovation opportunities. Instead, it succeeds in bridging technological discontinuities because the firm is flexible, re-evaluates relationships providing innovation information, and learns over time.

Thus, the point of analytical and theoretical departure should not be whether the 'correct' SI exists to help a firm or group of firms deal with radical technological change. Instead, analysis should be on the dynamics of interaction. At the firm level, this means how and why the firm develops internal routines and competencies to access and evaluate external information about technical and market developments.

The importance of interaction for size of firm The second area of conclusion relates seeing innovation as an interactive process to the questions about the size of firm. Our starting assumption has been that information does not flow in and of itself throughout a system of innovation. Based on modern innovation theory, McKelvey (1996c, p. 34) argues, 'Instead of a flow, knowledge and technology transfer must be seen as an active and interactive process which involves the multiple creation of knowledge and technology. Actors have to actively participate in the process and create knowledge specific to their needs and intended uses'.

This perspective indicates, therefore, that the traditionally identified tradeoff between doing R&D in-house versus accessing knowledge externally needs to be left behind. That view is based on an idea of the flow of knowledge from point A to B, and, the only question is where best to obtain knowledge. Instead, seeing innovation as an interactive process means that both parties have to be active in developing and interpreting information about innovation opportunities.[21]

The most recent part of the Ericsson case clearly indicates that it both had to have extensive in-house R&D and also had to monitor and access information about external developments. Ericsson has developed significant own technical competencies, and, increasingly, market competencies, but has also developed extensive relationships with others. Ericsson decided it needed a very high level of search capabilities; the problem is, of course, the high costs and hence financing of this strategy over time. The problem is deciding how much is enough – or too little, or too much.

Thus, what we have shown is that a large firm can bridge a major technological discontinuity through a strategy of extensive in-house R&D as well as extensive external relationships. The firm actively created and leveraged relationships important for innovation in order to strengthen its own position. In retrospect, it is possible to see how the type of relationships in the SI which were vital changed over time, forcing the firm both to react and act. The example of EIS indicates the difficulties of dealing with uncertainty, and of finding ways of monitoring potential technical and market changes.

The comparison with Motorola points to an interesting question for further research, namely how much, and what type of innovative activities should be done in the firm and what obtained through innovative relationships with others. Motorola had much of the knowledge related to components in-house, but it did not have all three elements of a mobile telecommunication system in-house. Ericsson did, as does Nokia. We are not suggesting these are the only reasons for their success. However, it does raise interesting future research questions about what, exactly, has to be kept in-house in large firms to deal with technological discontinuities.

Mechanisms for creation, inheritance and selection of knowledge about innovation opportunities Based on the dynamic and interactive view of innovation, the three levels of evolutionary change – generation of novelty, inheritance (transmission/retention) and selection – are relevant for focusing our attention only on those aspects of an SI that are relevant for helping (or hindering) a firm, or group of firms, to innovate (McKelvey 1997).

In the Ericsson case, generation of novelty initially occurred through in-house entrepreneurs who identified innovation opportunities but who were backed up by a strong interest from Televerket, which both opened a market but also developed technical standards. Later on, a combination of internal firm search activities in R&D as well as monitoring and accessing external developments has helped the firm identify novelty generated elsewhere, a behaviour strongly spurred by rapid market development.

Inheritance has come through knowledge (in the sense of abstract knowledge and of experience with ways of doing things), through firm routines, and through institutionalized relationships with others. Technical knowledge and routines related to designing switches came from existing business while knowledge about radio communication came from a small part of the corporation, which reoriented towards civilian activities. In the 1980s, Ericsson was oriented mainly towards the 'paperless office' as top management backed that area, but there was still room and resources to develop the expanding market of cellular telecommunication.

Selection has been touched upon at three levels in relation to Ericsson – namely within the firm, in the market, and in the SI. Selection within the firm has to do with what types of product lines and/or R&D projects to develop and it occurs a priori to market whereas selection in the market determines *ex post* which products become profitable.[22] Selection in the SI has been discussed mostly in terms of the interactions between firm and SI, namely what types of information are developed and whether and how this helps the firm identify and act upon innovation opportunities.

Obviously, what is particularly interesting for analysing firms in relation to SIs is not these three evolutionary mechanisms by themselves but how they interact. Thus, the discussion of inheritance should include not only a static analysis of what is inherited at a particular point in time, but also a dynamic one because much of what has been inherited at one point may have to be forgotten, or changed, in order to make room for novelty. Moreover, the set of what can be selected is influenced both by what has been inherited as well as by the novelty generated. Hence, another challenge for the future is to specify the dynamics at the three levels more clearly.

Although these three topics have been discussed both for the Ericsson case and in general terms, nothing has been said here to argue that Ericsson's strategy and its dynamic relationship with SI are the only ways to succeed. In fact, the firm's success in mobile telecommunication has been contrasted with its failure in information systems.

Finally, despite the risk of repeating ourselves, there are a few main points which challenge some existing ideas in the literature and should hence be stressed again. What has been important for Ericsson's success in mobile telecommunication is not that it was 'plugged into' the 'correct' SI. Nor did management identify what turned out to be the most profitable innovation opportunity. Instead, success has been based on a flexible, interactive process of gathering and evaluating information in a firm and in an SI to try to learn about innovation opportunities in a world of limited information and true uncertainty.

NOTES

* We are grateful to our colleagues of the Systems of Innovation Research Program (SIRP) at the University of Linköping for their comments on this chapter.
The empirical data of this chapter was collected during a targeted socio-economic research programme of the European Commission called Innovation Systems and European Integration (ISE).

1. Ehrnberg and Sjöberg (1995, p. 93) define a technological discontinuity as 'a large change in the set of technological knowledge necessary to design and produce a product, often resulting in a new or greatly improved product'. While they emphasize production knowledge, our definition is that the technological shift may be either for production processes or for products.
2. The life-cycle model has stimulated much discussion and criticism, as well as complementary and (partially) contradictory models. Hidefjäll (1997) gives detailed empirical evidence of the multitude of firms and firm strategy as well as lack of dominant design in the pace-maker industry.
3. Klepper and Graddy (1990) argue that even without a stabilization in dominant design, product innovations by small firms will decline over time because of overall decline of average costs (hence prices and profits) in the industry. Despite this, small firms continue to innovate, even though large firms will have the advantage.
4. At the level of national economic leadership and very long time frame, Freeman and Perez (1988) take a step further to argue about the necessity of changing institutions to take advantage of radical technical change, or shifts in techno-economic paradigms. These are quite rare but they not only open new opportunities for innovation throughout all sectors but also require changes in institutions to take advantage of those opportunities.
5. McKelvey (1991) makes this point when comparing different approaches to national systems of innovation.
6. The current classic distinction is between innovation and invention. Technical novelties which remain hidden in their inventors' garages or in R&D departments or which are mentioned in patents but never used, produced or sold are technical

inventions. Technical novelties include a combination of knowledge, techniques (ways of doing things) and technologies (things). However, inventions only become innovations when they are sold or used to make marketable products. As conceptualized here, an innovation must have some degree of technical novelty and some exposure to market selection.

7. As an initial understanding, economic opportunities can be thought of as 'market pull' and technical and scientific opportunities can be thought of as 'technology push'.
8. At the same time, the empirical evidence clearly indicates that new firms have potentials to exploit innovation opportunities. The evidence is different in different high-tech sectors, where in some sectors, like software and computing, start-ups have become serious competitors to large firms whereas in others like biotechnology for pharmaceuticals the two types of firms develop relationships. So far, much research on SI or networks of relationships for innovations has been done on high-tech sectors, but we know less about low- and medium-tech sectors.
9. The empirical material is based largely on McKelvey et al. (1997).
10. However, it would be extremely interesting to analyse the dynamic relationships between services and goods in mobile phones. After all, Edquist et al. (1998) postulate that R&D-intensive areas with a high interaction between services and goods provision are those with the greatest potential for employment growth. Thus, one could study the dynamics of the relationship between operators or providers of the service and those providing the hardware.
11. (McKelvey et al. 1997, ch. 2, tables 2.1 and 2.2). See also Fransman (1995) for an analysis of Japanese computer and communication firms in an international context.
12. However, there are indications of changing trends in 1998. The new CEO, Sven Christer Nilsson has reorganized the firm and part of the head office is moving to London.
13. McKelvey et al. (1997, ch. 2 tables 2.1 and 2.2, and ch. 9).
14. More processing power meant both that it could handle a larger load than was expected and that the quality of speech and ability to handle functions (like roaming and hand-over) were high. Furthermore, AXE is built on modules, which increase flexibility in applications for customers and in up-dating services (McKelvey et al. 1997, ch. 4).
15. Mölleryd (1996, chs 4, 5, 6) argues that there were three phases – local, national and international. The argument here differs from his because we do not define changes in terms of one or the other of these geographical levels because, for example, Ericsson's international expansion also explicitly has a national component. However, we do agree that there was a more national phase during the late 1970s and early 1980s.
16. In the case of the first Nordic standards NMT 450, the initial forecast demand was 50 000 subscribers by 1990. In reality, by 1992, it had approximately 250 000 and its replacement system already had more than 350 000 subscribers (Hultén and Mölleryd, 1993, pp. 4, 7).
17. For an in-depth analysis of Nokia, see Palmberg and Lemola (1998).
18. The year 1998 has, however, been a dramatic period for Ericsson stocks, which rocketed during the first half of the year, but then plummeted by 50 per cent between mid-July and October. This is due both to financial actors' interpretation of how well Ericsson can take advantage of future possibilities towards the internet as well as the global crisis, which hits areas like Asia and Russia which were previously seen as having a great potential for growth.
19. Although Televerket was also the only PTT to play an active role in developing the European GSM standard, standard setting has largely been taken over by international standards bodies, with much discussions with firms (see also Lindmark 1995).
20. In 1998, Ericsson invested 12.5 per cent of its turnover in R&D, while Nokia invested 8.7 per cent (*Ny Teknik* 1998).
21. Thus, in-house R&D was more than 'absorptive capacity' in the sense Cohen and Levinthal (1990) meant of having some in-house R&D in order to understand what is going on externally.
22. See McKelvey (1998b) for the argument that R&D is a form of pre-market selection.

REFERENCES

Abernathy, W.J. and J.M. Utterback (1978), 'Patterns of industrial innovation', *Technology Review*, **80**, 41–47.

Ahlbom, H. (1992), 'Ericsson gör allt – än så Länge' (Ericsson makes everything – so far), *Veckans Affärer*, no. 43, 21 October, 52, 54.

America's Network (1996), 'Wonders of the world phone: L.M. Ericsson's planned development of the world phone', **100**(22), 46–7.

Audretsch, D. (1997), 'Technological regimes, industrial demography and the evolution of industrial structures', *Industrial and Corporate Change*, **6**(1), 49–82.

Carlsson, B. (ed.) (1995), *Technological Systems and Economic Performance: The Case of Factory Automation*, Dordrecht: Kluwer Academic.

Caroll, G. and M. Hannan (eds) (1995), *Organizations in Industry: Strategy, Structure and Selection*, Oxford: Oxford University Press.

Cohen, W. and D. Levinthal (1990), 'Absorptive capacity: a new perspective on learning and innovation', *Administrative Science Quarterly*, **35**, 128–52.

Cooke, P., G. Etxebarria and M.G. Uranga (1997), 'Regional systems of innovation: institutional and organisational dimensions', *Research Policy*, **26**, 475–91.

Dosi, G., C. Freeman, R. Nelson, G. Silverberg and L. Soete (eds) (1988), *Technical Change and Economic Theory*, London: Pinter.

Dosi, G., F. Malerba, O. Marsili and L. Orsenigo (1997), 'Industrial structures and dynamics: evidence, interpretations and puzzles', *Industrial and Corporate Change*, **6**(1), 3–24.

Dosi, G., K. Pavitt and L. Soete (1990), *The Economics of Technical Change and International Trade*, London: Harvester Wheatsheaf.

Edquist, C. (1997a), 'Systems of innovation approaches – their emergence and characteristics', in C. Edquist (ed.) (1997b), pp. 1–35

Edquist, C. (ed.) (1997b), *Systems of Innovation – Technologies, Institutions, and Organizations*, London: Pinter.

Edquist, C., L. Hommen and M. McKelvey (1998), 'Product vs. process innovation: implications for employment', in J. Michie and A. Reati, *Employment, Technology, and Economic Needs*, Cheltenham: Edward Elgar, pp. 128–52.

Edquist, C. and M. McKelvey (1998), 'High R&D intensity without high tech products: a Swedish paradox?', in K. Nielsen and B. Johnson (eds), *Evolution of Institutions, Organizations, and Technology*, Cheltenham: Edward Elgar, pp. 131–52.

Edquist, C. and F. Texier (1996), 'The growth pattern of Swedish industry 1975–1991', in Osmo Kuusi (ed.), *Innovation Systems and Competitiveness*, Helsinki: Taloustieto Oy Publisher, pp. 103–22.

Ehrnberg, E. and S. Jacobsson (1997), 'Technological discontinuities and incumbents performance: an analytical framework', in Edquist (ed.) (1997b), pp. 318–36.

Ehrnberg, E. and N. Sjöberg (1995), 'Technological discontinuities, competition, and firm performance', *Technology Analysis and Strategic Management*, **7**(1), 93–107.

Fransman, M. (1995), *Japan's Computer and Communications Industry: The Evolution of Industrial Giants and Global Competitiveness*, Oxford: Oxford University Press.

Freeman, C. and C. Perez (1988), 'Structural crises of adjustment: business cycles and investment behavior', in Dosi et al. (eds) (1988), pp. 38–66.

Hagedoorn, J. and J. Schakenraad (1990), 'Inter-firm partnerships and cooperative strategies in core technologies', in C. Freeman and L. Soete (eds), *New Explorations in the Economics of Technological Change*, London: Pinter, pp. 3–37.

Håkansson, Håkan (1989), *Corporate Technological Behaviour: Cooperation and Networks*, London: Routledge.
Hidefjäll, P. (1997), 'The pace of innovation: pattems of innovation in the cardiac pacemaker industry', PhD dissertation, Department of Technology and Social Change, Linköping University.
Hultén, S. and B. Mölleryd (1993), 'Cellular telephony in retrospect – foundations of success', *Tele*, **1**, 2–7.
International Management (1994), 'Sweden's mobile leader', May, pp. 24–5, 27–9.
Hodgson, G. (1988), *Economics and Institutions*, Cambridge: Polity Press
Johnson, B. (1992), 'Institutional learning', in Lundvall (ed.) (1992), pp. 23–42.
Klepper, S. and E. Graddy (1990), 'The evolution of new industries and the determination of market structure', *Rand Journal of Economics*, **21**, 27–44.
L.M. Ericsson (LME) (1996), *Annual Report.*
L.M. Ericsson (LME) (1998), *Ericsson Interim Report*, October.
Lindmark, S. (1995), 'The history of the future – an investigation of the evolution of mobile telephony', Licenciate dissertation, Chalmers University, Sweden.
Lundqvist, Å. (1997), Former CEO of Ericsson Radio System, interview at his private residence, Södertälje, 10 April.
Lundvall, B.-Å (ed.) (1992), *National Systems of Innovation: Towards a Theory of Innovation and Interactive Learning*, London: Pinter.
Mäkitalo, Ö. (1997), Director of Telia Research Laboratories, interview at Telia Research Laboratories, Stockholm, 7 May.
Malerba, F., S. Breschi and F. Lissoni (1997), 'Patterns of technological entry and exit: evidence from France, Germany, Italy and the UK', in ISE on CDrom, edited by F. Texier (1998), ISBN 91-7219-266-6.
Malerba, F. and L. Orsenigo (1997), 'Technological regimes and sectoral patterns of innovative activities', *Industrial and Corporate Change*, **6**(1), 83–118.
McKelvey, M. (1991), 'How do national systems of innovation differ? A critical analysis of Porter, Freeman, Lundvall and Nelson', in G. Hodgson and E. Screpanti (eds), *Rethinking Economics: Markets, Technology and Economic Evolution*, Aldershot: Edward Elgar, pp. 117–37.
McKelvey, M. (1996a), *Evolutionary Innovations: The Business of Biotechnology*, Oxford: Oxford University Press.
McKelvey, M. (1996b), 'Technological discontinuities in genetic engineering in pharmaceuticals? Firm jumps and lock-in in systems of innovation', *Technology Analysis and Strategic Management*, **82**. 107–16.
McKelvey, M. (1996c), 'Redefining transfer in biotechnology and software: multiple creation of knowledge and issues of ownership', in A. Inzelt and R. Coenen (eds), *Knowledge, Technology Transfer and Forecasting*, Amsterdam: Kluwer Academic Publishers, pp. 33–47.
McKelvey, M. (1997), 'Using evolutionary theory to define systems of innovation', in Edquist (ed.) (1997b), pp. 200–222.
McKelvey, M. (1998a), 'Evolutionary innovations: learning, entrepreneurship, and the dynamics of the firm', *Journal of Evolutionary Economics*, **8**, 157–75.
McKelvey, M. (1998b), 'R&D as pre-market selection: of uncertainty and its management', in C. Green and G. Eliasson (eds), *Microfoundation of Economic Growth: A Schumpeterian Perspective*, Ann Arbor, MI: University of Michigan Press, pp. 188–211.
McKelvey, M., F. Texier and H. Alm (1997), 'The dynamics of high tech industry: Swedish firms developing mobile telecommunication systems', in ISE on CDrom, edited by F. Texier (1998), ISBN 91-7219-266-6.

Meurling, J. and R. Jeans (1994), *The Mobile Phone Book, the Invention of the Mobile Phone Industry*, London: Communications Week International.

Mölleryd, B. (1996), *Så Byggdes en Världsindustri – Entreprenörskappets Belydelse för Svensk Mobiltelefoni* (This is how a world industry was built – the importance of entrepreneurship for Swedish mobile telephony), Licenciate Dissertation, EFI (Economic Research Institute) Research Report, Stockholm School of Economics, Sweden.

Nelson, R. (1991), 'Why do firms differ, and how does it matter?', *Strategic Management Journal*, **12**, 61–74.

Nelson, R. (ed.) (1993), *National Systems of Innovation: Case Studies*, Oxford: Oxford University Press.

Nelson, R. and S. Winter (1982), *An Evolutionary Theory of Economic Change*, Cambridge, MA: Belknap Press of Harvard University Press.

Ny Teknik (1997), 'Han bestämmer Sveriges framtid: Ramqvist and Ericsson' (He decides about Sweden's future: Ramqvist and Ericsson), 15/5, No. 20, pp. 24–31.

Ny Teknik (1998), 'Nokia vinner telejättarnas Finnkamp' (Nokia is winning the Nordic competition between telecommunications giants), 8/10, No. 41, p. 24

Organization for Economic Cooperation and Development (OECD) (1997). *Communications Outlook 1997*, Vol. 1, Paris: OECD.

Palmberg, C. and T. Lemola (1998), 'Nokia as a related diversifier – Nokia's entry into mobile phone technologies and markets', in ISE on CDrom, edited by F. Texier (1998), ISBN 91-7219-266-6.

Powell, W., K. Koput and L. Smith-Doerr (1996), 'Interorganizational collaboration and the locus of innovation: networks of learning in biotechnology', *Administrative Science Quarterly*, **41**, 116–45.

Saxenian, A. (1994), *Regional Advantage: Culture and Competition in Silicon Valley and Route 128*, Cambridge, MA: Harvard University Press.

Skidé, U. (1994), 'Jan Uddenfeldt i spetsen för mobiltelefonin' (Jan Uddenfeldt at the forefront of mobile telephony), *Elektroniktidningen*, No. 1, 13–14.

Storper, M. (1995), 'The resurgence of regional economies, ten years later: the region as a nexus of untraded interdependencies', *European Urban and Regional Studies*, **2**, 191–221.

Teece, D., R. Rumelt, G. Dosi and S. Winter (1994), 'Understanding corporate coherence: theory and evidence', *Journal of Economic Behavior and Organization*, **23**, 1–30.

Tetzeli, Rick (1997), 'And now for the next trick', *Fortune*, 28 April, 122–30.

Texier, F., C. Edquist and N. Widmark (1997), 'The East Gothia regional system of innovation: a descriptive prestudy', Working Paper No. 199, Department of Technology and Social Change, University of Linköping.

Tushman, M. and P. Anderson (1986), 'Technological discontinuities and organizational environments', *Administrative Science Quarterly*, **31**, 439–65.

Utterback, J. and F. Suarez (1993), 'Innovation, competition and industry structure', *Research Policy*, **22**, 1–21.

9. The properties of routines: tools of decision making and modes of coordination

Bénédicte Reynaud

1 INTRODUCTION

Since the collapse of the Walrasian paradigm in the 1970s, there has been a real acceleration in the development of concepts other than the price system aimed at analysing market coordination. Heterodox economic theories rework notions already in use, while the more orthodox currents are drawing on concepts hitherto foreign to their approach. Economists now dispose of an impressive panoply of coordination mechanisms: institutions, conventions, norms, rules, plans, 'scripts', procedures, routines,[1] and so on.

Each theory tends to privilege one of these concepts and ignore the others. The choice depends on the level of analysis (micro, macro, individual, organizational), the domains being studied (the forms of organization of firms, long-term evolutionary dynamics, the stability of regimes of accumulation) and the kinds of links traditionally maintained with other disciplines (such as philosophy, history or law). Convention, in the philosopher David Lewis's sense, is the cornerstone of the work of Leibenstein (1982) on the determination of the level of effort in the workplace, and is the inspiration for the economics of convention (Dupuy et al. 1989). The concept of institution, issued from the 'Old Institutional Economics' of which Veblen (1919) was one of the founders, occupies a central place in the 'New Institutional Economics' (Williamson 1975, 1995) and in the Theory of Regulation (Aglietta 1976; Boyer 1978, 1986). Finally, the notion of routine which is at the confluence of artificial intelligence and cognitive psychology (Newell and Simon 1972), is one of the three pillars, along with selection and learning, of Evolutionist Theory[2] (Nelson and Winter 1982; Dosi et al. 1988).

This chapter focuses on one of these mechanisms: routines. Two remarks should be made concerning its intellectual context. First, the notion of routine was originally developed in artificial intelligence (AI). We are indebted above all to Simon, but also to Hayek (1973) – two Nobel laureates in economics – for having introduced it in the social sciences: Simon is at once a theorist of organizations[3] and a specialist in artificial intelligence and cognitive psychology.[4] As for Hayek, he is one of the founders of cybernetics. Second, there is an abundant literature on routines, due especially to the development of Evolutionist Theory, for which it is an important device linking the micro, and macro levels.

In order to bring out the properties of routines, we have chosen to compare them with two concepts that are quite close, at least in AI: algorithms and rules. In Sections 2 and 3, we shall show that a routine is a 'pragmatic tool' of decision making. To do this, we shall first undertake a survey of the literature (Section 2), then a comparison between routines and algorithms, which will allow us to characterize the routine as a situated transformation device, dependent on context (Section 3). However, in economics, the analysis of routines as pragmatic decision tools only takes on real interest and importance to the extent that they contribute to collective coordination. Sections 3 and 4 will therefore be devoted to the role of routines in micro dynamics. We shall identify the links between routines and rules (Section 3) in order to analyse their respective roles in coordination (Section 4).

2 ROUTINE AS A DECISION-MAKING DEVICE

2.1 The Four Notions of Routine in the Literature

A reading of the principal texts on the topic (Cyert and March 1963; Nelson and Winter 1982; Dosi et al. 1992; Cohen and Bacdayan 1994; Cohen et al. 1996,[5] and so on) reveals two possible points of entry into the notion of routine.

In the first, a routine is considered to be a pattern of behaviour to the extent that certain authors such as Egidi adopt the expression 'routinized behaviours' and reserve the term routine for use in computation theory.[6] But there remain two ambiguities with this conception. The first, stemming from the biological inspiration of the theory, has to do with the level at which the concept applies. Does it apply to individuals, organizations, or both (in which case one has to spell out the passage from one to the other)? Thus, for Nelson and Winter (1982) routines may, at certain times, 'refer to a repetitive pattern of activity in an entire organization, [or] to an individual skill' (p. 97). Rare are those who, like Cohen (1987) or Cohen and Bacdayan (1994) reserve the notion of routine for either individuals or organizations.

The second ambiguity has to do with the characterization of patterns of behaviour or of related concepts such as: (i) 'patterns of regular and predictable behaviors' (Nelson and Winter 1982); 'patterns of interactions, which represent efficient solutions to particular problems' (Dosi et al. 1992, p. 191); (iii) 'behaviors guided by norms' (Dosi et al. 1997), (iv) 'a way of doing things' or, more precisely, 'a relatively complex pattern of behavior ... functioning as a recognizable unit in a relatively automatic fashion' (Winter 1986, p. 165); (v) 'sequences of patterns of actions that lead to the realization of a final goal (Egidi 1998, p. 1); (vi) 'established patterns of organizational action': 'By organizational routine we mean patterned sequences of learned behavior involving multiple actors who are linked by relations of communication and/or authority' (Cohen and Bacdayan 1994, p. 555).

The ambiguity stems from an overly encompassing definition of the notion. The fact that a routine is an action pattern does not mean that all action patterns are routines. Aware of this difficulty, the leading experts on the question (Cohen, Dosi, Egidi, Marengo, Warglien and Winter) met at Santa Fe in August 1995[7] in order to try to specify what should go in which category (see Section 2.2).

The second point of entry defines a routine as a capacity to learn at distinct levels. This is what the opposition between static and dynamic routines tries to translate. 'Static routines embody the capacity to replicate certain previously performed tasks. Needless to say, routines are never entirely static, because with repetition routines can be constantly improved.' Learning curves concretely display the operation of this type of routine. Dynamic routines, on the other hand, 'are explicitly directed at learning' (Dosi et al. 1992, p. 192).

We propose to combine the distinctions – individual/organization and static/dynamic – to express four senses of the notion of routine. Thus, to each square in Table 9.1 corresponds the concept that seems to us to best characterize the pair identified. Below each concept we refer to the theory that seems most closely associated with it. Note that we have not sought to give a unified academic vision of the concept of routine.

Table 9.1 The many senses of the term 'routine'

Type of behaviour → Learning capacity ↓	Individual	Organizational
Static	(I) 'skill' or 'routines in a narrow sense' Nelson and Winter (1982) Cohen and Bacdayan (1994) Winter (Santa Fe, 1995)	(II) 'standard operating procedures' or 'rules of thumb' Cyert and March (1963) Nelson and Winter (1982) Winter (Santa Fe, 1995)
Dynamic	(III) 'search capability' 'heuristics' Dosi and Egidi (1991) Egidi (1992), Nelson (1995)	(IV) 'meta-routines' Dosi et al. (1990), (1992) Winter (Santa Fe, 1995)

- *Square (I)* Static routines, applied to individuals, will be called skills. A skill is a capacity to execute the same task repeatedly and derives from a kind of 'satisficing'. We make use of Cohen and Bacdayan (1994), who explicitly reserve the term 'skill' for the routines of individuals, and above all the chapter of Nelson and Winter (1982) who, following Cyert and March (1963), posit the 'satisficing' rationality of routines and devote an important chapter to skills. Nelson and Winter emphasize the automatic quality of routines and compare them to computer programs. This comparison also features in Stinchcombe (1990): 'The parts of an individual's skill which are completely routinized are the parts that he or she does not have to think about – once a routine is switched on in the worker's mind, it goes on the end without further consultation of the higher faculties' (p. 63, cited by Cohen 1991, p. 135). Skills have another property: they are a form of tacit knowledge. As Polanyi stresses 'We know more than we can tell' (1967, p. 4). Skills are, thus, a set of rules that are not known as such by the person who follows them.
- *Square (II)* Static routines at the level of organizations are the 'standard operating procedures' or 'rules of thumb' defined initially by Cyert and March (1963, p. 101) and afterwards by Nelson and Winter (1982, p. 17). These are decision rules adopted by the firm;

they are quite simple and can be carried out on the basis of minimal information. They constitute 'the memory of an organization' (1963, p. 101). Their definition is applied word for word to routines by Nelson and Winter (1982, p. 99) and has been adopted in their wake by everyone else.[8]

- *Square (III)* Dynamic routines operating at the individual level are referred to by the term 'search capability'. What this means is that routines are part of a organization's capacity to resolve new problems. In this category are found the 'search behaviours' analysed by Nelson and Winter (1982), and the 'deliberative processes' or 'search procedures' developed by Egidi (1992, p. 154); (see also Nelson 1995, p. 68 and Dosi and Egidi (1991)). Search capability is the capacity to advance through process of trial and error. In this, it is quite distinct from 'satisficing' by acquisition of skills, which is not oriented towards innovation. This search capability is based on heuristics, in other words on concepts and dispositions that provide an orientation and a common structure for treating similar problems (Winter in Cohen et al. 1996, p. 663).
- *Square (IV)* Dynamic patterns of behaviour, operative at the level of the organization, are called 'meta-routines'. By this is meant that they represent a capacity for changing routines (Dosi et al. 1990, 1992).

2.2 The Santa Fe Compromise (1995)

The efforts of the Santa Fe working group culminated in a proposal for a common definition of the notion of routine. It is an 'executable *capability* for repeated performance in some *context* that has been *learned* by an organization in response to *selective pressures*' (Cohen et al. 1996, p. 683). Four points are important: aptitude, the role of context, learning and selection: (i) aptitude is characterized by the capacity to generate an action, to guide or direct an action sequence; (ii) execution is possible only in an organizational context that is considered as a form of external memory or of representation of portions of routines; (iii) the emphasis placed on learning implies the possibility, but not the certainty, of the tacit and automatic character of the routine; (iv) routines are the outcome of a selection process.

The routine is an action pattern, but clearly not all action patterns are routines: witness heuristics, 'rules of thumb.' The 'Santa Fe group' has not succeeded in stipulating with sufficient precision criteria that would have made it possible to establish a distinction between routines and other action patterns.

2.3 The Contrast Between Routines and Procedures

We define a routine as a transformation mechanism aimed at obtaining a particular result. Routines are situated mechanisms that are embedded in particular contexts. They are applicable when individuals, lacking full knowledge of the world, must restrict themselves to local exploration in response to problems. Here we depart from Egidi, whose stipulation that '"Routine" is here a synonym of "not completely specified procedure"' (1992, p. 170, note 4), might lead one to think that a complete specification were possible. We also depart from Dosi et al. for whom it is only the complexity of individual behaviour that stands in the way of a codification of routines (1990, p. 243; 1992, p. 192).

The concept that stands in most pertinent contrast to individual routines is that of *procedure* or *algorithm*.[9] Routines and procedures are decision devices in a broad sense: they are used in the decision process, rather implicit as in the case of routine. We adopt the usual definition of a procedure or algorithm as 'a finite list of instructions before being followed in a given order. In following the instructions step by step, one should arrive at a result after a finite number of steps. The result should be reproducible in an infinitude of individual cases which are treated in the same manner' Lassègue 1994, p. 49. A computer program is an example of this.

We differentiate these notions by the nature of the reasoning required to apply them. This may be of a cognitive type: calculation, selection, search for an algorithm and so on (hypothesis H1), or of a situated type: the context is such that there is little room to manoeuvre and the application is overdetermined (hypothesis H2). The first type of reasoning is required by procedures, the second by routines. H2 implies that agents are 'pseudo-reactive', which is to say that amongst all the options theoretically available to them, very few are applicable. In the case of H1, in contrast, agents are said to be 'cognitive'.

Procedures and routines differ in other respects. The decisive reasoning with regard to future action is performed at different points in time: before the execution of a procedure and in the course of application of a routine. Procedures are explicit and codified, routines are tacit. The execution of a procedure does not require any interpretation because one is in the domain of syntax and of the calculable. The kind of rationality at work is procedural, *à la* Simon (1976). Use of the results of procedures require that they be interpreted. On the other hand, application of routines demands only a minimum of interpretation because they operate within a domain predelimited by an interpretation. That is why routines are part of adaptive rationality, in the sense of Cyert and March (1963). Table 9.2 summarizes these differences.

Table 9.2 Procedures and individual routines

Procedure	Routine
Procedure or algorithm: a finite list of instructions to be followed in a given order and leading to a reproducible result	A routine is a transformation mechanism
Non context-bound, since all possibilities are explored	Context-bound, since exploration is local
The solution is guaranteed, either with a certain probability (probabilistic algorithms) or with certainty (deterministic algorithms)	*A solution* is not guaranteed by the application of a routine
The problem *was explored systematically*	The problem *is explored pragmatically*: it is in applying the routines that the solution may be found. The solution is not 'found' in advance
Explicit, codified character of the procedure	Tacit, non-codified character of the routine
Carried out automatically (absence of interpretation, interpretation necessary afterwards)	Carried out automatically (minimal interpretation)
Procedural rationality (Simon 1976)	Adaptive rationality (Cyert and March 1963)

3 ROUTINES, RULES AND MICRO DYNAMICS

The analysis of routines as pragmatic tools of decision making is ultimately of importance to economics only in so far as they contribute to collective coordination.

3.1 Relationships Between Routines and Rules

Before dissecting the relations between rules and routines, we first need to spell out more precisely our conception of what rules are. Rules are permanent relations between a hypothesis or an antecedent and its consequence. They must be abstract, hypothetical and permanent.[10] Their abstract character stems from the fact that the hypothesis refers to a type of situation and not to particular events or persons. Formally, rules have the logical structure 'condition–action': they trigger the action to be performed when a condition is given. I use the AI perspective in order to find a common formal structure between rules.

Here my approach differs from that of researchers who restrict rules to the meaning of condition–action. I would underline this point because, contrary to a fairly widespread idea, rules do not always dictate behaviours. Rules are often frameworks for requiring some kind of interpretation, the nature of which depends on how the rule is specified.[11] Thus, this is the reason why I shall distinguish two subcategories: the 'rule-ready-to-use' and the 'interpretative rule' or 'rule-to-be-interpreted'. Rules put into play routinized behaviours in the former case, and interpretative behaviours, in the latter case.

In our perspective, a routine is a pragmatic means for the resolution of a problem to which the rule gives a theoretical, abstract and general answer. Rules form the background of routines. One cannot adopt routines without having rules present behind the scenes. Routines are rule-based behaviours.[12] The distinction drawn by Argyris and Schön (1974) between the 'theory-in-use' and the 'espoused theory' of the members of the organization is used to capture the difference between rules and routines. I shall establish a parallel with Argyris and Schön's distinction between the theory-in-use and the espoused theory:

> When someone is asked how he would behave under certain circumstances, the answer he usually gives is his espoused theory of action for that situation. This is the theory of action to which he gives allegiance and which, upon request, he communicates to others. However, the theory that actually governs his actions is theory-in-use, which may or may not be compatible with his espoused theory; furthermore, the individual may or may not be aware of the incompatibility of the two theories. (Argyris and Schön 1974, p. 7)

Thus, routines are opposed to conscious rules (which are the 'espoused theory'). More precisely, routines fill out what is not specified by the rule, the latter being by nature incomplete since it is general. In this sense, routines are virtual. They are (implicit) rules, incomplete and situated. This is the one element that constitutes the dynamic aspect of routines, namely that they take form only as the action unfolds.

To follow a rule, routines must therefore be adopted. But the rule is not the stipulation of routines. Take the following example: the worker who repairs an item solves well-known problems with familiar rules. He/she applies the appropriate rules to given conditional situations following an appropriate sequence. Note that he/she does not have in mind necessarily one deterministic sequence of actions to be performed: this would require a huge amount of memory to be occupied, because the conditions for the execution of any routine can vary slightly and the action sequence fitting these conditions may vary in consequence. Hence his/her goal-oriented activity is driven by a reasonably restricted set of rules. These rules synthethize the relevant interactions between the conditions of the work and the appropriate action. Therefore the worker's mental activity consists in recognition of patterns which trigger the appropriate action: a sequence of action is produced, the same in relation to the same conditions, and as a consequence the behaviour appears as routinized. But this is only half true, because the worker is supposed to have the ability to change the rules if it is required by radical changes in the problems to be solved. In such processes, routines are akin to interpreted rules. In this sense, routines may be understood as 'pragmatic interpretative uses'. That is the reason why I say that the behaviour looks routinized. This point highlights the proximity between following rules and adopting routines.

We contend that routines suspend the indeterminacy created by the rule. The situation in question is one in which it is impossible for any individual to know whether his/her interpretation of another individual's action is indeed what the other wanted to indicate by following the rule. Routines are a sort of setting-aside of the other's intentionality. Because of their automatic quality they allow members of an organization engaged in a collective action to bypass questions as to the motives of the others. Thus, problems of coordination, linked to the need for guarantees as to the intentions of others, are suspended (if not eliminated).

3.2 Organizational Routines Constitutive of Active Collectives

The literature employs the term 'organizational routine' to designate a pattern of collective behaviour specific to firms. The principal question that then arises is the following: what is the foundation of such collective behaviour?

Some theorists, like Nelson and Winter (1982), seem to avoid this question. Inspired by evolutionism in biology, they compare the routines of firms to the genes of living systems. Without setting any limits on the metaphor, they treat the firm as a 'quasi-subject',[13] ascribing to it characteristics hitherto reserved for individuals. Thus, for Nelson and Winter, organizational routines are the extensions to the firm of the properties of the routines of individuals. Nelson writes, for instance, that 'the term "routine" connotes behaviour that is conducted without much explicit thinking about it, as habits and customs' (1995, p. 68).

I contend that organizational routines are constitutive of active collectives. Rules are different: they exist only in their application and, hence, fashion only virtual collectives. Routines, in spite of their tacit and quasi-invisible character, contribute to making the collective 'visible' by giving it a sort of identity. To develop this hypothesis we make use of the advances made by some theorists on the issue.

First, Dosi et al. analyse routines as patterns of interaction internal to the behaviour of a group (1992, p. 192). They depend on each member of the group: (i) having precise knowledge of his/her work; (ii) correctly interpreting messages from others; (iii) being capable of making the 'right' response in the full range of existing routines without unnecessary reflection (the response should by nearly immediate). The response of each agent will indicate to others what they should do.[14] Routines are, thus, more or less stable patterns of human action which ensure the consistency of individual decisions. According to this perspective, routines only exist in the interaction of individuals.

Second, routines are a tacit form of knowledge stored in the 'memory of the organization'. Many authors claim this: for example, Nelson and Winter (1982) and Marengo (1992). But they have merely shifted the problem to the constitution of this 'collective memory'. Can one apply to an organization properties specific to individual subjects, as Marengo proposes? 'Routines constitute the organizational memory: the organization "remembers" a routine by applying it, in the same way that individuals "remember" their skills by exercising them' (Marengo 1992, p. 1). The question remains open, but we hypothesize the following. Organizational memory is not just the transposition of individual memory to the level of the collective, because the organizations have an interactive memory (organizational memory could, however, be an 'emergent property' of the interaction between the routines of individuals).

Third, routines give an identity to the organization for at least two reasons: (i) they operate to produce consistency (Marengo 1992, p. 1), and (ii) being tacit, they are not transmitted outside of the organization. Dosi et al. (1992) compare them to the 'specific assets' of Williamson.

4 CONCLUDING REMARKS

Starting from a survey of the literature on routine, we reach three conclusions.

1. A reading of the principal texts on the topic reveals two possible points of entry into the notion of routine: in the first, a routine is considered to be a pattern of behaviour; the second point of entry defines a routine as a capacity to learn at distinct levels. In each case, we show the difficulties of the notion. These arise mainly from a lack of distinction between two levels: individual versus organizational.

2. Therefore, we propose the following definition of routine, which has three characteristics: first, a routine, is a transformation mechanism aimed at obtaining a particular result. Routines are situated mechanisms that are embedded in particular contexts. Second, they are applicable when individuals, lacking full knowledge of the world, must restrict themselves to local exploration in response to problems. Finally, a routine is a pragmatic means for the resolution of a problem.
3. In so far as routine is considered as a pragmatic means for the resolution of a problem, it becomes easier to compare it to rules which gives a theoretical, abstract and general answer. Rules form the background of routines. Thus, routines are opposed to conscious rules. More precisely, routines fill out what is not specified by the rule, the latter being by nature incomplete since it is general. In this sense, routines are virtual. They are (implicit) rules, incomplete and situated. This is the one element that constitutes the dynamic aspect of routines, namely that they take form only as the action unfolds. In this perspective, we contend that routines may suspend the lack of determination created by rules.

NOTES

* These reflections on routines and rules have benefited from discussions with Richard Bradley, Giovanni Dosi, Massimo Egidi, Jean Lassègue, Pierre Livet, Luigi Marengo, Bart Nooteboom and Paolo Saviotti. They have my thanks. However, I alone am responsible for the ideas expressed.

1. Winter (1986, p. 165) allows that the generic term 'routine' can be rendered as: 'decision rule, technique, skill, standard operating procedure, management practice, policy, strategy, information system, program, script and organizational forms', even if he adopts a more restrict sense of it (see below).
2. We are not thinking here only of the evolutionist theory of the firm, but also of the analysis of the conditions for stability and growth in long-term economic dynamics.
3. See March and Simon (1958).
4. See Newell and Simon (1972).
5. This is a working group that met at Santa Fe in 1995. See Section 2.2 and the following note.
6. Egidi in Cohen et al. (1996, p. 687).
7. This group's report was published in *Industrial Corporate Change* (Cohen et al. 1996).
8. See among others Marengo (1992, p. 1).
9. We consider the latter to be synonymous. In that point we agree with Egidi (1992) who writes 'I use the word "procedure" in a precise sense of algorithm, which (by Church's thesis) can be represented by means of a Turing Machine and mechanically executed' (p. 170, note 4).
10. This is what makes it possible to distinguish rules from decisions, which are concrete, categorical, and non-permanent. The author has developed these notions in Reynaud (1992, ch. 2).
11. Most often, the way in which the rule is specified depends on the context: the nature of the organization, the power relations which exist in it, the goal of its managers and those of its members, the strategies of the individuals who put the rule into effect, etc.

12. See Egidi in Cohen et al. (1996, p. 687).
13. This expression is Jean-Pierre Dupuy's (1992, p. 15 and 1994, pp. 174–7).
14. See also Hodgson (1991 pp. 122 ff.)

BIBLIOGRAPHY

Aglietta, M. (1976), *Régulation et crises du capitalisme*, Paris, Calmann-Lévy, 2nd edn 1997, Paris: Odile Jacob.

Argyris C. and D. Schön (1974), *Theory in Practice*, San Francisco, CA: Jossey-Bass.

Argyris, C. and D. Schön (1978), *Organizational Learning*, Reading, MA: Addison-Wesley.

Billaudot, B. (1995), 'Formes institutionnelles et macroéconomie', in R. Boyer and Y. Saillard (eds), *Théorie de la régulation. L'état des savoirs*, Paris: La Découverte, pp. 215–24.

Boyer, R. (1978), 'Les salaires en longue période', *Economie et Statistique*, **103**, September, 27–57.

Boyer, R. (1986), *La théorie de la régulation: une analyse critique*, Paris: La Découverte.

Cohen, M. (1987), 'Adaptation of organizational routines', Workshop on Organizational Science, Massachusetts Institute of Technology, 10–12 June.

Cohen, M. (1991), 'Individual learning and organizational routine: emerging connections', *Organizational Science*, **2**(1), 135–9.

Cohen, M. and P. Bacdayan (1994), 'Organizational routines are stored as procedural memory: evidence from a laboratory study', *Organizational Science*, **5**, November, 554–68.

Cohen, M., R. Burkhart, G. Dosi, L. Marengo, M. Warglien and S. Winter (1996), 'Routines and other recurring actions patterns of organizations: contemporary research issues', *Industrial and Corporate Change*, **5**(3), 653–98.

Coriat, B. and G. Dosi (1995), 'Evolutionnisme et régulation: différences et convergences', in R. Boyer and Y. Saillard (eds), *Théorie de la régulation. L'état des savoirs*, Paris: La Découverte, pp. 500–509.

Coriat, B. and O. Weinstein (1995), *Les nouvelles théories de l'entreprise*, Paris: Le Livre de Poche, Hachette.

Cyert, R.M. and J. March (1963), *A Behavioural Theory of the Firm*, Englewood Cliffs, NJ: Prentice-Hall.

Dosi, G. (1992), 'Industrial organization, competitivenes and growth', in *Revue d'économie industrielle*, **59**(1), 27–45.

Dosi, G. and M. Egidi (1991), 'Substantive and procedural uncertainty', *Journal of Evolutionary Economics*, **1**, 145–68.

Dosi, G., C. Freeman, R. Nelson, G. Silverberg and L. Soete (1988), *Technical Change and Economic Theory*, London: Francis Pinter.

Dosi, G. and F. Malerba (1996), 'Organization learning and institutional embeddedness: an introduction to the diverse evolutionary paths of modern corporations', in G. Dosi and F. Malerba (eds), *Organization and Strategy in the Evolution of the Enterprise*, London: Macmillan.

Dosi, G. and L. Marengo (1994), 'Some elements of an evolutionary theory of organizational competences', in R.W. England (ed.), *Evolutionary Concepts in Contemporary Economics*, Ann Arbor, MI: Michigan University Press, pp. 157–78.

Dosi, G., L. Marengo, A. Bassanini and M. Valente (1997), 'Norms as emergent properties of adaptive learning', in B. Reynaud (ed.), *Les Limites de la rationalité. Les figures du collectif*, vol 2, Paris: Ed. La Découverte, pp. 45–64.

Dosi, G. and J. Metcalfe (1991), 'Approches de l'irréversibilité en économie', in R. Boyer, B. Chavance and O. Godard (eds), *Les figures de l'irréversibilité*, Paris: EHESS, pp. 37–68.

Dosi, G. and R. Nelson (1994), 'An introduction to evolutionary theories in economics', *Journal of Evolutionary Economics*, **4**, 173–84.

Dosi, G., D. Teece and S. Winter (1990), 'Les frontières des entreprises: vers une théorie de la cohérence de la grande entreprise', *Revue d'économie industrielle*, **51**(1), 238–54.

Dosi, G., D. Teece and S. Winter (1992), 'Toward a theory of corporate coherence: preliminary remarks', in G. Dosi, R. Giannetti and P.A. Toninelli (eds), *Technology, and Enterprise in a Historical Perspective*, Oxford: Clarendon Press pp. 184–211.

Dupuy, J.- P. (1992), *Introduction aux sciences sociales. Logique des phénomènes collectifs*, Paris: Ellipses.

Dupuy, J.- P. (1994), *Aux origines des sciences cognitives*, Paris: La Découverte.

Dupuy, J.- P., F. Eymard-Duvernay, O. Favereau, A. Orlean, R. Salais and L. Thevenot (1989), 'L'économie des conventions', *Revue Economique*, **40**.

Egidi, M. (1992), 'Organizational learning, problem solving and the division of labour', in M. Egidi and R. Marris (eds), *Economics, Bounded Rationality and the Cognitive Revolution*, Aldershot: Edward Elgar, pp. 148–73.

Egidi, M. (1996), 'Routines, hierarchies of problems, procedural behaviour: some evidence from experiments', in K. Arrow and the International Economic Association (eds), *The Rational Foundations of Economic Behaviour*, London: Macmillan and New York: St Martins Press.

Egidi, M. and A. Narduzzo (1996), 'The emergence of path-dependent behaviors in cooperative contexts', *International Journal of Industrial Organization*, **5**, 677–709.

Favereau, O. (1995), 'Apprentissage collectif et coordination par les règles: application à la théorie du salaire', in N. Lazaric and J.M. Monnier (eds), *Coordination économique et apprentissage des firmes*, Paris: Economica, pp. 23–38.

Hayek F.A. (1973), *Law, Legislation and Liberty. Rules and Order*, vol. 1, French translation, Paris: PUF, 1980.

Hodgson, G. (1991), 'Evolution and intention in economic theory', in Saviotti and Metcalfe (eds) (1991), pp. 108–32.

Hodgson, G. (1994), 'Natural selection and economic evolution', in G. Hodgson, W. Samuels and M. Tool (eds), *The Elgar Companion to Institutional and Evolutionary Economics*, Aldershot: Edward Elgar, pp. 113–17.

Kavassalis, P. (1995), 'La stabilité dans le mouvement. Economie et évolution des technologies de la télévision', Thèse de Sciences des organisations, Université de Paris IX Dauphine, ch. 2, pp. 28–59.

Lassègue J. (1994), 'L'intelligence artificielle et la question du continu', Thèse de Doctorat (Philosophie), Paris X-Nanterre, ch. 2, 'Présentation classique de la notion de calculabilité', pp. 49–87.

Leibenstein, H. (1982), 'The prisoners' dilemma and the invisible hand: an analysis', *American Economic Review*, **72**, 92–7.

Lewis, D.K. (1969), *Convention: A Philosophical Study*, Cambridge, MA: Harvard University Press.

March, J. (1991), 'Rationalité limitée, ambuïté et ingniérie des choix', in *Décisions et Organisations*, Paris: Ed. d'Organisations, pp. 133–61.

March, J. and H.A. Simon (1958), *Organizations*, New York: John Wiley & Sons.

Marengo, L. (1992), 'Structure, competence and learning in an adaptive model of the firm', *Papers on Economics and Evolution*, no. 9203.

Mosconi, J. (1989), 'La constitution de la théorie des automates', Thèse de Doctorat d'Etat (Histoire et Philosophie des Sciences), Université de Paris 1.

Nelson, R. (1994), 'Routines', in G. Hodgson, W. Samuels and M. Tool (eds), *The Elgar Companion to Institutional and Evolutionary Economics*, Aldershot: Edward Elgar, vol. 2, pp. 249–53.

Nelson, R. (1995), 'Recent evolutionary theorizing about economic change', *Journal of Economic Literature*, **33**, March, pp. 48–90.

Nelson, R. and S. Winter (1982), *An Evolutionary Theory of Economic Change*, Cambridge, MA: Harvard University Press.

Newell, A. and H. Simon (1972), *Human Problem Solving*, Englewood Cliffs, NJ: Prentice-Hall.

Polanyi, M. (1967), *The Tacit Dimension*, Garden City, NY: Doubleday Anchor.

Reynaud, B. (1992), *Le salaire, la règle et le marché*, Paris: Christian Bourgois.

Reynaud, B. (1996), 'Types of rules, interpretation and collective dynamics: reflections on the introduction of a salary rule in a maintenance workshop', *Industrial Corporate Change*, **5**(3), 699–721.

Saviotti, P. and J. Metcalfe (eds) (1991), *Evolutionary Theories of Economic and Technological Change*, Cambridge, MA: Harvard Academic Publishers.

Simon, H.A. (1976), 'From substantive to procedural rationality', in S.J. Latsis (ed.), *Method and Appraisal in Economics*, Cambridge: Cambridge University Press, pp. 129–48.

Stinchcome, A. (1990), *Information and Organization*, Berkeley: University of California Press.

Veblen, T. (1919), *The Place of Science in Modern Civilization and Other Essays*, New York: Huebsch.

Williamson, O.E. (1975), *Markets and Hierarchies: Analysis and Antitrust Implications*, New York: Free Press, Macmillan.

Williamson, O.E. (1995), 'Hierarchies, markets and power in the economy: an economic perspective', *Industrial and Corporate Change*, **4**(1), 21–49.

Winter, S. (1986), 'The research program of the behavioral theory of the firm: orthodox critique and evolutionary perspective', in B. Gilad and S. Kaish (eds), *Handbook of Behavioral Economics*, vol. A, London: JAI Press, pp. 151–88.

Winter, S. (1995), 'Four Rs profitability: rents, resources, routines and replication', in C. Montgomery (ed.), *Resource-based and Evolutionary Theories of the Firm: Toward a Synthesis*, Hingham, MA: Kluwer Academic Publishers, pp. 147–77.

10. Persistence and change of economic institutions: a social-cognitive approach

Silke R. Stahl-Rolf

1 INTRODUCTION

In the course of the last few decades the functional explanation of social and economic institutions has received growing attention in economics, namely in New Institutional Economics. Building on the pioneering works of Harold Demsetz (1967) and Oliver Williamson (1985) a whole school has developed which tries to functionally describe the existence of social institutions as for example property rights arrangements (Dahlman 1980). These works have offered many interesting insights. However, they failed to cope with a problem that has become relevant with reference to the process of transition in Central and Eastern Europe: human behaviour and social and economic institutions do not react directly to a change in the formal conditions of economic action, that is, an induced change of transaction costs. Instead, institutions exhibit a kind of inertia and only change, if ever, with a lag. In the literature this hesitancy to respond has been explained by referring to cultural continuities and a biased perception of reality (Tversky and Kahneman 1986; Denzau and North 1994). Recently, it has become obvious that economic costs are not the only determinants of institutions, and humans are not calculators that are able to deduce behaviour directly from the respective cost structures. On the contrary, there is an enormous influence from cultural and historical factors that exert a certain impact on the subjective perception of costs and benefits.

Nevertheless, approaches that have tried to explain institutional inertia or persistence by referring to a limited human cognitive capacity to process information were rather vague and not able to shed light on the process by which the socio-cultural environment – of which economic institutions are an integrative part – influences human behaviour.

Whereas it is rather difficult to find contributions to this problem in the economic literature, psychology and especially social psychology have developed a range of instruments that can be used to explain the persistence as well as the change of economic institutions. By doing this the endogenous development of institutions rather than their function lies at the core of the explanation. Thus, the approach presented here is in the tradition of the old institutionalists whose aim was to explain the institutional development by cumulative causation (Veblen 1898/1990) or, using today's language, by referring to the notion of path dependency.

By building on this tradition, a further problem of many approaches (see Granovetter 1985, pp. 482ff.), the reduction of human behaviour exclusively to cognitional limits, psychological laws or underlying social structure, will be avoided. It is generally accepted that the human capability to judge and decide is limited because of these factors. Nevertheless, many decisions, especially those that are irreversible and have consequences which are decisive for an individual's future, will induce individuals to decide as rationally as possible, that is to take into account as much economic information as possible. Thus, if a theory of such decision making has to be developed, social limitations as well as a certain degree of rationality have to be taken into consideration. This means that aspects of observational learning, especially those relevant for the passing on of values and traditional practices, have to be combined with cognitive learning, that is, the conscious evaluation of behavioural alternatives and the search for new possibilities of action.

In this chapter I shall try to close this gap. In Section 2 some methodological problems concerning the possibility of integrating social psychological models into economics will be discussed. In the following section the stages of the transmission process by which cultural factors influence human behaviour and economic institutions will be presented in detail. Section 4 is devoted to the question of how the stages of the transmission mechanism are interlinked. The explanation of the interlinkage will be based on the findings of the theory of social learning and on balance theory, a theory of cognitive dissonance that has been developed by the psychologist Fritz Heider. Starting from the results of the preceding sections, I shall formulate some hypotheses about the forces behind the persistence of economic institutions and especially about the factors relevant for institutional change. The chapter closes with a review of possible applications of the model.

2 METHODOLOGICAL PRELIMINARIES: ON THE INTEGRATION OF (SOCIAL-) PSYCHOLOGICAL APPROACHES INTO ECONOMICS

2.1 (Social) Psychology and Methodological Individualism

As has already been outlined in the introduction, the approach I shall present in this chapter will be based on elements stemming from psychology and social psychology. It is social psychology devoted to the investigation of the influence of the social environment on social behaviour (Witte 1994, p. 4) which is in the sociological tradition, thus in a tradition that is concerned with the investigation of *collective* phenomena. Also, psychology sees processes within individuals against a background of the social environment. Thus, it is appropriate to ask whether social psychology can be integrated into economics at all, since social psychology considers collective processes while economics is based on the central paradigm of methodological individualism. This question can be answered from two perspectives.

On the one hand, the problem of the possible persistence of institutions which are inefficient in terms of an economic rationality characterized by costs and benefits makes it easy to recognize that there must be other factors influencing (economic) institutions as well. If these factors are social ones it is no longer possible to maintain the fiction of an individual acting independently. Yet, even if social factors are integrated into the analysis, the individual remains the central unit of investigation and this is precisely the perspective of (social) psychology as well: individual human behaviour is seen against the background of its social embeddedness. Thus, social psychology is not contradictory to economic theory. On the contrary, it is helpful in making explicit that individual behaviour is always socially embedded and that an analysis on the individual level always has to take into account the social environment; for example, the existing culture which is represented by a collective of individuals or the social structure which describes the reciprocal power and dependency relationships among individuals.

On the other hand, there are an increasing number of economists working in the field of psychology who fall back upon the models of behaviourism in which humans are assumed to react to environmental stimuli as automata. Cognitive processes, anticipation or creativity play no role within the framework of these models: 'A person does not act upon the world, the world acts upon him' (Skinner 1971, p. 212). One has to ask whether a theory that considers individual behaviour but reduces it to reactions to stimuli can be labelled 'individualistic' at all, as all behaviour is explained by the environment that produces the stimuli. Thus, an individualistic theory must contain some element of intentionality and provide real behavioural alternatives if it wants to offer a true alternative to the narrow-minded neoclassical models. From this perspective, cognitive psychology supplies much better possibilities for modelling the individual with all his/her capabilities and limitations within the framework of methodological individualism. The individual is attributed a degree of freedom that is offered neither by behaviourism nor by neoclassical economics. Without question this freedom is limited by individual shortcomings and widespread conformism, yet it seems as if human discretionary powers are too large to reduce them to behaviourism.

If one looks at the possibility of integrating (social) psychological ideas into economics from these two perspectives of the relevance of the individual and the importance of attributing a degree of discretionary freedom to the individual, indeed, they shed a new light on the old paradigm of methodological individualism and show that social factors have to be integrated into economic models because this is in accordance with methodological individualism.

Economics is enriched and gains explanatory power if it is open to psychological considerations as well. This had already been expressed at the turn of the century by the predecessors of today's institutionalists – Gustav Schmoller (1901, p. 107) and Thorstein Veblen (1898/1990, p. 75). They considered psychology, while still in its infancy, to be the key to all social sciences and thus economics as well. Today, with psychology being much more developed than a hundred years ago, one can only endorse that opinion.

2.2 Institutions and Norms in Economics and Other Social Sciences

This chapter aims to describe an essential mechanism leading to the emergence of institutions. While the central elements of this mechanism stem from (social) psychology, the notion of 'institution' chosen here has its origin in economics. There is a whole host of definitions in the literature (for example, see Furubotn and Richter 1991). For the purpose of this chapter I shall use a notion that has been introduced in the literature by Ulrich Witt (1989). According to this notion, institutions are behavioural regularities that can be observed in a population of individuals if these individuals are confronted with the same decision situation.

Yet, in most other social sciences, for example, sociology, institutions are associated with something totally different – for example, in everyday language, an institution is seen as an organizational entity (Opp 1982,

p. 332). Nevertheless, there is a concept in these social sciences as well that comes quite near to the economic notion of an institution. This is the social norm. What relationship exists between social institutions and social norms? Thibaut and Kelley (1959, p. 134) describe social norms as 'rules about behaviour', meaning rules that govern behaviour.

These rules may exist either explicitly in the form of laws or implicitly in the form of customs or shared values. This implies that an institution is the expression of a social norm in the sense of Thibaut and Kefley or, vice versa, if there is an institution there must be a corresponding underlying social norm as well. This means that institutions are sufficient conditions for the existence of social norms. However, it does not necessarily follow that if there is a social norm then there is also an institution. To explain this it is necessary to distinguish between implicit and explicit social norms. While the first are not codified and their existence can only be derived from the behaviour resulting from them, the latter can exist without a corresponding behaviour. This means that institutions are the expressions of social norms, be they explicit or implicit, but not all norms will find their expression in economic institutions.

Thus, there is a certain interdependency between social norms and institutions, and concepts referring to social norms can be treated as equivalent to concepts referring to institutions if the norms are not codified and implicit ones. Therefore, all the investigations in sociology or social psychology dealing with (implicit) norms can be consulted in order to analyse economic institutions.

3 THE ELEMENTS OF THE TRANSMISSION MECHANISM

The mechanism that governs human behaviour and that leads to the emergence of institutions which themselves react upon culture is depicted in Figure 10.1. As shown in this figure, the process takes place on the micro level as well as on the macro level. Culture and its personal representation are phenomena on the macro level while its mental representation and the resulting governing of behaviour are phenomena

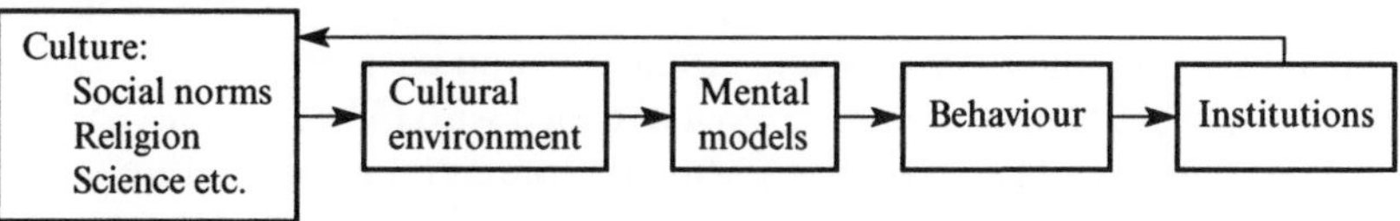

Figure 10.1 The elements of the basic mechanism

on the micro level. The development of a social institution out of a behavioural regularity, again, is a problem of the macro level.

3.1 Culture, its Personal Representation and the Role of Institutions

Persistence and change of economic institutions will be explained in this chapter by showing that cultural traditions have a decisive influence on the emergence and development of economic institutions. Thus, it is necessary to make explicit what is actually understood by 'culture', how culture is represented in society and what relationships exist between culture and institutions.

There is a whole host of possibilities to make explicit the concept of culture. One definition from cultural anthropology is 'the complex whole which includes knowledge, beliefs, art, morals, law, custom, and other capabilities and habits *acquired* by man as a member of society' (Taylor 1871, italics added). Another definition stems from Werner Sombart, who identifies culture with its integral parts, namely state, law, economic organization, religion, art, philosophy, science and language (1929, p.2).

It becomes clear that culture does not relate only to the fine arts. Nor is it all that has been formally codified. The crucial difference between the definition of culture used in this chapter and Sombart's notion of culture is that culture is always more than the formal law or the formal constitution of a state. Culture is the result of formal as well as of informal norms. This means it is not the formal codification of a culture alone but the totality of its formal and informal components that can be seen as the expression of a specific culture. Because of the underlying social norm, institutions are one element of culture and cultural change, whether through creative processes of individuals, or through change in other elements of culture, results in cultural change.

Taylor's definition addresses an important aspect: culture is transmitted by learning. It does not suffice that it is stored in books. Each generation must learn it anew and live according to it. Culture cannot exist and be transmitted without the representatives of culture, that is, each single individual in a population. Yet, not every individual contributes to the representation of culture equally. Rather, the different individuals will embody different aspects of culture and with differing intensity. Thus, in most societies parents play an important role in the transmission of the rules of social interaction and basic attitudes concerning the way challenges are perceived and coped with (McClelland 1961).

The totality of the respective personal representation of culture forms the cultural environment to which an individual is exposed. After his/her own initial socialization each individual becomes the representative of at least one aspect of culture and at the same time is permanently exposed to all the other representatives of culture. Thus the individual is permanently exposed to these 'models' (Bandura 1976, pp. 30ff.). Culture is *socially* learned with the help of models. The nature of this learning process and the mental representation of what is learned is the subject of Section 4.

3.2 Mental Representation of the Cultural Environment and Behaviour

It is one function of culture to allow people to classify the world, to perceive it in a structured way or, as Douglass North puts it, 'culture provides a (. . .) conceptual framework for encoding and interpreting the information that the senses are presenting to the brain' (1990, p. 37). Thus, culture helps to reduce uncertainty and offers a device to facilitate decisions by influencing the development of mental models with the help of which reality can be interpreted.

What is a mental model and what function does it have? The notion of a mental model or frame is one of the central concepts of cognitive science (Johnson-Laird 1983). It is based on the idea that humans do not necessarily react to environmental stimuli in a uniform and predictable way. Such predictable reactions may be the case with reflexes vital for life but perceived information is usually processed before an individual reacts to it with a certain behaviour.

Interestingly, the development of information science and theories of artificial intelligence has made us aware of the vast amount of sensory perceptions that humans are constantly exposed to and how many data have to be processed. It became evident that the processing of every sensory perception and accordingly a rational decision being made based on all these perceptions is not possible at all. Rather it has proved to be advantageous in the course of human evolution for humans to develop equipped with devices in order to facilitate perception and decision making which allow them to come to a decision based on experience and observational learning.

Mental models are the symbolic representation of the environment. It is the human brain which transforms the mechanical, chemical, auditory and visual stimuli into conscious perceptions. Humans principally need models in order to perceive the world, and the way they perceive the world largely depends on these models: 'our view of the world is causally dependent both on the way the world is and on the way we are. (. . .) All our knowledge of the world depends on our ability to construct models of it' (Johnson-Laird 1983, p. 402).

The specific nature of the models depends on whether it refers to the physical or the conceptual (that is, abstract) environment. Whereas the physical environment is perceived directly and represented pictorially, the abstract environment created by humans (for example, the rules of human interaction) is rather perceived discursively. Its representation is based not on pictures but on language and organized according to the criteria of structural logic. Thus, for example, certain groups are attributed certain characteristics and between the groups exist relationships. As will be shown later, these relationships can also be found in models of cognitive dissonance and referring to them it is possible to explain phenomena of institutional inertia on the level of cognitive models as well as on the level of individual behaviour.

A subgroup of mental models are the so-called 'shared mental models' (Denzau and North 1994), models which can be found among a group of individuals in a rather uniform way and which allow them to see the world and act similarly. From the concept of culture presented in this chapter one can derive the reason for the existence of shared mental models: although individuals form their mental models in the course of their personal history they are exposed to similar environmental influences if they grow up and live in the same cultural milieu. Thus, their behaviour will also show regularities.

Social groups differ very much from one other concerning the degree of correspondence of mental models (Rogers and Shoemaker 1971, p. 214). If a group of individuals is open, that is, if it has much contact with the outside, the mental models one can find in this group will show a much greater variance than those found in groups that are self-referential. As will be shown in Section 4, the variance of mental models is of great importance for the generation of new ideas and the adoption of new practices within a group.

3.3 Economic Institutions as Behavioural Regularities

While mental models and the behaviour resulting from them are phenomena on the micro level, the emergence of an institution is a problem of the macro level. Under what conditions will a sufficient number of individuals behave similarly so that a behavioural regularity can be observed and thus allow one to conclude that there must be an underlying social norm?

The answer to this question can be derived from what has been said concerning (shared) mental models. Due to the fact that individuals are exposed to similar cultural influences they share mental models, which induce them to act similarly. Thus, it is usually not the case that a group of individuals decides to adopt and enforce a new mode of behaviour because it is related to lower transaction costs. Rather, a new rule that emerges has been used by one or only a few individuals first because *they* considered it to be advantageous *and* because the new mode of behaviour was somehow in accordance with their view of the world and thus also with the mental models of their social environment.

At this point the two components of human learning mentioned earlier are clearly given expression. Within the framework of culturally determined models individuals are able to decide according to their own estimation and to a certain extent they are able to overcome existing mental models and cultural standards. A new institution will emerge if the new mode of behaviour spreads throughout the population of individuals because others consider it to be advantageous as well. Sometimes it is necessary that the mental models of the individuals change in order that a new mode of behaviour can be adopted.

Property rights, the way property relations are organized in a society, play an important and central role in institutional economics. A simple example of the way individual behaviour influences the social order, in this case the economic order and thus the cultural environment, are systems of land use and the corresponding respect for property in rural societies. Each individual can behave in a way that he or she either respects or violates social norms. For instance, in agricultural societies the individual has the possibility either to look after the cattle and make sure that they stay in the meadow or not to look after them with the result that they eat and trample on the neighbour's wheat. The reason for such carelessness could be that there are mental models which render the decision between 'mine' and 'yours' difficult, as in the case of religious beliefs forestalling the possibility of attaining property by postulating that land can be God's property only.

It is obvious that in such a case no property rights structures will develop that are based on personal property but rather where the differentiation between 'mine' and 'yours' is not necessary, in other words, a property rights structure based upon collective ideas and in which decisions concerning the use of land are taken collectively. At the same time this collective property rights structure strengthens the religious norms that led to its development, which will make it even more difficult to overcome this form of societal organization.

Nevertheless, it is extremely interesting to understand how it is possible to move out of such a social lock-in situation. This is the aim of the following section, in which as a first step the transmission process between the elements of the basic mechanism (Figure 10.1) will be explained. As a next step it will be shown where one could localize sources of institutional change.

4 THE FUNCTIONING OF THE TRANSMISSION MECHANISM

4.1 Learning of Culture

The central question of this section is how individuals learn and internalize culture, that is, how the cultural environment influences the mental models of an individual. Two established theories from social psychology, the theory of social-cognitive learning (Bandura 1976) and the theory of the balance of cognitive and social systems (Heider 1958; Cartwright and Harary 1956) will be used in order to answer this question.

By including both of the above theories, persistence *and* change of economic institutions can be taken into account: while the theory of social learning refers to the learning and strengthening of mental models, balance theory deals with the congruence between old and new mental models and thereby with the question of stability of mental models. Yet, especially if aspects of social learning are taken into consideration as well, it is also possible to explain why and under what circumstances individuals are willing to adopt new mental models.

Culture and learning are closely related in this hypothesis. As has been shown already, culture is represented by the individuals in a society and central to the hypothesis of social learning theory is that humans generally do not learn by trial and error or by being conditioned, but by adopting values, attitudes and behaviour of other persons who are called *models* in the language of the theory of social learning. Humans observe models and adopt their attitudes and modes of behaviour. It is not necessary that they themselves exhibit the behaviour of these models in order for the behaviour to be reinforced. Rather, they can observe that the model and his/her behaviour is reinforced.

Albert Bandura distinguishes four phases in observational learning (1986, p. 52): attentional processes, retention processes, production processes, and motivational processes. Because the phases of production and motivational processes become relevant after a time lag only and that of retention processes refers to the formation of mental models which have already been discussed in the previous section, this section deals only with the first phase.

The phase of attentional processes is of special importance, as in this phase it will be settled whether and with what intensity a model will be perceived. This depends on the modelling stimuli on the one hand and on the observer attributes on the other. Modelling stimuli determine the way a model is perceived. One component of the modelling stimuli is the clarity with which a model presents himself/herself to the environment. Thus, if a model consciously tries to propagate a certain value, the modelling stimulus will be much stronger than in a case where a model merely abides by this value. Also of great relevance are the power, prestige and status characteristics of a model (Webster 1975, pp. 152ff.). Another important factor is affective valence, that is, the emotional relationship between model and observer. In contrast to these modelling stimuli which characterize the model, observer attributes describe the learning situation of the observer, as in, for example, his or her willingness to learn or the existence of mental models which somehow relate to what can be learned by observation.

It is possible to conceptualize both the modelling stimuli as well as the observer attributes within the framework of the theory of the balance of cognitive and social systems. This theory has been developed by the social psychologist Fritz Heider and has been formalized in the language of graph theory by Cartwright and Harary. The theory resembles Leon Festinger's theory of cognitive dissonance (1957). Yet, there is a decisive difference between the two theories, as a result of which the balance theory is much more appropriate to explain the phenomena of persistence and change in economic institutions: while Festinger's theory of cognitive dissonance starts from the hypothesis that dissonances occur *after* decisions only, balance theory is able to describe situations that precede and lead to decisions.

Heider tried to understand under what conditions cognitive structures change. He conceptualized cognitive models as relational structures in which cognitive elements are conceived as objects between which relationships exist. There are two kinds of relationships between these entitles: L-relationships denote emotional ties; R-relationships denote unity as for example neighbours, possession or causality. These relationships can be either positive ('likes', 'possesses') or negative ('disgusts', 'does not possess').

This situation is balanced if *in the case of a dyad* all relationships between the entities are either only positive or only negative. *In the case of a triad* the situation is balanced if either all relationships are positive or two relationships are positive and one is negative. It becomes interesting if the situation is not balanced. Then the individual is exposed to tensions which can also be interpreted as *psychological costs*. Heider's central hypothesis is that 'if no balanced state exists, then forces toward this state will arise. Either the dynamic characteristics [L-relations] will change, or the unit relations will be changed through action or cognitive reorganization' (1958, p. 25). In Figure 10.2 this correlation is depicted schematically:

Larger relationships than only triads can be analysed if the hypothesis is formalized in the language of graph theory. Then the relationships are depicted in the form of a 'directed line *x*' which has either a positive or a negative sign:

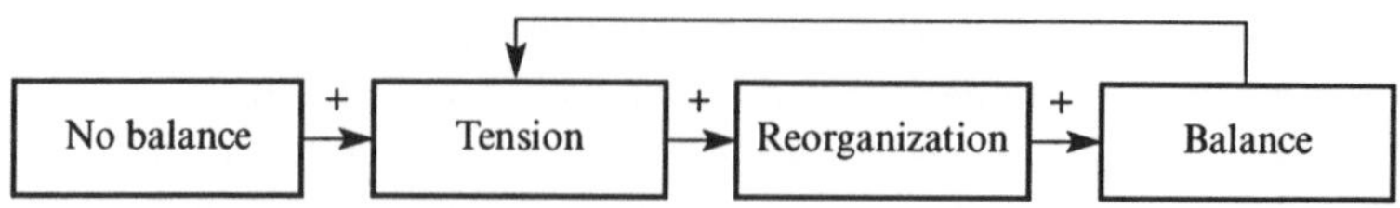

Figure 10.2 Heider's central hypothesis

$c(x) = 1$ if the relation is positive
$c(x) = -1$ if the relation is negative.

A graph consists of cycles. These are sequences of directed lines which have a starting and an ending point. If the direction of lines is of no importance one speaks of semicycles (S). The lines of the semicycle have been attributed a certain value via the function $c(x)$ and thus it is possible to determine the value of the semicycle as the product of the value of the lines:

$$w(S) = \prod_{i-1}^{n} c(x_i).$$

To what extent the complex relationship is balanced is described by the degree of balance B which is defined as the number of positive semicycles $|P|$ divided by the number of all semicycles

$$|S|: B = \frac{|P|}{|S|}.$$

If the number of positive semicycles coincides with the number of all semicycles then $B = 1$; if, by contrast, in the limiting case there is no single positive semicycle, then $B = 0$. The more the degree of balance B approaches 1 the smaller is the probability that the complex relationship will be reorganized; the nearer it is to one, the greater is this probability because of the associated growing tensions or psychic costs, respectively. Thus, the probability of change $p(R)$ is given by $p(R) = 1–B$.

Figure 10.3 shows a complex relationship characterized by the fact that one observer attribute, the existence of a mental model that represents an attitude, possibly incompatible with an attitude that is to be learned, is of decisive importance. It is possible to analyse this situation with the help of the range of instruments just presented. Let us assume that up to now individual p only interacted with model o1 who holds attitude x1. Both model o1 and attitude x1 can be conceptualized as cognitive elements. p learned this attitude by observing o1 and formed a corresponding mental

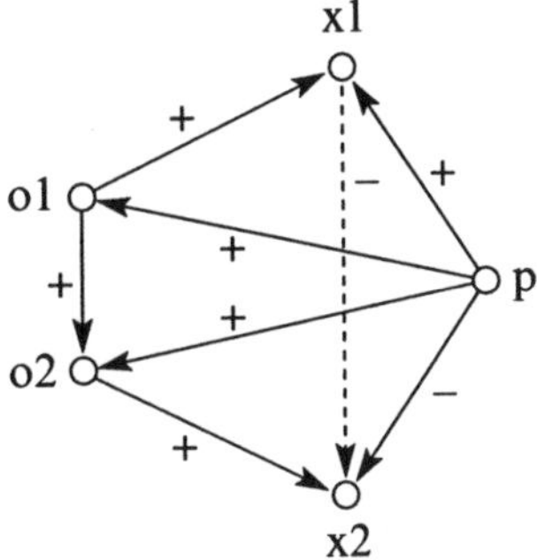

Figure 10.3 Complex relationship

model according to which attitude x1 is not compatible with attitude x2 which is thus rejected by individual p. If individual p gets to know o2 and feels attracted to o2 as well, he/she has a problem if o2 supports attitude x2. Obviously, the situation is not balanced.

This can be seen as well if one determines the corresponding degree of balance

$$B = \frac{|P|}{|S|} = \frac{4}{8} = \frac{1}{2} .$$

In order to reduce tensions and thus psychic costs something in the cognitive structure has to be changed. Thus, it is possible that p negates the positive relation to o1. Or he/she may also negate the incompatibility of xl and x2, that is, reorganize the relevant mental models.

The central question then is in which way the mental structure is likely to be reorganized. In the theory of cognitive dissonance (Festinger 1957) several possibilities are mentioned: adding a new cognitive element (addition), eliminating a cognitive element (subtraction) or changing value judgements (substitution). If the relationships between the elements of the mental model are of equal strength, it would not be possible to make predictions about what will happen. Yet, as an extension to the original model by Heider it is possible to assign values to the relations between the elements of the cognitive model. This can be formalized in a graph-theoretic context as well by attributing not only 1 and –1 to the directed lines but specifying them according to their relative strength. This is depicted in Figure 10.4. There p can identify himself or herself quite well with attitude xl and x2 between which there is a

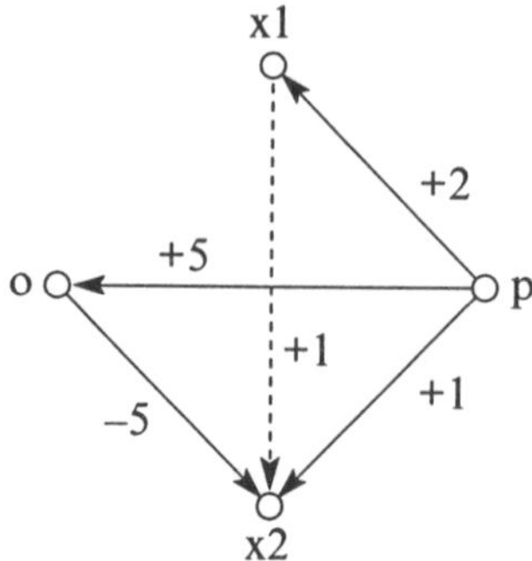

Figure 10.4 Graph with assigned values

positive relationship. Thus, she/he is in a state of cognitive balance when he/she meets o who enjoys status and respect, i.e. is a good model, and therefore experiences high regard. As p vehemently rejects x2 a negative semicycle is added and p is in a state of cognitive imbalance which he/she tries to reduce. He/she could do that by negating the positive relation to o (substitution). Yet, as p feels strongly attracted by o this is rather improbable. Another possibility would be to change the valuation of the relationship to x2. This seems to be more probable as this relationship is comparatively weak. What becomes obvious is that existing cognitive structures are much more likely to change if there are only weak ties between their elements whereas they are more stable if the relationships are stronger. This might be the case if the relationships are firmly established, have seldom been challenged and resulting behaviour led to positive feedbacks. This very often is the case if societies are culturally homogeneous rather than heterogeneous.

Thus, considering the results of this subsection, it is possible to model how 'culture enters people's heads' and thereby to describe the transmission mechanism from the macro level to the micro level with the help of the combination of the theory of social learning on the one hand and the theory of the balance of cognitive and social systems on the other. In the next subsection we shall discuss how a mental model on the micro level is transformed into behaviour.

4.2 The Determinants of Behaviour

On the micro level (the level of individual behaviour), the phenomenon of persistence and change appears just as at the macro level (the level of culture and institutions). Only if individuals alter their behaviour does an institution also change, and the inertia of individual behaviour is the basis of institutional inertia.

Individual behaviour is influenced by underlying mental models. Yet, it is not rigidly determined by these models. It will be shown in this subsection that a wide range of degrees of freedom exist which, apart from behavioural persistence, leave room for novelty and thus behavioural change.

Behavioural persistence has several causes. It can result from routinized behaviour that is only questioned if the behaviour leads to poor results, that is, a certain threshold value is reached after which any behaviour would be questioned (Simon 1957). In this case two things might happen: people might start to actively seek for new behavioural alternatives or they lower their threshold and start being content with 'less' compared to the previous situation. In both cases cognitive reorganizations must take place and within a specific cultural framework. Under different cultural and historical settings the kind of reorganization that in all probability occurs should differ. Moreover, in both situations persistence of behaviour might occur, but in the case of a lowering of the threshold the persistence is much stronger.

It is the function of mental models to structure the set of possible modes of behaviour. Individuals usually do not perceive all possible modes of behaviour but only those which seem to suggest themselves because they have realized them already or because in their social reference group it is customary to behave accordingly. Thus, the set of possible modes of behaviour is restricted from the beginning in order to reduce the complexity of the decision situation and the reduction of complexity follows rules which have been shaped by own past action and mental models which have been acquired by social learning.

Apart from this there is another way in which mental models influence human decisions: by adopting devices which judge behaviour according to a system of values that an individual has developed in the course of his/her personal history and by the internalization of group norms. This mechanism also precedes the choice situation and ensures that behaviour continues to be oriented towards a certain system of values which, by this interaction, is at the same time reinforced.

Yet, it is by no means assured that only those modes of behaviour are chosen that are in accord with *all* mental models of an individual. Often behaviour is so complex and contains so many elements that several mental models are activated by the choice of a mode of behaviour. Mental models that have been only latent so far are activated together and it is only in facing a new problem that their incompatibility might be revealed. While the individual's behaviour is reinforced by one mental model it might be punished by the other. The individual is exposed to tensions which he/she tries to avoid. Either the individual completely abstains from the tension-generating behaviour or he/she tries to resolve the supposed contradiction between the two mental models, for instance, by modifying them. Thus, new mental models develop which pave the way for the change in individual modes of behaviour. Activating behaviour on the basis of mental models might lead to endogenous change of the mental structure and lead to further changes of behaviour.

Moreover, it is possible that an individual may behave in a way that is in open contradiction to his/her system of values. The reasons for this are manifold. For instance it is possible that in a certain social context individuals are literally forced or feel forced to exhibit a certain behaviour that is in open contradiction to their system of values and the corresponding mental models. Yet, behaviour that contradicts existing mental models can also be due to changed incentives. If up to now the behaviour had been adapted to the environmental conditions and the conditions change, this can lead to new modes of behaviour under certain circumstances. For this to happen it is first of all necessary that the new mode of behaviour is perceived. Quite often this is not the case and then old modes of behaviour which are not adapted to the new environmental conditions will be retained.

If the impulse that comes from the new environmental conditions is so strong that the individual has no other possibility than to occupy him-/herself with the new situation and the resulting new behavioural possibilities then one can assume that the decision whether to adopt a new mode of behaviour or not results from the consideration of costs and benefits. The costs include psychic costs and result from the tensions that arise because the new mode of behaviour is not in accordance with existing mental models. The amount of these costs varies with the importance of the mental models that have been violated and with the number of mental models that are not in accordance with the new mode of behaviour. The costs will rise if there is a fundamental and generic model that induces the individuals to disapprove of the new mode of behaviour even more. For example, mental models that represent conformism or collectivism.

The expected benefit that results from the adoption of a new mode of behaviour has manifold reasons. For instance, benefits may result from reduced transaction costs, the individual is better adapted to the changed environment or he/she is able to realize rents due to his/her pioneering activities. Moreover, the expected benefit of a behavioural change will depend on how self-confident the individual is and how strongly he/she believes in his/her abilities to change the current situation. McClelland (1961) described this belief in one's own abilities as 'achievement motivation' and conjectured that it is culturally determined.

One can apply these findings to the theory of economic development. How is it possible to explain that some societies experienced early industrialization while others were latecomers? How is it possible that some latecomers develop extremely rapidly and are just about to overtake some early industrialized countries? One hypothesis might be that in the countries that industrialized early people shared mental models that did not severely punish deviant behaviour and the adoption of new kinds of technologies, whereas this was discouraged by the people's mental models in other countries. Morover, in these countries individual mental models might have been so strong, that changes, whether endogenous or induced, seldom occurred and no resulting deviant behaviour can be observed.

4.3 The Emergence of Institutions as a Collective Phenomenon

The last step of the transmission mechanism is the emergence of institutions. In this chapter, institutions have been defined as behavioural regularities between individuals exposed to a similar situation. Thus, the central question here is under what conditions does an institution develop and become stable considering that it stems ultimately from individual behaviour. As individual behaviour is influenced by mental models which are shaped not only by a person's individual history but also by the group-specific culture to which he/she is exposed, culture is one factor behind the sustaining force of existing institutions. Especially if the cultural setting is very homogeneous and therefore the relations within mental models are very strong, endogenous change occurs rather seldom and induced change would have difficulty gaining a foothold since in such societies mental structures tend to sustain existing mental models, whereas possible innovations, whether they are new cognitive elements, changed relationships between elements or innovative behaviour, are eliminated.

This strongly formative nature of culture with respect to individual behaviour is not surprising. Therefore, it is important to try to understand how new institutions could emerge if individual behavioural innovation is limited by culture. How is it possible to explain that culturally determined institutions change? If one wants to answer this question it is necessary to distinguish between two cases: the conscious adoption of a social institution by the group on the one hand and the gradual spread of a new mode of behaviour which leads to the development of a social institution on the other.

The reason for the first case is that either this mode of behaviour is authoritarianly enforced or all individuals (possibly tacitly) agree to adopt a new mode of behaviour. Authoritarian decisions can be found in modern as well as in traditional societies. Yet, whether an authoritarian decision can be enforced or not largely depends on whether the observed group is modern or traditional. The precondition for the enforcement of authoritarian decisions is power, which in modern societies is derived from achievement, success or military strength, whereas in traditional societies power is based on factors which are outside a person's sphere of influence, for example, rules that ascribe power to certain families. As a number of institutional innovations interfere with these traditions and thus the legitimization of power could be questioned, it does not seem likely that in traditional societies an innovation is authoritarianly enforced. By contrast, an institutional innovation that has been successfully enforced in a modern society is able to strengthen the existing power.

It is rare that individuals explicitly agree upon the adoption of a new mode of behaviour and thus a new social institution. Also a tacit agreement is not usual but it is possible: if within a group a drastic change in environmental conditions leads to a uniform rise in the benefit of a new mode of behaviour that exceeds the associated costs, then the new mode of behaviour appears to be attractive to all members of the group at the same time and will emerge spontaneously. A precondition for this to happen is that mental models in a group do not differ significantly. Moreover, intense communication on the basis of the shared mental model makes the common adoption of a new institution much easier. Thus, especially small traditional societies in which mental models coincide to a considerable extent are able to demonstrate such a kind of institutional change. Yet, as change in these societies is rare, there are considerable difficulties in the way of change.

Therefore, the second case, the gradual spread of an institution, has considerably more relevance than the sudden and spontaneous adoption of an institutional innovation. In the second case the emergence of a new institution can be conceived of as a diffusion problem: a behavioural innovation spreads until one can speak of an institutional innovation. The diffusion process can be conceptualized as a dynamic process in which the probability of adopting a new behavioural alternative depends on the number of adopters. This can be explained by the cognitive phenomenon that the more an existing mental structure is challenged the more probable it is that it will actually change, that is, the relative strength of existing relations between mental models decreases.

More formally, the frequency $F_{t+1}(x)$ of adoption of the behavioural alternative x in $t + 1$ depends on the frequency with which x was chosen in t: $F_{t+1}(x) = r[F_t(x)]$ where r expresses the functional relationship. Graphically this dependence can be depicted as in Figure 10.5. In this figure the functional relationship leads to a diffusion process that until $F_t^*(x)$ is not self-sustaining as in each time period $r[F_t(x)]$ is smaller than

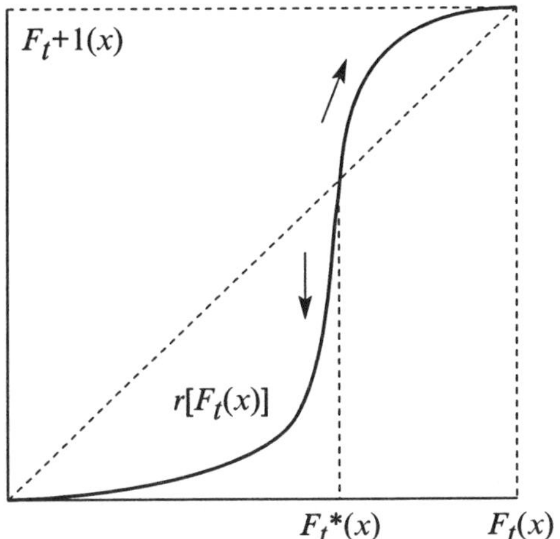

Figure 10.5 Frequency-dependent diffusion of an institutional innovation

$F_t(x)$. Only after $F_t^*(x)$ is reached does the situation reverse and a self-sustaining diffusion process starts. Thus, two questions become relevant: why does a behaviour spread after the critical mass of $F_t^*(x)$ is reached and how is it possible to overcome the critical mass?

Referring to the insights from the theory of social learning and cognitive representation the answer to the first question is not difficult: individuals learn in the process of social learning. They will tend to reject mental elements and thus possible behavioural innovations if they are presented by a minority only. This phenomenon is extremely immanent if the society is traditional and homogeneous, whereas in heterogeneous groups there are always some who are willing to be early adopters and facilitate the diffusion of a behavioural alternative in such a way that the problem of critical mass might lose its relevance.

Yet, in many cases, critical mass does play a role and has to be overcome. There are two ways to realize this. Changing the valuation of the behavioural alternative in general, with the result that $r[F_t(x)]$ shifts onwards and the critical mass moves to the left or disappears completely. This can be reached by changing the subjective perception of the costs and benefits of an institutional alternative which might have its cause in changes in the related cost structures or social learning accompanied by changing mental models that lead to a new evaluation of the behavioural alternatives.

If, as a first step, it is not possible to eliminate the problem of critical mass completely, as a second step a possible critical mass needs to be overcome. If one assumes that individuals are not perceived with equal strength one gets an idea how a critical mass could be overcome. Some individuals are extremely good models and count as a number of 'typical' individuals. If these models adopt a new behavioural alternative it is possible that their weight is enough to induce a spontaneous diffusion of a behavioural innovation. Thus, in modern societies it is especially the symbolic models, that is, the media, which exhibit such a function, whereas in traditional societies in which the influence of symbolic models is considerably smaller, direct modelling plays a much greater role. The model will be extraordinarily influential if he/she holds a power position or is equipped with status characteristics. Thus external diffusion agents are usually less successful than internal ones. The influence will be even stronger if the model deliberately propagates the new behavioural alternative. He/she might try to show the advantages of the new mode of behaviour and make the others believe that it is in accordance with the shared mental models. Moreover, the spread of the knowledge about the new mode of behaviour will be quicker the greater the cohesion of a group, that is, the stronger the ties of the members of the group and the more intensively they communicate.

Yet, there are a number of factors acting as obstacles for the spread of a new mode of behaviour and thus the emergence of a social institution. One problem is that often those individuals who are innovative are social outsiders. They are not so well integrated into the community and thus the social pressure they experience is comparatively low. Especially, the cost of behavioural alternatives are not so much dependent on the judgement of the other individuals and thus do not so much depend on the fact that shared mental models are violated. But as has been explained above, outsiders do not serve as good models. The better-integrated members of a group will probably ignore the innovating outsider. The result is that in a number of cases behavioural innovations do not spread within a group. In such a case, older institutions which, objectively seen, may be inefficient, can persist. There are perspectives for change but it will take time and might be initiated from the outside.

5 CONCLUSION

This chapter has shown how the emergence of economic institutions can be modelled within a framework characterized by ideas taken from social psychology and cognitive science. Apart from the methodological question of the integrability of social psychology into economics it could also be demonstrated that the chosen perspective can bring new insights into the process that leads to the development of social institutions.

This approach sheds new light on problems of institutional change in the transition countries in Middle and Eastern Europe and in many developing countries as well: persistence of institutions appears due to the persistence of mental models in people's heads *and* inappropriate incentives. Thus, the presented model integrates ideas which have been developed in institutional economics and social psychological theories.

Whether inappropriate incentives or shared mental models which are inadequate are responsible for the fact that in some countries institutional change either proceeds very slowly or goes in a direction that is not desired has to be analysed empirically. In order to do this it appears to be necessary to investigate not only the current situation but also the historical background. If it can be shown historically that certain mental models have been persistent in the past, one has to assume that they influence today's institutional development as well and thus institutional persistence is not so much a problem of incentives but rather a problem 'within people's heads'.

BIBLIOGRAPHY

Bandura, Albert, J. Grusec and F.L. Menlove (1967), 'Some social determinants of self-monitoring reinforcement systems', *Journal of Personality and Social Psychology*, **5**, 449–55.

Bandura, Albert (1976), *Lernen am Modell. Ansätze zu einer sozialkognitiven Lerntheorie* (Observational learning: a social-cognitive theory of learning), Stuttgart: Klett-Cotta.

Bandura, Albert (1986), *Social Foundations of Thought and Action. A Social Cognitive Theory*, Englewood Cliffs, NJ: Prentice-Hall.

Cartwright, Dorwin and Frank Harary (1956), 'Structural balance: a generalization of Heider's Theory', *Psychological Review*, **63**, 277–93.

Dahlman, Carl (1980), *The Open Field System and Beyond*, Cambridge: Cambridge University Press.

Demsetz, Harold (1967), 'Toward a theory of property rights', *American Economic Review*, **57**, 347–59.

Denzau, Arthur and Douglass C. North (1994), 'Shared mental models. Ideologies and institutions', *Kyklos*, **47**, 3–31.

Eisenstadt, S.N. (1992), 'The order maintaining and order transforming dimensions of culture', in Richard Münch and Neil Smelser (eds), *Theory of Culture*, Berkeley, University of California Press, pp. 64–87.

Festinger, Leon (1957), *A Theory of Cognitive Dissonance*, Stanford: Stanford University Press.

Furubotn, Eirik G. and Rudolf Richter (1991), 'The New Institutional Economics: an assessment', in Eirik G. Furubotn and Rudolf Richter (eds), *The New Institutional Economics*, Tübingen: J.C.B. Mohr, pp. 1–32.

Gardner, Howard (1987), *The Mind's New Science*, New York: Basic Books.

Granovetter, Mark (1985), 'Economic action and social structure: the problem of embeddedness', *American Journal of Sociology*, **9**, 481–510.

Heider, Fritz (1958), *The Psychology of Interpersonal Relations*, New York: John Wiley & Sons.

Homans, George C. (1975), 'What do we mean by social "structure"?', in P.M. Blau (ed.), *Approaches to the Study of Social Structure*, New York: Free Press, pp. 42–56

Johnson-Laird, Philip N. (1983), *Mental Models*, Cambridge: Cambridge University Press.

McClelland, David (1961), *The Achieving Society*, Princeton: Van Nostrand.

Meyer, Willi (1982), 'The research programme of economics and the relevance of psychology', *British Journal of Social Psychology*, **21**, 81–91.

North, Douglass C. (1990), *Institutions, Institutional Change and Economic Performance*, Cambridge: Cambridge University Press.

Opp, Karl-Dieter (1981), 'The economic theory of social norms ('property rights') and the role of social structures and institutions', *Archives for Philosophy of Law and Social Philosophy*, **67**, 344–60.

Opp, Karl-Dieter (1982), 'The evolutionary emergence of norms', *British Journal of Social Psychology*, **21**, 139–49.

Opp, Karl-Dieter (1983), *Die Entstehung sozialer Normen* (The evolution of social norms), Tübingen: J.C.B. Mohr.

Opp, Karl-Dieter (1984), 'Balance theory: progress and stagnation of a social psychological theory', *Philosophy of the Social Sciences*, **14**, 27–49.

Peterson, Richard A. (1979), 'Revitalizing the culture concept', *Annual Review of Sociology*, 137–66.

Rogers, Everett M. and F. Floyd Shoemaker (1971), *Communication of Innovations. A Cross-cultural Approach*, New York: Free Press.

Schmoller, Gustav (1901), *Grundriß der allgemeinen Volkswirtschaftslehre* (Foundations of economics), Leipzig: Duticker & Humblot.

Simon, Herbert A. (1957), 'Theories of decision-making in economics and behavioural sciences', *American Economic Review*, **49**(1), 77–107.

Skinner, Burrhus Frederic (1971), *Beyond Freedom and Dignity*, New York: Knopf.

Sombart, Werner (1929), 'Economic theory and economic history', *Economic History Review*, **2**, 1–19.

Stephan, Ekkehard (1990), *Zur logischen Struktur psychologischer Theorien*, Berlin, Heidelberg et al.: Springer.

Taylor, E. (1871), *Primitive Culture*, London: John Murray.

Thibaut, John W. and Harold H. Kelley (1959), *The Social Psychology of Groups*, New York: John Wiley & Sons.

Tversky, Amos and Daniel Kahneman (1986), 'Rational choice and the framing of decisions', *Journal of Business*, **59**, 251–78.

Veblen, Thorstein (1898/1990), 'Why is economics not an evolutionary science?', reprinted in Thorstein Veblen, *The Place of Science in Modem Civilization and Other Essays*, New Brunswick, NJ, pp. 60–83.

Webster, Murray (1975), *Actions and Actors. Principles of Social Psychology*, Cambridge: Winthrop.

Williamson, Oliver (1985), *The Economic Institutions of Capitalism. Firms, Markets, Relational Contracting*, New York: The Free Press.

Winter, Sidney G. (1971), 'Satisficing, selection, and the innovating remnant', *Quarterly Journal of Economics*, **85**, 237–61.

Witt, Ulrich (1989), 'The evolution of economic institutions as a propagation process', *Public Choice*, **62**, 155–72.

Witte, Erich H. (1994), *Lehrbuch Sozialpsychologie* (Social psychology), Munich: Beltz.

Index